Algorithmic Capital

Resistance in the Age of Artificial Intelligence

Jonathan Martineau and
Jonathan Durand Folco

Translated by David Broder

VERSO
London • New York

obvia

Fonds de recherche
Société et culture
Québec

SODEC
Québec

Social Sciences and Humanities Research Council of Canada
Conseil de recherches en sciences humaines du Canada
Canada

English-language edition first published by Verso 2026
First published as *Le capital algorithmique : Accumulation, pouvoir et résistance à l'ère de l'intelligence artificielle*

The manufacturer's authorized representative in the EU for product safety (GPSR) is LOGOS EUROPE, 9 rue Nicolas Poussin, 17000, La Rochelle, France
contact@logoseurope.eu

1 3 5 7 9 10 8 6 4 2

Verso
UK: 6 Meard Street, London W1F 0EG
US: 207 East 32nd Street, New York, NY 10016
versobooks.com

Verso is the imprint of New Left Books

ISBN-13: 978-1-80429-900-5
ISBN-13: 978-1-80429-902-9 (US EBK)
ISBN-13: 978-1-80429-901-2 (UK EBK)

British Library Cataloguing in Publication Data
A catalogue record for this book is available from the British Library

Library of Congress Cataloging-in-Publication Data
A catalog record for this book is available from the Library of Congress

Typeset in Minion Pro by MJ & N Gavan, Truro, Cornwall
Printed in the UK by CPI Group (UK) Ltd, Croydon, CR0 4YY

Algorithmic Capital

Contents

Acknowledgments

The authors would like to thank the following people and groups, who, since we embarked on this research project, have at one point or another made significant contributions to our work, whether academically, publication-wise, or through moral support:

David Murray, Barbara Caretta-Debays, Élodie Comtois, and the entire team at Écosociété.

David Broder, Sebastian Budgen, Nick Walther, Gillian Robin, Steven Hiatt, and the entire team at Verso.

Milan Martineau-Ulman, Maïka Sondarjee, Aïna Sondarjee-Folco, Philomène Pellerin, Lysandre Champagne, Gavin Walker, Céline Castets-Renard, Alan Sears, David McNally, Martin Deron, Gabriel Monette, Vladimir de Thézier, Myriam Lavoie-Moore, Sue Ferguson, Martin Gibert, Yves-Marie Abraham, Ambre Fourrier, and Jeremi Léveillé.

Colleagues, students, and staff from the Liberal Arts College at Concordia University; colleagues, students, and staff of the École d'Innovation sociale Élizabeth Bruyère at St Paul's University; members of the Center for Research on Innovations and Social Transformations (CRITS); Polémos Collective; members of the Research Collective on Initiatives, Transformations and Institutions of the Commons (CRITIC); and members of the Center for Interdisciplinary Research on Time, Technology, and Capitalism (CIRTTC).

Thesis 1
We need a critical theory of algorithms

Thought itself is already a sign of resistance, the effort to keep oneself from being deceived any longer.

—Max Horkheimer

For better or worse, the honeymoon with digital technology and its promises is now long in the past. At the turn of the millennium, the rapid embrace of new technologies raised hopes in new forms of sharing, creativity, engagement with the world, and self-expression. Today, we are instead in an age of digital disillusionment, faced with fake news, mass surveillance, polarized online debates, the rampant monetization of personal data, militarized AI, large-scale electoral manipulation, social isolation, cyber-addiction, threats of massive job losses through automation, and much more. Dreams of an ever freer, more interconnected, humane, and efficient world are turning into nightmares worthy of *Black Mirror*.

The great wave of contemporary technological innovation carries with it both utopia and dystopia. It promises constant improvement and optimization of life, but also heralds a grim authoritarian world dominated by machines and Big Tech. The "connected freedom" dear to Silicon Valley walks hand in hand with the foul specter of Big Brother. Today, the rapid development of Big Data, algorithms, and other appendages of "artificial intelligence" embodies this contradictory dream, at once attractive and threatening.

The founder of the World Economic Forum, Klaus Schwab, coined the term "Fourth Industrial Revolution" to describe this convergence of technological innovations, including Big Data, the widespread use of algorithms, the Internet of Things, blockchains, and the accelerated development of robotics and AI.[1] This unprecedented technological revolution, which some describe as "the greatest transformation in history,"[2] is accompanied by narratives that oscillate between epic and tragedy, the technophile embrace of the new and its technophobic rejection. These issues are no longer limited to the online world, formerly known as "cyberspace." After all, the world of algorithms now permeates the physical and "real" world, with the arrival of autonomous cars, personal assistants, online objects, household robots, military drones, smart homes, smart cities, and conversational robots. The separation between the "real" and online worlds is by now obsolete, and society as a whole is transforming before our eyes.

These social changes are rushing toward us in an accelerated, chaotic flow, with no apparent unity between them. What connects such diverse phenomena as the rise of social media, the extraction of rare earth metals, new forms of online entertainment, the fragmentation of our schedules, hyperconnected forms of socialization, the fortunes of Bezos, Musk, and Zuckerberg, US–China tensions over Huawei and TikTok, or the ubiquitous spread of screens? What if we also work in the Brexit vote, teleworking, emotional robots, ChatGPT, the vagaries of cryptocurrencies, military conflicts in Congo, insurance discounts for "good behavior," polarized public debate, targeted political assassinations by drones, the climate crisis, binge-watching, suicides at Foxconn, the semiconductor shortage, influencers, Nvidia's record market capitalization, or same-day delivery? Social change is accelerating, Hartmut Rosa tells us,[3] but there is something unprecedented about the speed at which social, economic, political, and cultural changes are developing in our present era. Some of these developments will affect us close at hand, and others from afar. But it is important to see that they are all symptoms of a huge social transformation set in motion by one particular historical development, often associated with the rise of AI, but not reducible to it. To identify this development, we must place all these phenomena in their sociohistorical context, since otherwise they are bound to remain inexplicable, disconnected, opaque, and even intimidating.

1 Klaus Schwab, "The Fourth Industrial Revolution," *Foreign Affairs*, December 2015.

2 Kai-Fu Lee, *AI Superpowers: China, Silicon Valley, and the New World Order*, Boston: Houghton Mifflin Harcourt, 2018.

3 Hartmut Rosa, *Social Acceleration: A New Theory of Modernity*, New York: Columbia University Press, 2013, and *Alienation and Acceleration: Towards a Critical Theory of Late-Modern Temporality*, Malmö: NSU Press, 2012.

Should we be afraid of "artificial intelligence"?

Before going any further, it is worth taking a critical look at the murky expression "artificial intelligence," which lends itself to various misinterpretations and ambiguities. One of the pioneers in the field, John McCarthy, defines it as "the science and engineering of making intelligent machines, especially intelligent computer programs."[4] This vague and circular definition suggests that machines can be "intelligent"; but what exactly does that mean? Another well-known scientist in this field, Marvin Minsky, thought of AI as machines displaying skills we would associate with human intelligence: "Even though we don't yet understand how brains perform many mental skills, we can still work toward making machines that do the same or similar things. 'Artificial intelligence' is simply the name we give to that research."[5] This definition refers to a set of theories and techniques aimed at creating systems capable of imitating human cognitive abilities: visual recognition algorithms, automatic language processing, automated decision-making systems, and the like.

Everyone agrees that computer systems can *simulate* processes related to human intelligence—but "can machines think?," as mathematician Alan Turing asked in a famous 1950 article.[6] Here, we will not dive into a philosophical debate about what it means to judge, think, believe, see, recognize, or decide. But we certainly find it problematic to attribute mental states, knowledge, or forms of consciousness to machines that are essentially sophisticated computing processes. This tendency toward anthropomorphism—in this case, attributing human qualities to objects or machines—brings with it a host of worries, anxieties, exaggerations, and scenarios that remain far from reality, while at the same time feeding the hype around AI.[7]

This is especially true given that all currently existing technologies belong to the category of *weak AI* (also known as narrow AI). That is, these are algorithms trained to perform specific tasks: playing chess, filtering spam emails, selecting content on a news feed, recommending

4 John McCarthy, "What Is Artificial Intelligence?," 2007, available at www-formal.stanford.edu.

5 Marvin Minsky, *The Age of Intelligent Machines: Thoughts About Artificial Intelligence*, universelle-automation.de/1991_Boston.pdf.

6 Alan Turing, "Computing Machinery and Intelligence," *Mind*, no. 236, 1950, pp. 433–60.

7 See the on-point discussion of AI anthropomorphism in chapter 2 of Emily Bender and Alex Hanna, *The AI Con: How to Fight Big Tech's Hype and Create the Future We Want*, New York: Harper, 2025.

products on a website, driving an autonomous vehicle, generating plausible text, and so forth. These are essentially sets of instructions designed to solve a particular problem, vast data processing systems or "artificial neural networks," without soul or consciousness.

In contrast, *strong AI*, also known as artificial general intelligence (AGI), is considered capable of performing a multitude of tasks, predicting and acting in ways far superior to humans. Philosopher Nick Bostrom uses the term "superintelligence" to describe this hypothetical machine that could eventually replace, dominate, or even eradicate human beings.[8] Investor Ian Hogarth, founder of the venture capital fund Plural, echoes this sentiment in the gravest tones:

> A three-letter acronym doesn't capture the enormity of what AGI would represent, so I will refer to it as what is: God-like AI. A superintelligent computer that learns and develops autonomously, that understands its environment without the need for supervision and that can transform the world around it. To be clear, we are not here yet. But the nature of the technology means it is exceptionally difficult to predict exactly when we will get there. God-like AI could be a force beyond our control or understanding, and one that could usher in the obsolescence or destruction of the human race.[9]

The existential risks associated with superintelligence—reminiscent of Skynet in the *Terminator* series (from 1984 onward), or the machines that took control of the world in *The Matrix* (1999)—are the subject of a great deal of technical and philosophical speculation. The problem of alignment, that is, the challenge of ensuring that AI produces results that are consistent with its designers' goals or values (rather than unpredictable ones that conflict with the interests of human beings), becomes critical when it comes to programming autonomous machines that far exceed human capabilities.

No one knows yet whether AGI will ever see the light of day—and, if so, when. Futurist Ray Kurzweil predicts that this pivotal moment, which he calls the "technological singularity," will occur around 2045.[10] Other experts, such as Erik J. Larson, remind us more soberly that the transition

8 Nick Bostrom, *Superintelligence: Paths, Dangers, Strategies*, Oxford: Oxford University Press, 2014.

9 Ian Hogarth, "We Must Slow Down the Race to God-like AI," *Financial Times Magazine*, April 13, 2023.

10 Ray Kurzweil, *The Singularity Is Near: When Humans Transcend Biology*, New York: Viking, 2005.

from weak AI to AGI would represent a qualitative leap impossible to deduce from current technological development. It is not enough to continue developing weak AI for it to eventually become AGI. Rather, it would involve qualitatively different technologies, and, for a programmer like Larson, the scenario of superintelligence suggested by those like Bostrom is more a product of a fertile imagination than the *telos* of current technologies.[11]

Fears over the imminent arrival of AGI have been reignited by the hype surrounding ChatGPT.[12] Many people claim that ChatGPT talks to them, responds to them, makes choices and analyses, and thinks. They are, in fact, anthropomorphizing this algorithmic system. In June 2022, Google engineer Blake Lemoine even made headlines by claiming that the conversational robot (chatbot) LaMDA was "sentient," that is, capable of thinking, feeling emotions, and being aware of its situation.[13] However, it is important to remember that a machine like ChatGPT is only an algorithmic completion tool; it predicts the next word, sentence, or token based on a complex statistical calculation using a huge data set made up of human-generated texts extracted from blogs, Wikipedia articles, novels, comments on social media, and many other sources. The ability to analyze a prompt and then generate sophisticated text is surely quite a technical feat—but this does not mean that the machine is "conscious" or "sentient."

It seems that many people, amazed or terrified by ChatGPT's spectacular capabilities, are fooled by the idea that AI has a real "mind," when in fact it is merely regurgitating altered versions of texts produced by the human mind. As science fiction author Arthur C. Clarke puts it, "Any sufficiently advanced technology is indistinguishable from magic."[14] In behavioral psychology, the "mirror test" is designed to determine whether an animal can show self-awareness by recognizing its reflection in a mirror. Many people today keep failing this test when they forget that algorithms excel at reflecting our own ideas, and as Shannon Vallor puts it, are only mirrors that reproduce the past.[15] The automatically generated

11 Erik J. Larson, *The Myth of Artificial Intelligence: Why Computers Can't Think the Way We Do*, Cambridge, MA: Belknap Press, 2021, pp. 33–43.

12 Alex Blake, "GPT-5 Could Soon Change the World in One Incredible Way," *Digital Trends*, March 28, 2023.

13 Nitasha Tiku, "The Google Engineer Who Thinks the Company's AI Has Come to Life," *Washington Post*, June 11, 2022.

14 Arthur C. Clarke, *Profiles of the Future: An Inquiry into the Limits of the Possible*, New York: Harper & Row, 1973.

15 Shannon Vallor, *The AI Mirror: How to Reclaim Our Humanity in an Age of Machine Thinking*, Oxford: Oxford University Press, 2024.

texts, concepts, stories, and images are in fact reconstructions of elements created by millions of human beings around the world. "Knowing this," James Vincent observes, "we *should* be able to recognize ourselves in our new machine mirrors, but instead, it seems like more than a few people are convinced they've spotted another form of life."[16]

To avoid falling into anthropomorphism or fetishizing or attributing magical powers to machines, some suggest abandoning the misleading and ambiguous metaphor of artificial intelligence altogether. For example, for researcher Evgeny Morozov this "term belongs to the same scrapheap of history that includes 'iron curtain,' 'domino theory' and 'Sputnik moment.' It survived the end of the Cold War because of its allure for science fiction enthusiasts and investors."[17] Morozov contends that AI is neither entirely intelligent nor entirely artificial. First, intelligence cannot be reduced to statistical calculation and pattern matching, because it also includes the abilities to generalize, deduce, and improvise, which have a strong social, symbolic, emotional, and physical dimension. Intelligence presupposes having a body and being surrounded by peers. Yet neither are these systems purely mechanical or artificial, because they "draw their strength from the work of real humans: artists, musicians, programmers and writers whose creative and professional output is now appropriated in the name of saving civilisation. At best, this is 'non-artificial intelligence.'"[18] In the same vein, Emily Tucker argues that the label "artificial intelligence" distracts public attention from the real issues surrounding the advent of algorithmic systems. The threat of humans falling under domination is not simply posed by the specter of a menacing superintelligence. Rather, it comes from real social, economic, and political problems that largely result from the activities of companies that develop algorithms, control them, and sometimes use them against us.[19]

But even if we realize that the term "artificial intelligence" has become a buzzword, a catch-all term, or a dubious metaphor, this leads to another question: What should we replace it with? Tucker suggests various ways of better formulating our ideas to avoid anthropomorphism and remain more in touch with reality: "Be as specific as possible about what the technology in question is and how it works," she urges.[20] This is the spirit

16 James Vincent, "Introducing the AI Mirror Test, Which Very Smart People Keep Failing," *The Verge*, February 17, 2023, theverge.com.

17 Evgeny Morozov, "The Problem with Artificial Intelligence? It's Neither Artificial nor Intelligent," *The Guardian*, March 30, 2023.

18 Ibid.

19 Emily Tucker, "Artifice and Intelligence," *Tech Policy Press*, March 17, 2022.

20 Ibid.

in which Bender and Hanna propose to refer to large language models (LLMs) as "synthetic text extruding machines."[21] We should also identify the companies that produce these technologies and avoid attributing undue agency to machines—remembering that it is always humans who program and use them.

The word "smart"—used in all sorts of contexts to refer to such innovations as "smart cities" or "smart watches"—is a good example of marketing speak that spreads confusion among the general public. Italian computer scientist Stefano Quintarelli even proposes replacing the term "artificial intelligence" with the acronym SALAMI, which stands for "Systematic Approaches to Learning Algorithms and Machine Inferences." The advantage of the term SALAMI is that it allows us to circumvent anthropomorphic language and dispel the magical aura surrounding algorithms:

> Now we have redefined the name, will we still support the idea that SALAMI will develop some form of consciou[s]ness?
> Will SALAMI have emotions?
> Can SALAMI acquire a "personality" similar to humans'?
> Will SALAMI ultimately overcome human limitations and develop a self superior to humans?
> Can you possibly fall in love with a SALAMI?
> Can we suddenly perceive a sense of how all these far flung (unrealistic) predictions look somewhat ridiculous?[22]

But this poses a problem. Can we completely dispense with the terms "artificial intelligence," "machine learning," "emotional computing," or "facial recognition" and replace them with the term SALAMI when these buzzwords are already so widely used—everywhere from scholarly literature to public debate and even everyday talk? While it may be more accurate to replace the sentence "the Los Angeles Police Department is using AI to fight crime" with "the Los Angeles Police Department has signed a contract with Palantir to use data-driven risk prediction software," what is gained in accuracy here is lost in clunkier phrasing. This could have the perverse effect of sowing doubt in the reader's mind: Are we actually talking about AI? In short, since the term "artificial

---

21 Bender and Hanna, *The AI Con*.

22 Stefano Quinterelli, "Let's Forget the Term AI: Let's Call Them Systematic Approaches to Learning Algorithms and Machine Inferences (SALAMI)," Quinta's Weblog, November 24, 2019, blog.quintarelli.it.

intelligence" has become so established, we prefer to use it—albeit sparingly—for the sake of simplicity and to avoid misunderstandings. This, even while we remember that AI is neither entirely intelligent nor entirely artificial. We also favor the use of the adjective "algorithmic," which has the advantage of being less anthropomorphic and, as we shall see, more accurate.

Although semantic debates are important, our higher ambition in this book is to understand and critique the thing in question. We see developing a *critical theory of algorithms* as especially important in this regard. And this is what we set out to do, across the various theses of this book. To this end, we need to avoid two kinds of traps that could quickly ensnare us in an uncritical approach: technological determinism and the mirage of an ethical, responsible AI.

The blinders of technological determinism

The technological determinist approach is ubiquitous in mainstream discourse, both in the media and in the corporate world, around new technologies that claim to "build a better world." Technological determinism conceives of technical change as an endogenous, autonomous, and inevitable process able to transform large sections of society: "The hand-mill gives you society with the feudal lord; the steam-mill society with the industrial capitalist," to borrow Marx's caricature.[23] Generally, technological determinism goes hand in hand with a *techno-optimistic* attitude focused on the beneficial and potentially salvific effects of technological innovations. Sometimes it is accompanied by a cult of personality surrounding "great innovators" such as Elon Musk or Steve Jobs. Freedom and social progress are to be achieved through the great epic of human technical ingenuity. For example, Jeremy Rifkin conceives of the Internet of Things as a great "intelligent infrastructure" that would facilitate ecological transition, the eclipse of capitalism, and the emergence of a zero-marginal-cost society:

> If the steam engine freed human beings from feudal bondage to pursue material self-interest in the capitalist marketplace, the Internet of Things frees human beings from the market economy to pursue nonmaterial shared interests on the Collaborative Commons.[24]

---

23 Karl Marx, "The Poverty of Philosophy," text from marxists.org.

24 Jeremy Rifkin, *The Zero Marginal Cost Society: The Internet of Things, the Collaborative Commons, and the Eclipse of Capitalism*, New York: St. Martin's Press, 2014.

For thinkers like Rifkin, technology is the main determining force behind historical development and social progress.

Nevertheless, the idea of technological development as an autonomous force can also lead to a *technopessimistic* stance. The "megamachine" analyzed by Lewis Mumford, Jacques Ellul's "technological system," and Bernard Charbonneau's "industrial totalitarianism" are, in their different ways, classic examples of this concern over the power of technology, which ultimately tends to dispossess and subjugate human beings. More recently, Éric Sadin has associated AI with the arrival of a "radical anti-humanism," an "automated invisible hand," and an "algorithmic Leviathan." Other authors see the convergence of Big Data, "dataveillance," and neoliberalism as a dangerous cocktail that may potentially lead to the suppression of political freedom, the replacement of humans by algorithms, and the dissolution of society by cybernetic ideology.[25]

The media hype surrounding AI, added to Silicon Valley's own triumphalist discourse, is thus whipping up a moral panic that sees radical criticism of the digital world and the "great refusal" as the best means of resisting technological encroachment on human life. In these two faces of technological determinism, technical progress appears as an overarching, reified force over which we have virtually no control; we must either adapt to it, or else reject the "digitally administered world" outright.[26]

However, we think it essential to move beyond both naïve optimism and visceral pessimism. Individuals are never *totally* subjugated or alienated, but they may be dominated to a greater or lesser degree. Despite the various obstacles and constraints they face, human beings are able

25 Lewis Mumford, *Technics and Civilization*, New York: Harcourt, Brace, 1934; Jacques Ellul, *The Technological Society*, New York: Vintage, 1964; Bernard Charbonneau, *Le totalitarisme industriel*, Paris: L'Échappée, 2019; Éric Sadin, *L'intelligence artificielle ou l'enjeu du siècle: Anatomie d'un antihumanisme radical*, Paris: L'Échappée, 2018; Pierre Henrichon, *Big Data: Faut-il avoir peur de son nombre?*; *Cybernétique, dataveillance, néolibéralisme, des armes contre la société*, Montreal: Écosociété, 2020. While often read as a promotion of a certain form of technological determinism, Bernard Stiegler's works on technology are far more complex and nuanced than this suggests and deserve greater attention. See, among others, Bernard Stiegler, *Technics and Time, 1: The Fault of Epimetheus*, Stanford, CA: Stanford University Press, 1998; and Stiegler, *Automatic Society, Volume 1: The Future of Work*, Cambridge: Polity, 2016.

26 Éric Sadin, *La silicolonisation du monde: L'irrésistible expansion du libéralisme numérique*, Paris: L'Échappée, 2016; Maxime Ouellet et al., "Big Data et quantification de soi: La gouvernementalité algorithmique dans le monde numériquement administré," *Canadian Journal of Communication* 40, no. 4, November 2015, pp. 597–613; Maxime Ouellet, *La révolution culturelle du capital: Le capitalisme cybernétique dans la société globale de l'information*, Montreal: Écosociété, 2016.

to mold their worlds, according to the resources, strategies, and power relations that exist within a given sociohistorical context. Contemporary technology may appear as a vast, reified, automated, and rigid complex. It nonetheless remains dependent on changing, ever-shifting, and ultimately historical social relations. If we want to advance a critical analysis of algorithms, we must avoid one-sided approaches that tend to erase the complex and contradictory social dynamics that influence new technologies and their development.

It would also be misleading to see technology in general, and algorithms more particularly, as mere tools that can be used for good or ill depending on the circumstances. Digital applications, robots, algorithms, and platforms are always designed to produce certain effects, and to direct actions and guide user behavior toward specific objectives. Moreover, technological innovations do not evolve in a vacuum, supervised by experts in laboratories, but within a socioeconomic system whose imperatives direct and select the technologies that can generate profits, productivity gains, and conquer new market shares. Placing technology in its proper sociohistorical context is thus a first step toward a critical theory that avoids technological determinism.

The ethical mirage of responsible AI

This comprehensive reading is precisely what seems to escape the proponents of AI ethics. This rapidly growing field of study presents itself as a benevolent and reasonable exercise in reflexivity faced with the rise of algorithms. A highly fashionable approach, it essentially seeks to determine "how to program virtuous robots."[27] While this literature is surely interesting, it skips a crucial step and glosses over an even more fundamental critical question: Do we need robots, and if so, at what cost?[28] Ethics as applied to algorithms does not primarily aim to assess the social and moral impacts of AI in a broad sense, but stops at proposing an ethic of programming suited to promoting the "responsible development" of algorithmic technologies. This approach, above all, focuses on the interactions between individuals and machines, without sufficient attention given to sociohistorical context. Building public trust in intelligent machines seems to be its fundamental concern.

27 Martin Gibert, *Faire la morale aux robots: Une introduction à l'éthique des algorithmes*, Montreal: Atelier 10, 2020.

28 We will return to these broader ethical issues in Thesis 19.

Ethicists do, of course, mention the biases of algorithms that can reproduce certain forms of racism or sexism for example. However, in so doing they do not seek to pose critical questions about capitalism or to understand how algorithms reinforce patriarchy, racism, or colonialism from a global and systemic perspective. Their rather more modest goal is to limit the algorithms' discriminatory effect. In short, this is about a better programming of effective and "inclusive" robots, with a touch of morality added at the right moment to prevent robots from acting inappropriately. A good example is ChatGPT, the conversational robot programmed to avoid sexist, racist, and hateful comments. To this end, OpenAI has contracted another company, Sama, to hire thousands of Kenyan workers paid less than $2 an hour to identify large quantities of hateful texts, so as to train the chatbot not to use such language.[29] Such is the hidden face of "ethical" algorithms that rely on the exploitation of precarious and fragmented human labor, largely outsourced to the Global South.

It could surely be argued that this ethics of algorithms is only a subfield of a broader reflection on the ethics of AI, which would properly reckon with the social, economic, political, and ecological consequences of these innovations. In Quebec, this broader approach is illustrated by the "Montréal Declaration for a Responsible Development of Artificial Intelligence," produced through a participatory process and signed by a group of experts in AI and ethics. This declaration proposes ten principles to guide the development of AI in the service of humanity: well-being, respect for autonomy, protection of privacy and intimacy, solidarity, democratic participation, equity, diversity inclusion, caution, responsibility, and sustainable development.

This upstanding list of values implicitly recognizes that the development of AI could be (if it is not already) detrimental to well-being, social justice, democracy, and all the other principles. Indeed, as the declaration points out, algorithmic machines "can restrict the choices of individuals and groups, lower living standards, disrupt the organization of labor and the job market, influence politics, clash with fundamental rights, exacerbate social and economic inequalities, and affect ecosystems, [and] the climate."[30] All of this is true, but a narrow ethical response

29 Billy Perrigo, "Exclusive: OpenAI Used Kenyan Workers on Less Than $2 Per Hour to Make ChatGPT Less Toxic," *Time*, January 18, 2023. We will devote Theses 5 and 6 to the question of labor, and Thesis 12 to international relations.

30 Yoshua Bengio et al., "The Montréal Declaration for Responsible Development of Artificial Intelligence," 2018, montrealdeclaration-responsibleai.com.

to these complex issues falls well short of what is needed. The aim of such declarations is not to question the social and economic causes of these problems, but to provide an "ethical compass" to guide the development of AI. The key goal of building public trust is explicit here, aiming to "lay the foundation for cultivating social trust toward artificially intelligent systems."

This scramble to find virtuous robots and responsible AI is accompanied by the worldwide spread of codes of ethics, though none of them are binding. Already in 2019, Marietje Schaake, a member of the European Parliament, noted that there were more than 128 lists of ethical standards for AI in Europe alone, generally aimed at achieving a "broad consensus" on principles that do not really disrupt the status quo.[31] The field of AI ethics has not slowed its output of principles and declarations since then. Clearly, no binding mechanisms can be found in the documents produced by the tech giants' PR operations, such as the 2016 Partnership on AI to Benefit People and Society, which includes Amazon, Meta, Alphabet, Microsoft, IBM, and Apple. Moreover, while they set out general principles, they are largely silent about how these are to be put into practice, as the AI industry has no formal governance framework or professional order that could ensure compliance.[32] Furthermore, most of these ethical statements from the Global North rarely represent the voices of the Global South, and thus fail to include some of the groups and populations most affected by the development of AI.[33]

As things stand, we can hardly expect AI ethics to deliver convincing results. The main pitfall of responsible development strategies is that they go no further than a moral approach, based on how things ought to be. Philosophers, scientists, and experts thus prescribe moral principles that ignore the material and social dynamics that guide the actual development of AI. But since when have statements of principle, corporate social responsibility, and articles of faith in sustainable development and self-regulation done anything to curb rising social inequalities, the hollowing out of democracy, and environmental crisis? Ethical thinking creates standards that individuals, programmers, and companies are meant to comply with, as if Big Tech had the choice to act morally rather

31 Marietje Schaake, "What Principles Not to Disrupt: On AI and Regulation," *Medium*, November 5, 2019.

32 Kate Crawford, *Atlas of AI: Power, Politics, and the Planetary Costs of Artificial Intelligence*, New Haven, CT: Yale University Press, 2021.

33 Anna Jobin, Marcello Iena, and Effy Vayena, "The Global Landscape of AI Ethics Guidelines," *Nature Machine Intelligence*, no. 1, 2019, pp. 389–99.

than respond to the economic imperatives that decide corporate strategies and courses of action.

Admittedly, several voices linked to responsible AI have begun to speak up about the need to move beyond a purely ethical approach and to call for government regulation of the development of algorithms. The open letter "Pause AI Giant Experiments," cosigned by leading figures in industry and research (including Yoshua Bengio and Elon Musk), put forward the idea of a six-month pause in the development of algorithmic models superior to GPT-4. It even contended that, if necessary, "governments should step in and institute a moratorium."[34] But while the letter sounds the alarm, it ends on an optimistic, even candid note: "Humanity can enjoy a flourishing future with AI. Having succeeded in creating powerful AI systems, we can now enjoy an 'AI summer' in which we reap the rewards, engineer these systems for the clear benefit of all, and give society a chance to adapt." It is clear, then, that the call to slow down the development of these machines is only temporary. The idea is simply to allow society to "adapt" to this technological progress, so that these machines can then be rolled out on a larger scale.

Two other examples illustrate the limitations of responsible AI when this approach is confined to the industry's own chosen parameters. One of the individuals working to bring about change from within this field is the computer scientist Timnit Gebru. Along with another prominent tech researcher, Joy Buolamwini, Gebru studied how facial recognition systems discriminate against black women, who are 35 percent less likely to be recognized than white men.[35] During her studies at Stanford University, she cofounded the organization Black in AI to increase the representation of black people in the field, then completed a postdoc at Microsoft's FATE (Fairness, Accountability, Transparency, and Ethics in AI) lab before being hired by Google in 2018 to join a team dedicated to AI ethics. In December 2020, she was dismissed by Google for refusing to remove her name from an article identifying the pitfalls and dangers of "large language models," because the company did not want to be associated with the publication.[36]

---

34 "Pause Giant AI Experiments: An Open Letter," Future of Life Institute, March 22, 2023.

35 Steve Lohr, "Facial Recognition Is Accurate, if You're a White Guy," *New York Times*, February 9, 2018.

36 Emily Bender, Timnit Gebru, Angelina McMillan-Major, and Shmargaret Shmitchell, "On the Dangers of Stochastic Parrots: Can Language Models Be Too Big?," *Proceedings of the 2021 ACM Conference on Fairness, Accountability, and Transparency*, March 2021, pp. 610–23.

Ethics is thus tolerated only so long as it does not overly disrupt the industry's own interests. Despite her dismissal, Gebru received the support of 1,200 Google employees and 4,300 members of the academic community and civil society organizations, becoming one of Silicon Valley's most prominent dissidents. By choosing the path of independent research, with a much freer, more radical, and militant discourse, Gebru was recognized as one of the ten greatest scientists of 2021 according to *Nature*, and as one of *Time* magazine's 100 most influential people of 2022.[37] Gebru calls for moving beyond a narrow approach to AI ethics:

> What I've realized is that we can talk about the ethics and fairness of AI all we want, but if our institutions don't allow for this kind of work to take place, then it won't. At the end of the day, this needs to be about institutional and structural change.[38]

The second example is OpenAI, a company at the forefront of generative AI with the development of GPT, a large language model on which ChatGPT is based. OpenAI was founded in December 2015 as a nonprofit by Sam Altman, Elon Musk, Peter Thiel, and Reid Hoffman. Based on values of transparency, openness, and concern for the common good, this organization's announced mission was to put general artificial intelligence at the service of all humanity. Originally, its goal was to act as a socially responsible company, in a spirit of open and disinterested innovation, in contrast to the financial interests of the digital giants. Following Musk's departure in 2018 (after he had attempted to take control of the company), and certain difficulties related to the exorbitant costs of training the algorithms, in March 2019 its leaders decided to turn the nonprofit into a private, for-profit company to attract more investment. This worked handsomely, with Microsoft injecting $1 billion in 2019 in order to take its technologies to market, followed by another $10 billion in 2023 after the early signs of ChatGPT's phenomenal success. As of February 2026, OpenAI's estimated market value was at $730 billion, and ChatGPT broke the record for the fastest-growing application in history, after reaching 100 million users before it was even two months old.[39]

---

37 Safiya Noble, "Timnit Gebru: The 100 Most Influential People of 2022," *Time*, May 23, 2022.

38 Dylan Walsh, "Timnit Gebru: Ethical AI Requires Institutional and Structural Change," Stanford University, Human-Centered Artificial Intelligence, May 26, 2022, hai.stanford.edu.

39 By comparison, TikTok took nine months to reach 100 million users, Instagram took two and a half years, and Twitter took five years. See Dan Milmo, "ChatGPT

This commercial success comes at some cost in terms of ethics and transparency. OpenAI has a deeply opaque internal culture,[40] and it refuses to share technical details related to the architecture of its algorithmic models and the data on which they are trained.[41] Here is a textbook case of mission drift—an organization that has strayed from its social mission due to its quest for financial resources, competitive pressures, and other economic imperatives. Meanwhile, one of its main partners for a while, Microsoft, decided to dismiss its entire responsible AI team in March 2023. This aimed to limit internal resistance and obstacles that might hamper its ability to finish first in the AI race by integrating OpenAI technologies into its Office suite and Bing search engine.[42] This is, again, a clear example of ethics being sacrificed on the altar of maximizing profits.

For now, the high-tech sector is no different from others, and the imperative of profitability generally takes precedence over ethics. In general, corporate social responsibility can apply as long as it remains sufficiently profitable for shareholders, and only for that long. While technological determinism conceives of technical progress as a mechanical, necessary, and inevitable process, algorithmic ethics overestimates the freedom and the moral potential of the for-profit companies that are developing these technologies on a mass scale. A critical theory of algorithms must take care to avoid these two pitfalls.

Problem-solving or critical theory?

To better understand the contrast between mainstream approaches to AI and a critical theory of algorithms, it is worth turning to the work of Robert Cox. He distinguishes between two types of theories defined by their respective purposes. The first approach, which he calls problem-solving theory, seeks to solve specific problems posed by a particular perspective that it takes as its starting point. The second, more reflexive

Reaches 100 Million Users Two Months After Launch," *The Guardian*, February 3, 2023.

40 For a fascinating investigation into the inner workings and corporate culture of OpenAI, see Karen Hao, *Empire of AI: Dreams and Nightmares in Sam Altman's OpenAI*, London: Penguin, 2025.

41 Tiernan Ray, "With GPT-4, OpenAI Opts for Secrecy Versus Disclosure," *ZDNet*, March 16, 2023.

42 Stéphanie Bascou, "Pourquoi Microsoft s'est délesté de son équipe dédiée à l'IA responsable," 01net, March 14, 2023, 01net.com.

approach is the one taken by critical theory. It questions social phenomena in a comprehensive manner, paving the way for the construction of an alternative social order:

> Critical theory, unlike problem-solving theory, does not take institutions and social and power relations for granted but calls them into question by concerning itself with their origins and how and whether they might be in the process of changing. It is directed towards an appraisal of the very framework for action, or problematic, which problem-solving theory accepts as its parameters. Critical theory is directed to the social and political complex as a whole rather than to the separate parts … Whereas the problem-solving approach leads to further analytical sub-division and limitation of the issue to be dealt with, the critical approach leads towards the construction of a larger picture of the whole of which the initially contemplated part is just one component, and seeks to understand the processes of change in which both parts and whole are involved.[43]

Of course, we should avoid being too quick to judge the superiority of one approach over another: Each has different objectives, with its own advantages and disadvantages. The strength of the problem-solving approach lies in its ability to clearly define or identify the parameters of a problem, refine the analysis, or limit the number of variables.[44] Still, it has the disadvantage of being restricted to a particular sphere of activity, implicitly assuming that the current social world is fixed, and it potentially serves—intentionally or not—the interests that benefit from the existing order. This analytical approach, which can be applied in a multitude of practical ways, should not be rejected wholesale, since it is often useful to isolate a phenomenon from its context in order to solve a particular problem. Still, its limitations should be borne in mind. Critical theory, for its part, is historical and comprehensive, highlighting the assumptions and constraints of the existing social order and allowing us to imagine alternatives to it. Its scope is thus both theoretical and practical: Critical theory aims to understand the world and change it. On the other hand, it may seem to lack precision at times, as it must "continually adjust its concepts to the changing object it seeks to understand

---

43 Robert W. Cox, "Social Forces, States and World Orders: Beyond International Relations Theory," *Millennium* 10, no. 2, June 1981, pp. 128–9.

44 Ibid., p. 129.

and explain."[45] It must also be interdisciplinary, which requires effort and sometimes difficult compromises between the methodologies and literature from different disciplines.

Cox thus offers a relatively clear, simple, and elegant description of the specificity of critical theories, regardless of their particular tradition, whether Marxism, the Frankfurt School, feminist theories, intersectionality, or decolonial thought. This is also reminiscent of Max Horkheimer's distinction between traditional theory and critical theory.[46] In his account, traditional theories are rooted in the specialization of the sciences, focusing on a single type of object, with an analytical approach that abstracts from society and adopts the disinterested stance of the logical and impartial observer. Critical theory instead focuses on historical totality, social praxis, and its contradictions, analyzing these through an interdisciplinary approach in the social sciences, recognizing that theory is always linked to a sociohistorical context that remains subject to change. Furthermore, it makes a clear commitment to human emancipation.

How do these two opposing visions play out today in the abundant literature on algorithms? The problem-solving approach generally analyzes the risks of algorithms by abstracting from the socioeconomic system in order to focus on specific issues amenable to technical solution. Such an "ethics of parameterization" that aims to "code morality" can certainly be useful for improving the functioning of various kinds of robots, machines, and algorithms. But must philosophy necessarily serve to set better parameters for search engines, recommendation systems, personal assistants, autonomous cars, sex robots, surveillance cameras, financial algorithms, and automated recruitment software? In the thirteenth century, Thomas Aquinas contended that philosophy should be the servant of theology; today, it seems that ethics is becoming the servant of technology.

There is thus an urgent need to develop a *critical theory of algorithms*, an interdisciplinary approach that aims to understand the multiple ramifications of algorithmic logic, its various devices, and its ideological dimension. This means drawing on the social sciences and philosophy to uncover the power relations that are connected to these new technologies and to understand the political economy that produces them. The social and political issues raised by the rapid rise of algorithms should

45 Ibid.

46 Max Horkheimer, "Traditional and Critical Theory," in *Critical Theory: Selected Essays*, New York: Continuum, 1972.

not be analyzed in isolation, but as phenomena linked to the institutional, economic, and cultural structures of the dominant social order. A central feature of critical theory is that it refuses to adopt the impartial and detached gaze of the supposedly "neutral" observer. Critical theory is *normative*, and it reveals relations of domination by taking the side of those who suffer the harshest repercussions of these technical innovations. Researcher Kate Crawford agrees:

> To understand what is at stake, we must focus less on ethics and more on power. AI is invariably designed to amplify and reproduce the forms of power it has been deployed to optimize. Countering that requires centering the interests of the communities most affected. Instead of glorifying company founders, venture capitalists, and technical visionaries, we should begin with the lived experiences of those who are disempowered, discriminated against, and harmed by AI systems. When someone says, "AI ethics," we should assess the labor conditions for miners, contractors, and crowdworkers. When we hear "optimization," we should ask if these are tools for the inhumane treatment of immigrants. When there is applause for "large-scale automation," we should remember the resulting carbon footprint at a time when the planet is already under extreme stress.[47]

Of course, we are not advocating the complete abandonment of AI ethics. It surely is preferable to give direction to the programming of algorithms rather than count on their blind development. But ethics extends far beyond the issues of computer programming: It is a matter of reflecting on the general conditions for living well, whether in terms of practices, social interactions, institutions, or ways of life. Beyond wanting to program "good robots," this means knowing what place digital technology, algorithms, and robots should occupy in our lives.

This is why, in our opinion, ethics calls for critical theory. We must question the imperatives and mechanisms that govern the development of algorithms, which are closely linked to the reconfiguration of a capitalism that is already using these new tools to strengthen its hold on our lives. The traditional approach to AI ethics takes the social relations and institutions of capitalism for granted and focuses on problems posed by specific algorithms. Critical theory must develop a broader vision

47 Crawford, *Atlas of AI*, pp. 224–5.

of the historical totality of which intelligent machines are only components. It is necessary to understand the processes of sociotechnical change in which the whole and its parts are caught up. That is why we need a critical theory of algorithms. But can such a critical theory be constructed without a theory of capitalism? We will address this question in Thesis 2.

Thesis 2

A theory of algorithms needs a theory of capitalism

I have striven not to laugh at human actions, not to weep at them, nor to hate them, but to understand them.

—Baruch Spinoza

It would be an understatement to say that today there is an abundant literature on artificial intelligence and algorithms. However, most analyses of AI's impacts address different areas in isolation—from privacy to work, politics, the environment, education, and so on. Too few studies systematically address the complex relationships between the capitalist mode of production and algorithms.[1] Let us therefore start from a hypothesis: We cannot understand the rapid development of algorithms without understanding the reordering of capitalism, and vice versa.

To grasp the social, political, and economic implications of algorithmic machines properly, these innovations need to be situated in the context of the transformation of capitalism in the first quarter of the twenty-first century. We suggest that the harvesting of Big Data and the rapid rollout of artificial intelligence are accompanied by a significant transformation of capitalism, similar to the Industrial Revolution that propelled the empire of capital in the nineteenth century. We have entered a new stage of capitalism, based on a new logic of accumulation and a specific form of power, which we call algorithmic capital. This term

1 On the other hand, a burgeoning critical literature on the political economy of Big Tech is significantly adding to our understanding of these phenomena. We will come back to and draw on such contributions in subsequent theses in this book.

refers to a process with many dimensions: It is at once a formal logic, a dynamic of accumulation, a social relation, and an original form of power based on algorithms. Our aim here is to conceptualize how the logic of capital accumulation is converging with the increased use of new algorithmic tools. In contrast, the term "algorithmic capital*ism*" refers to the social formation taken as a whole—to the historical articulation of a mode of production and an institutional context, what Nancy Fraser calls an "institutionalized social order."[2] We can therefore distinguish between the capitalist society in which we live and the specific social relation that expresses itself in manifold different ways.

Clearly, it is also important to understand what an algorithm is. We will return to this question in more detail in Thesis 4. For now, we need only point out that in its simplest form, an algorithm is a sequence of operations that solves a particular problem. One oft-cited example is a recipe for preparing a dish in the kitchen: an ordered series of steps to follow that lead to the desired result. It is thus a clear and precise procedure for accomplishing a certain task. Most mechanical actions lend themselves well to this type of analytical breakdown, but so do a growing number of complex cognitive tasks. In computer science, an algorithm performs a set of calculations at very high speed using massive amounts of data, incorporating huge quantities of information. On that basis, the algorithm can prioritize the information, select what most interests us or what we will prefer, and perform a multitude of tasks.[3] It is this type of algorithm we discuss in this book, not those found in cookbooks!

Generally, theoretical reflections on algorithms begin with basic classification exercises. For example, some authors seek to distinguish between them by function: calculating, categorizing, communicating, modeling, predicting, and so on.[4] Others, such as Dominique Cardon, instead suggest classifying them into four main families of digital calculations. Algorithms can be used to: (1) measure audiences and rank the *popularity* of products for advertising purposes; (2) give a ranking to websites' *authority* (like Google's PageRank); (3) measure the *reputation* of individuals or companies on social media (such as by counting likes on social media); or (4) *predict* behavior by analyzing individuals' digital

---

2 Nancy Fraser and Rahel Jaeggi, *Capitalism: A Conversation in Critical Theory*, Cambridge: Polity Press, 2018.

3 Dominique Cardon, *À quoi rêvent les algorithmes: Nos vies à l'heure des big data*, Paris: Seuil, 2015, p. 7.

4 Serge Abiteboul and Gilles Dowek, *Le temps des algorithmes*, Paris: Le Pommier, 2017.

trails (for example, Amazon recommendations). However, such analyses often merely list the diversity of forms and uses of particular algorithms without placing them in an overall context, let alone within the framework of a critical theory.

In Thesis 7, we will explain how recent technological innovations such as machine learning and deep learning have revolutionized the field of AI research while accelerating the use of these algorithms in economic and social life. Algorithmic machines have a wide variety of possible applications, and this makes it difficult to grasp algorithmic logic in a clear, unified, comprehensive, and consistent manner. Some authors contend that we have entered a "computational society," an "algorithmic society," or an "age of Big Data." These evocative but vague formulations obscure the specific nature of the socioeconomic system underlying these developments. To explain the origins, current forms and impacts of algorithms, most analyses focus on the promises and pitfalls of these new technologies, while ignoring the "elephant in the digital room": capitalism.[5]

Going beyond buzzwords

Capitalism's recent history has seen the rapid acceleration of technological development and the emergence of a globalized networked economy. This has also given rise to a wealth of literature rich in concepts aimed at understanding the changes in capitalism. The notions of information capitalism, digital capitalism, postindustrial society, the new attention economy, a knowledge economy, cognitive capitalism, communicational capitalism, attentional capitalism, mental capitalism, cybernetic capitalism, platform capitalism, or surveillance capitalism emphasize different dimensions of these developments.[6] This proliferation of newly coined

5 Robert McChesney, *Digital Disconnect: How Capitalism Is Turning the Internet Against Democracy*, New York: New Press, 2013.

6 Daniel Bell, *The Coming of Post-Industrial Society*, New York: Basic Books, 1999; Michael H. Goldhaber, "The Attention Economy and the Net," *First Monday*, April 1997; Dan Schiller, *Digital Capitalism: Networking the Global Market System*, Cambridge, MA: MIT Press, 2000; David Rooney, Greg Hearn, and Abraham Ninan, eds., *Handbook on the Knowledge Economy*, Cheltenham: Edward Elgar, 2005; Yann Moulier Boutang, *Le capitalisme cognitif: La nouvelle grande transformation*, Paris: Éditions Amsterdam, 2007; Jodi Dean, "Communicative Capitalism: Circulation and the Foreclosure of Politics," *Cultural Politics* 1, no. 1, March 2005, pp. 51–74; Yves Citton and Jonathan Crary, eds., *L'Économie de l'attention: Nouvel horizon du capitalisme?*, Paris: La Découverte, 2014; Georg Franck, "Capitalisme mental," *Multitudes* 54, no. 3, 2013, pp. 199–213; Maxime

terms might itself be a sign of a fundamental change in capitalism, even if it also risks fueling confusion about what is happening.

Among these terms, "digital capitalism" is the one that comes closest to our perspective. Still, in our view it is short of analytical sharpness, because it fails to identify the real heart of contemporary capitalism. Does it refer to the transition from industrial society to a "networked society" based on information technology?[7] Or are we talking more about the emergence of new Internet-related business strategies?[8] Is this new version of capitalism based on the predominant role of knowledge and "immaterial labor" in economic life, or is it based on the extension of market logic, enclosures, and privatization strategies into the digital world?[9] Should we understand digital capitalism as the result of the IT revolution that began in the 1970s, evolving in parallel with financialized, neoliberal, and globalized capitalism? Or is it a much more recent phenomenon, which emerged at the dawn of the twenty-first century, representing a new phase of capitalism beyond neoliberalism? This book's intention is not to add a new label, adjective, or buzzword to the pantheon of theories about capitalism, only for the sake of it. The key aim is to analyze the technical, social, and economic reconfigurations of a system that continues to revolutionize the instruments of production and social relations. In our view, the term "digital capitalism" already seems outdated in light of recent developments in AI and Big Data. Today, there is a reconfiguration of the fundamental social relationship that governs the accumulation of capital, and it is being reconfigured precisely around algorithmic technologies. This change requires us to adapt our critical concepts accordingly.

While the literature has produced a wide range of terms that emphasize different aspects of contemporary capitalism, the concept of "algorithmic capital" has several advantages in terms of interpreting the logic of the new regime of capitalist accumulation. For one, it allows us to place labor and production back at the center of the analysis, something

---

Ouellet, *La révolution culturelle du capital: Le capitalisme cybernétique dans la société globale de l'information*, Montreal: Écosociété, 2016; Nick Srnicek, *Platform Capitalism*, Cambridge: Polity, 2016; Shoshana Zuboff, *The Age of Surveillance Capitalism: The Fight for a Human Future at the New Frontier of Power*, London: Profile Books, 2019.

7 Manuel Castells, *The Rise of the Network Society*, Malden, MA: Wiley-Blackwell, 2012.

8 Michael E. Porter, "Strategy and the Internet," *Harvard Business Review*, March 1, 2001.

9 Antonio Negri and Carlo Vercellone, "Le rapport capital/travail dans le capitalisme cognitif," *Multitudes* 32, no. 1, 2008, pp. 39–50.

that several influential analyses neglect. It is also useful for identifying what, in our view, constitutes the real heart of the logic of accumulation in capitalism today and the social relation unfolding therein: the algorithm. Indeed, the algorithm is the structuring principle of the new capitalist accumulation regime, which builds on, rearticulates, and goes beyond financialized neoliberalism. In the following theses, we will analyze several trends that illustrate this. First, the algorithm is becoming the dominant mechanism for the allocation of digital labor.[10] Second, algorithmic accumulation is becoming the dominant mechanism for governing the production process. Third, algorithms and the data associated with them are becoming a central object of competition between capitalist enterprises. Fourth, algorithms are now mediating social relations, particularly through social networks and digital platforms. Fifth, algorithms now mediate access to information, culture, and collective memory. Sixth, algorithms generate revenue by participating in the production of goods and services, or as a mechanism for extracting rent. Seventh, even beyond the realm of private companies, many organizations and social spheres, ranging from public authorities, police services, and military complexes to development NGOs, and across the fields of healthcare, education, transport, and public infrastructure, use algorithmic technologies to exercise their power and/or carry out their activities. This is why algorithms—much more than "the digital," "surveillance," or anything else—are at the heart of the new regime of accumulation: They mediate social relations, preside over socioeconomic (re)production, and spread their predictive logic throughout contemporary society.[11]

Thus, while the term "digital capitalism" can be used broadly to describe the many ways in which digital technologies influence capitalism and vice versa, this categorization remains too generic and abstract to fully describe the mechanisms and transformations at work in twenty-first-century societies. More specifically, the hypothesis of algorithmic capitalism suggests that there is a *break* between two phases

10 "Digital labor" is also reflected in the literal sense of the "work with the fingers" done by platform users. See our full discussion of the term in Thesis 5.

11 The term "algorithmic capitalism" is mentioned here and there in the literature. However, it has not yet been conceptually developed. See James H. Mittelman, "The Power of Algorithmic Capitalism," *International Critical Thought*, June 2022, pp. 1–22; Haytham Karar, "Algorithmic Capitalism and the Digital Divide in Sub-Saharan Africa," *Journal of Developing Societies* 35, no. 4, 2019, pp. 514–37; Michael A. Peters, "Algorithmic Capitalism in the Epoch of Digital Reason," *Fast Capitalism* 14, no. 1, 2017, pp. 65–74; Luciana Parisi, "Instrumental Reason, Algorithmic Capitalism, and the Incomputable," in Matteo Pasquinelli, ed., *Alleys of Your Mind: Augmented Intelligence and Its Traumas*, Lüneburg: Meson Press, 2015, pp. 125–38.

or moments of digital capitalism. While the period beginning in the late 1970s corresponds to the emergence of neoliberal, financialized, and globalized capitalism, with the arrival of new technologies such as the personal computer and mass Internet, a series of transformations in the first decade of the twenty-first century produced a tipping point. Social media, smartphones, the platform economy, Big Data, and the machine learning revolution converged with the global financial crisis of 2007–8 to accelerate the transition toward a reconfiguration of capitalism in which algorithmic logic plays a decisive role. We will return to this in detail in Thesis 8.

To avoid an overly mechanical and reductive conception of algorithmic capitalism, it is important to establish where our perspective sits within the abundant literature of critical theories of contemporary capitalism. There are a variety of positions and debates surrounding the complex relationships between algorithmic technologies and the capitalist system. To briefly map this field, we can distinguish three main areas of debate: AI as a tool of capital; the nature of capitalism; and the possibilities of liberation.

Is AI a tool of capital?

We should start by asking to what degree algorithms really are tools subject to a capitalist logic. We can first consider an uncritical approach holding that algorithms have their own reality, functions, and potential uses. On this reading, it would be possible to analyze and develop algorithms without reference to the capitalist mode of production and the specific imperatives of accumulation, modes of exploitation, and constraints that it imposes. This is indeed the dominant view of algorithms, widely shared across scholarly, industrial, and media circles. It encompasses mainstream theories of AI, AI ethics, and techno-optimistic positions that contend that Big Data, platforms, and machine learning can create a more efficient, freer, and more prosperous world. The techno-optimistic bestsellers of the last decade exemplify perspectives that take capitalist social and institutional relations as a given in any discussion of the economics of algorithms.[12] While the authors are

12 Viktor Mayer-Schönberger and Kenneth Cukier, *Big Data: A Revolution That Will Transform How We Live, Work, and Think*, Boston: Mariner Books/Houghton Mifflin Harcourt, 2014; Erik Brynjolfsson and Andrew McAfee, *The Second Machine Age: Work, Progress, and Prosperity in a Time of Brilliant Technologies*, New York: W. W. Norton, 2014.

willing to acknowledge the risks inherent in these technologies, the ultimate goal is to "harness" the algorithmic revolution so that it serves the common good or even allows capitalism to be reinvented in the age of Big Data.[13]

In contrast to these uncritical (and often procapitalist) positions, critical theories zero in on the fallacy of the idea that technology is independent, and instead see algorithms as a new tool in service of capital. The question of whether there are several possible uses for Big Data and AI is a matter of debate. But all critical approaches to capitalism share the observation that Big Tech companies largely control the development of these technologies. It is they who own the intellectual property rights, the research budgets, and the data centers, as well as the working time of scientists and of the lobbyists who pressure governments to prevent legislation harmful to their interests.[14]

In this book, we argue that algorithmic technologies are primarily or mainly developed as tools for capital accumulation, while also being used in a variety of spheres of social life with consequences that go far beyond this economic purpose. Big Data and algorithms cannot be entirely reduced to their profit-seeking function.[15] And yet they are appropriated, shaped, and developed according to a logic that responds to the imperatives of the capitalist mode of production, and they in turn contribute to changing this system.

We believe, as do others, that fears associated with the dangers of AI, robots, and algorithms are often rooted in anticipated threats related to the functioning of capitalism. As sci-fi writer Ted Chiang notes: "Most of our fears or anxieties about technology are best understood as fears or anxiety about how capitalism will use technology against us. And technology and capitalism have been so closely intertwined that it's hard to distinguish the two."[16] For example, some people fear that AI will one day dominate us, escape our grip, or even replace us. However, behind the doomsday scenario of an all-powerful Terminator-style AI,

---

13 Viktor Mayer-Schönberger and Thomas Ramge, *Reinventing Capitalism in the Age of Big Data*, New York: Basic Books, 2018.

14 See, for example, Nick Dyer-Witheford, Atle Mikkola Kjøsen, and James Steinhoff, *Inhuman Power: Artificial Intelligence and the Future of Capitalism*, London: Pluto Press, 2019.

15 "Big Data" refers to extremely large datasets that exceed the processing capacity of traditional data processing software. We discuss this technical term in more detail in Thesis 7.

16 Cited by Ezra Klein, "The Imminent Danger of A.I. Is One We're Not Talking About," *New York Times*, February 26, 2023.

or Bostrom's "superintelligence" posing an existential risk to humanity, lies the *current* domination of capitalist companies that extract our data, train their algorithms, and guide our behavior with their countless "intelligent machines." The problem is not AI itself, but the capitalist social relations that underpin it and largely determine its trajectory. This is compounded by another problem: Algorithmic capital is itself intertwined with other social relations such as patriarchy, colonialism, racism, and state domination. Some will say that AI (or some of its applications) must be abolished, and algorithmic machines sabotaged, in an outright rejection of technology. However, such reactions only scratch the surface of the problem.

Today, we face a dynamic similar to the one Marx observed during the Industrial Revolution, when machines were starting to replace human labor. The "Fourth Industrial Revolution" hailed by Klaus Schwab is reviving the dynamics of industrialization, now with the help of algorithmic machines and automated decision-making systems. The neo-Luddite reaction, which aims to destroy these machines, is quite normal and predictable. And yet it is important to realize that it is pointless to sit in fear of ChatGPT, robots, or algorithms. We need, instead, to question and change the whole set of social relations that produce and reproduce the domination that is today partly expressed through the power of algorithms. As Marx points out:

> The struggle between the capitalist and the wage-labourer starts with the existence of the capital-relation itself. It rages throughout the period of manufacture. But only since the introduction of machinery has the worker fought against the instrument of labour itself, capital's material mode of existence. He is in revolt against this particular form of the means of production because it is the material foundation of the capitalist mode of production ... It took both time and experience before the workers learnt to distinguish between machinery and its employment by capital, and therefore to transfer their attacks from the material instruments of production to the form of society which utilizes those instruments.[17]

But let us add here that the Luddite movement discussed by Marx was complex and had a subtle understanding of exploitation and social relations, and in the age of Big Tech we have many lessons to learn from

17 Karl Marx, *Capital,* Vol. I, London: Penguin, 1976, pp. 554–5.

their uprisings.[18] Moreover, we still need to develop an adequate analysis of this "form of society which utilizes those instruments," the better to combat it and to replace it with other, more egalitarian and democratic social relations.

But what is capitalism?

What exactly do we mean by capitalism? Is it simply a market economy where individuals come together to exchange goods and services? Is it better defined as an economic system based on private ownership of the means of production and the exploitation of labor by capitalists? Or is it a dynamic of accumulation based on the abstract logic of the self-valorization of value, which tends to transform capital into an "automated subject"? Any analysis of the exact role of algorithms is bound to vary greatly depending on what perspective we take on the nature of capitalism.

In readings that take for granted the institutions of capitalism, such as private enterprise and the market, analyses of the use of data and algorithms will generally see them through the prism of the potential ways in which they can be applied to improve business performance. Algorithms will be understood as "predictive machines" capable of performing various tasks: reducing production costs, improving work processes, optimizing value chains, enhancing customer experience, guiding strategic decisions, and so on.[19] Just as Marx, in striving to better understand capitalism, assiduously read the theorists of bourgeois political economy, here we will at times draw on the perspectives of the "bourgeois" technological economics of the last couple of decades to investigate the new and ingenious use of algorithmic technologies in the world of business and digital platforms.

Those adopting a critical perspective on capitalism, and who follow in the footsteps of classical Marxism, will tend to highlight the dynamics of private appropriation of new technologies, the relations of exploitation, and resistance from the oppressed classes. In this book, we will see that algorithmic capitalism is largely based on the predominant power of web giants and their exploitation of labor and human experience. We

18 Brian Merchant, *Blood in the Machine: The Origins of the Rebellion Against Big Tech*, New York: Little, Brown, 2023.

19 Ajay Agrawal, Joshua Gans, and Avi Goldfarb, *Prediction Machines: The Simple Economics of Artificial Intelligence*, Boston: Harvard Business Review Press, 2018.

will see that it fuels various tensions, inequalities, and contradictions in the social world, and that numerous fronts of resistance are emerging to combat its pernicious effects, such as movements for the protection of privacy, mobilizations against Uber and Airbnb, the sabotage of surveillance devices, and Data 4 Black Lives.

Some approaches, not necessarily Marxist in orientation, but closer to various forms of Internet activism and "hacktivism," also help us understand the digital world as a space of conflict and struggle. For example, in his book *Les batailles d'Internet*, Philippe de Grosbois emphasizes that the development of the Internet did not occur in a homogeneous or linear manner. Rather, he explains, it is the result of numerous actors, origins, and contradictory forces: cybernetic logic, the military system, the academic community, computer hackers, hippies, Silicon Valley entrepreneurs. This conception of the digital world helps us see it as a battlefield.[20]

That said, analyses focused solely on the "class struggle" and approaches inspired by classical Marxism tend to reduce digital capitalism to a clash between web giants (capitalists) and "users" or "click workers" who are exploited and/or have their data stolen. Accounts from more activist, digital commons, and hacktivist perspectives might also overlook the need for a careful conceptual analysis of the workings of capitalist *accumulation* and the *social relations* that structure capitalist society as a whole. This, too, needs closer examination.

This is what a third critical approach, inspired by Marxian critiques of value, seeks to do: to adopt a holistic view that considers capital as "a totality that takes the form of a quasi-objective structure of domination … a situation of social heteronomy that takes the form of a blind, uncontrolled and unthinking mode of development over which societies have no political control."[21] In such contributions, the analysis of capitalism focuses on the central categories of capitalist society, such as abstract labor, commodities, and value. The critique of capitalism as an "alienated social totality" thus implies a radical critique of value and of "fetishized social mediations," of the "communicational imaginary of cybernetic capitalism," and of the "cultural revolution [carried forth by] the Californian ideology," said to be leading us toward a "digitally

20 Philippe de Grosbois, *Les batailles d'Internet: Assauts et résistances à l'ère du capitalisme numérique*, Montreal: Écosociété, 2018, p. 30. See also the work of activist and writer Cory Doctorow, notably *The Internet Con: How to Seize the Means of Computation*, London: Verso, 2023.

21 Ouellet, *La révolution culturelle du capital*, p. 38.

administered world."[22] However, despite their theoretical prowess, these totalizing approaches often tend to sidestep the conflicts, resistances, and possibilities for emancipation that are also present in technical and economic development. This is something we will examine in detail in this book.

These two critical Marxist approaches—the one based on class conflict and the other on the totalizing logic of capital—thus each have their strengths and limits. Activist/hacktivist sociological approaches also make valuable contributions to understanding the digital world as a space of conflict but often lack theoretical depth. In this book, we will also analyze the multiple dimensions of algorithmic capital, which tends to colonize all spheres of human existence (economic, social, political, and cultural), showing how the dynamics of accumulation based on data extraction and the accelerated development of AI combine with the deployment of algorithmic power, with numerous ramifications. Capitalism cannot be reduced to a power struggle between antagonistic classes: It represents a *social totality*. However, we must also avoid the fatalistic trap of overarching and reifying approaches and analyze in detail the power relations and conflicts that are unfolding in the wake of the rise of algorithmic capital. A critical theory of algorithms must, therefore, articulate the two distinct but complementary logics of capital: "the *logic of capital* as a complete system" and "the *strategic logic of confrontation*."[23]

What are the possibilities of emancipation?

The different degrees of emphasis placed either on the logic of capital (as a totalizing dynamic) or on the strategy of confrontation (thus shining a light on the agency of various actors) imply different assessments of the possibilities for emancipation under algorithmic capitalism. Approaches centered on the class struggle and social conflict sometimes argue that the impact of AI is greatly exaggerated by capitalists, who aim to frighten the labor movement about the imminent dangers posed by automation. Astra Taylor, for example, has argued that doomsday narratives about the arrival of robots and mass technological unemployment are deliberate attempts to intimidate and discipline workers by wielding the ideological discourse of what she calls "*fauxtomation*."[24] There is no doubt

22 Ibid., p. 60.
23 Pierre Dardot and Christian Laval, *Marx, prénom: Karl*, Paris: Gallimard, 2012.
24 Astra Taylor, "The Automation Charade," *Logic*, no. 5, 2018.

that capitalists do use such rhetoric in this way. However, it would be a mistake to downplay the impact of AI on the work process, jobs, and entire sectors of the economy. Even if estimates of the unemployment due to be caused by automation vary widely (from 9 to 50 percent, depending on the methodologies used), we should recognize that algorithmic capital does have a tendency to replace humans with robots, or at least to supplement, monitor, and control them with sophisticated new algorithmic technologies. If it could achieve its logic in total form, algorithmic capital would aim to develop a fully automated form of capitalism.

At the opposite end of the spectrum, authors of the "accelerationist" current school of thought view the prospect of widespread automation in a generally positive light, provided it is combined with the socialization of the means of production.[25] Algorithmic technologies would thereby lay the foundations for a postcapitalist economy beyond labor.[26] Other movements, such as xenofeminism and theories of "cognitive capitalism," which we will discuss in Thesis 8, also emphasize the emancipatory potential of immaterial labor, collaborative practices, human/machine hybridization, and new technologies.[27] Following Lenin, who said that "Communism is Soviet power plus electrification," we could say that today techno-optimistic anticapitalist currents consider cybercommunism to mean "universal basic income plus artificial intelligence."[28] This is reminiscent of Trotsky's position on Taylorism: "bad in its capitalist use and good in its socialist use." The most developed vision of this accelerationist socialism can be found in Aaron Bastani and his vision of fully automated luxury communism, which proposes the utopia of a post-scarcity economy through the full development of algorithmic productive forces released from the constraints of capitalist logic. In this perspective, algorithmic capital is said to be digging its own grave, and this holds out the possibility of a fully automated world based on mining asteroids, producing synthetic food, and drastically reducing human labor.[29] This vision brings together techno-optimism

25 Nick Srnicek and Alex Williams, *Inventing the Future: Postcapitalism and a World Without Work*, London: Verso, 2015.

26 Paul Mason, *Postcapitalism: A Guide to Our Future*, London: Allen Lane, 2015.

27 Helen Hester, *Xenofeminism*, Cambridge, MA: Polity Press, 2018; Laboria Cuboniks (Collective), *The Xenofeminist Manifesto: A Politics for Alienation*, London: Verso, 2018; Michael Hardt and Antonio Negri, *Empire*, Cambridge, MA: Harvard University Press, 2000; Yann Moulier Boutang, *Le capitalisme cognitif: La nouvelle grande transformation*, Paris: Éditions Amsterdam, 2007.

28 Dyer-Witheford, Kjøsen, and Steinhoff, *Inhuman Power*, p. 6.

29 Aaron Bastani, *Fully Automated Luxury Communism*, London: Verso, 2020.

and postcapitalist ideals, as if Jeff Bezos or Elon Musk was suddenly donning a socialist cap.

That said, it seems rather naïve to believe that accelerated algorithmic automation is bad under Big Tech, but good under socialism. As we shall see, far from laying the foundations for a society beyond poverty and inequality, algorithmic capital automates various forms of domination, contributes to the overexploitation of natural resources, and drains our attention through its many addictive technologies. In other words, the ever-increasing extraction of data and the current development of algorithms are helping to lay the foundations for a fully automated *capitalism*, accelerating the increasing scarcity of time and natural resources. Contrary to the techno-optimistic dream of accelerationists, we are instead heading toward a fully automated scarcity capitalism.

Finally, it would be misleading to limit ourselves to the logic of capital while ignoring strategies of confrontation that could open up possibilities for emancipation beyond the horizon of automated capitalism. Unlike technoskeptic currents of thinking, which reject algorithmic technologies outright, at the end of the book we will discuss perspectives for resistance and anticapitalist strategies aimed at a potential collective reappropriation of data and algorithms. Far from locking ourselves into a teleological view of the progress of history, we will address the possibilities and limitations of algorithmic planning from the perspective of ecological self-limitation, social justice, and radical democracy. To avoid the pitfalls of both technophile and technophobic stances, we propose a critical *technosober* approach that allows us to analyze the mechanisms of algorithmic capital and identify courses of action that can help guide today's social struggles toward a world beyond algorithmic capital. We will continue this discussion in greater detail in Thesis 19.

Algorithmic capitalism as an institutionalized social order

What then is the added value, the particular angle, the originality of a book that claims to offer a new theoretical contribution to the understanding of the great issues of our time? We should say from the outset that we do not claim to be reinventing the wheel. Rather, this book adds but one brick to the edifice of a new critical theory of contemporary capitalism. As such, our research begins with a comparative reading of three major intellectuals who provide essential insights into the socioeconomic dynamics of our time: Nancy Fraser, Shoshana Zuboff, and Kate Crawford.

In recent years, Nancy Fraser has developed a comprehensive theory of capitalism. She grapples not only with a specific mode of economic production, but also with its relationship with nature, politics, and the labor of social reproduction on which the dynamics of capitalist accumulation are based.[30] This reconceptualization makes it possible to grasp the ways in which capitalism is intertwined with gender inequalities and the care crisis, the dynamics of expropriation and exploitation within racialized capitalism, and the complex dynamics that fuel the ecological crisis, which exacerbates racial, economic, and environmental inequalities. This broad conceptual framework, inspired by Marxist feminism, allows us to grasp the entanglement of multiple systems of domination (capitalism, patriarchy, racism) and to shed light on the "frontier struggles" present in different corners of the social world. We thus consider Fraser's critical perspective the best starting point for painting a systematic portrait of capitalist society, which will then allow us to appreciate how algorithmic technologies are reshaping it.

To this end, it is important that we define the contours of capitalism as a whole, as well as the different spheres that make up this system. For us, capitalism entails an *economic system* in which labor, knowledge, the control of attention, data extraction, and algorithms are used to *accumulate capital*. Most current analyses of the digital economy address specific and partial dimensions of this system. As a result, these analyses remain insufficient, and this gap itself calls for a general theory. On the economic level, it is necessary not only to address the emergence of platforms, surveillance, AI, digital labor, and the rise of new data markets, but also to integrate these elements into a dynamic analysis of the multiple moments and metamorphoses of value in the spheres of production and circulation. Algorithmic capital is thus a new way of producing, exchanging, and accumulating value through massive data extraction, the exploitation of digital labor, and the accelerated development of algorithmic machines.

In orthodox Marxism, capitalism is often represented as a mode of production composed of a base and a superstructure. The base includes the conditions of production (natural resources and transport and

30 Nancy Fraser, "Behind Marx's Hidden Abode," *New Left Review*, no. 86, March/April 2014, pp. 55–72; "Expropriation and Exploitation in Racialized Capitalism: A Reply to Michael Dawson," *Critical Historical Studies* 3, no. 1, 2016; "Contradictions of Capital and Care," *Dissent* 63, no. 4, 2016, pp. 30–7; Fraser and Jaeggi, *Capitalism*; "Climates of Capital," *New Left Review*, no. 127, January/February 2021, pp. 94–127; *Cannibal Capitalism: How Our System Is Devouring Democracy, Care, and the Planet—and What We Can Do About It*, London: Verso, 2022.

communication systems), the productive forces (labor, tools, machinery), and the relations of production (property rights, wage labor). Meanwhile, the superstructure includes the political, cultural, and ideological dimensions of a society: its laws, science, religion, morality, art, and so on. Conceptualizing things in this schematic way has the advantage of integrating different spheres of social life into one overall vision. However, this model has the disadvantage of sometimes giving excessive weight to the economic "base." So-called "materialist" analysis too often translates into a mechanistic or economistic conception of society. Even approaches that recognize a "dialectical relationship" between base and superstructure or the "relative autonomy" of different spheres often end up maintaining, explicitly or implicitly, that the social and ideological dimensions are *determined in the last instance* by the economic structure.[31]

In contrast to this orthodox approach, Nancy Fraser proposes viewing capitalism as an institutionalized social order. What does this mean? Fraser first defines capitalist production in terms of four fundamental characteristics: (1) private ownership of the means of production; (2) a labor market; (3) the dynamics of accumulation; (4) market allocation of factors of production and the social surplus. In this book, we will see how algorithmic capital is reshaping business models (with the hegemony of digital platforms), the forms of work (with digital labor and algorithmic labor), the dynamics of accumulation (through data extraction and algorithms), and the emergence of new markets (predictive product markets). However, this summary characterization of the capitalist economic system remains far short of what we need.

According to Fraser, economic production ultimately relies on the exploitation of three "noneconomic" spheres: nature, socially reproductive labor, and political power. These represent its *conditions of possibility*. So, rather than focus solely on the *capitalist economy*, we will here examine different spheres of *capitalist society* in the age of algorithms. Such a society is based on divisions between society and nature, production and reproduction, economy and politics, center and periphery, Global North and South, and others. A theory of capitalism as an institutionalized social order must, therefore, include an understanding of the complex relationships between commodity production, the environment, the domestic sphere, public institutions,

31 This is an expression coined by Louis Althusser, Étienne Balibar et al., *Reading Capital*, London: Verso Books, 2016 [1965].

and geopolitical dynamics. It allows us to articulate *exploitation* and *dispossession* or *extraction* within a broad critical-theoretical framework, which will prove crucial to our analysis of the imbrication of data extraction and exploitation of digital labor in the accumulation of algorithmic capital.

One advantage of Fraser's perspective is that it broadens the focus from solely the economic contradiction between capital and labor and highlights as well the multiple trends toward crisis in capitalist society, including on the social, democratic, and ecological levels. She refers to "the built-in tendency of a capitalist economy to destabilize its own 'non-economic' background conditions of possibility: social reproduction, public power, sustainable natures, and a habitable planet."[32] Indeed, the expansionist dynamic of capital feeds the overexploitation of natural resources and the production of greenhouse gases, disrupts practices of social reproduction (mostly carried out by women), increases political instability by intensifying inequalities, and reinforces North/South injustices through the dynamics of externalization, colonialism, and accumulation by dispossession. For Fraser, we face "not just a sectoral crisis, but one that encompasses every major aspect of our social order. So the problem isn't simply 'economic' but concerns problems from climate warming to deficient healthcare and the hollowing out of public authority."[33]

As we shall see, algorithmic capital seeks to resolve its multiple crises and contradictions through the intensive use of data and artificial intelligence. This may involve technological innovations to accelerate the ecological transition (energy efficiency, agriculture 4.0, self-driving cars, robots to clean the oceans), aid the tasks of domestic and social reproduction (smart homes, domestic service platforms, care robots to look after the elderly and children), or resolve crises of governance (through smart cities, the optimization of public administration, predictive policing, and surveillance systems), and a range of others. It is no accident, then, that "technological solutionism" and "dataism"[34] today present

32 Fraser and Jaeggi, *Capitalism*, p. 179.

33 Ibid., p. 3.

34 The term "dataism" was first used by David Brooks to describe an emerging philosophy according to which data analysis is the preferred key to grasping reality, enabling us to assess the past, present, and future. The term was then popularized by Yuval Noah Harari, who defines it as the "religion of data"—a technocentric worldview which is now overtaking humanism. See David Brooks, "The Philosophy of Data," *New York Times*, February 4, 2013, and Yuval Noah Harari, *Homo Deus: A Brief History of Tomorrow*, London: Vintage, 2017.

the creation of algorithms and the judicious use of data as the magical remedy for the countless challenges of our time.

Still, even in adopting a perspective that sees capitalism as an institutionalized social order, we should recognize that Fraser's theory also has a major blind spot: the question of technology. For us, the technological infrastructure represents another *condition of possibility* for capital, which Fraser does not fully theorize. Without transport and communication systems, without machines and technical tools, it seems difficult to imagine capitalism functioning, even in its agrarian, mercantile, and preindustrial variants. In other words, in her overall theory of capitalism Fraser does not centrally address the *general conditions of production*, including the digital and algorithmic technologies that are contributing to its metamorphosis today. She analyses the historical evolution of capitalism through its different stages or regimes of accumulation, up to the stage of neoliberal, financialized, and globalized capitalism, a periodization we think is in need of an update. To fill these gaps, we believe it is essential to complement Fraser's rich perspective by drawing on the crucial contributions of Shoshana Zuboff and Kate Crawford. These two authors take a penetrating look at the technological dimensions of the contemporary economy, which increasingly relies on surveillance and the extractive AI industry.

Surveillance capitalism as a global extractive industry

Shoshana Zuboff's book *The Age of Surveillance Capitalism* offers a striking analysis of the changes in the digital economy, which has seen the emergence of new business models based on the extraction of personal data and the development of algorithms focused on predicting future behavior.[35] While the technological development of computers and the Internet could have led to various socioeconomic configurations, a particular form of capitalism took hold in the early 2000s and quickly became hegemonic both in the economy and in multiple spheres of our lives. Here is how Zuboff defines this new face of contemporary capitalism:

35 Zuboff, *The Age of Surveillance Capitalism*.

> Surveillance capitalism unilaterally claims human experience as free raw material for translation into behavioral data. Although some of these data are applied to product or service improvement, the rest are declared as a proprietary *behavioral surplus*, fed into advanced manufacturing processes known as "machine intelligence," and fabricated into *prediction products* that anticipate what you will do now, soon, and later. Finally, these prediction products are traded in a new kind of marketplace for behavioral predictions that I call *behavioral futures markets*. Surveillance capitalists have grown immensely wealthy from these trading operations, for many companies are eager to lay bets on our future behavior.[36]

Zuboff provides a real gateway to understanding the fundamental transformation of capital at the dawn of the current century: The extraction and exploitation of data through algorithms is now at the heart of the process of capital accumulation. She also makes a second major contribution. According to her, surveillance capital cannot be reduced to a simple economic process. Rather, it gives rise to the emergence of a new and disturbing form of instrumentarian power, meaning: "*the instrumentation and instrumentalization of behavior for the purposes of modification, prediction, monetization, and control*."[37] Thus, the rolling out of algorithmic machines does not only serve to extract value through targeted advertising that anticipates our future behavior. Rather, these predictive machines also aim to influence, control, nudge,[38] and manipulate our daily conduct.

If this configuration of capitalism has such a hold over social relations and our way of relating to the world, it is because this new logic of power has spread across the different spheres of social, economic, and political life. This conception of surveillance capital as a sui generis form of power implies abandoning a narrowly economic vision in favor of a critical social analysis that pays attention to the emergence of new relationships of domination. In this book, we will return to the critical analysis of instrumentarian power suggested by Zuboff. We will show

36 Ibid., p. 25 (original emphasis).

37 Ibid., p. 472 (original emphasis).

38 The idea of the "nudge" is used in behavioral economics to describe mechanisms that make it possible to shape and change individuals' behavior without resorting to constraint or compulsion. As one account puts it, a nudge "is any aspect of the choice architecture that alters people's behavior in a predictable way without forbidding any options or significantly changing their economic incentives. To count as a mere nudge, the intervention must be easy and cheap to avoid." Richard H. Thaler and Cass R. Sunstein, *Nudge: Improving Decisions About Health, Wealth, and Happiness*, London: Penguin, 2008, p. 2.

the strengths but also the pitfalls of this approach, which seems to lack at times the subtlety required to dissect, in a more precise manner, the multiple ramifications of algorithmic power.

A further limitation of Zuboff's approach is that it places such great emphasis—perhaps too much—on the issue of surveillance and the appropriation of personal data. Her theory has the merit of dissecting the new logic of capital accumulation and the unprecedented mechanisms of algorithmic control of individuals. But it ultimately leads us to the "right to future time," the "right to sanctuary," and other demands that are mostly limited to the protection of privacy. Zuboff's theory does not really allow us to address the class, gender, and racial injustices that are amplified by algorithmic domination. She suggests a reframing of capitalism, highlighting the fundamental contradiction between data extraction and respect for privacy. A Marxian critique of the exploitation of labor by capital is thus replaced by a critique of the extraction/manipulation of individuals' private lives in their everyday existence. It is thus no coincidence that Zuboff's analytical framework ignores the role of labor, the extraction of natural resources, the state, and geopolitical issues. From our point of view, surveillance is just one of the many faces of algorithmic capital, and one that has much further-reaching consequences.[39]

This is why we believe it is essential to supplement this initial overview of Zuboff with a broader landscape of the many ramifications of AI in the contemporary world. We can get closer to this picture by way of Kate Crawford's rich and original approach. She proposes to go beyond a narrow analysis of AI that has by now become commonplace, which sees it as simple computing power—a tool such as Google's Gemini or Apple's personal assistant Siri, or even a deep learning computer process based on an artificial neural network that allows data to be classified for specific uses. Although these various descriptions are not untrue, they nevertheless conceal a complex infrastructure by focusing our attention on isolated, abstract, seemingly autonomous, rational, and disembodied technologies. For Crawford:

> Artificial intelligence is both embodied and material, made from natural resources, fuel, human labor, infrastructures, logistics, histories, and classifications. AI systems are not autonomous, rational, or able to discern anything without extensive, computationally intensive training with large datasets or predefined rules and rewards … And due to the capital

39 We come back to Zuboff's contribution in more critical detail in Thesis 4.

> required to build AI at scale and the ways of seeing that it optimizes[,] AI systems are ultimately designed to serve existing dominant interests. In this sense, artificial intelligence is a registry of power.[40]

For Crawford, algorithms cannot be properly understood from a strictly technical point of view that does not contextualize them within broader social structures and systems. Unlike Zuboff, who conceives of surveillance capital mainly in terms of the extraction of personal data from digital platforms and connected objects, Crawford takes a much broader view. She shows that AI relies on the widespread extraction of minerals, energy, human labor, data, emotions, and resources of our public institutions. This is why she proposes to conceive of AI from a materialist and holistic perspective, avoiding the pitfalls of an overly formal and idealistic vision, such as that often found in the field of AI ethics.

We will also draw on Crawford's conceptualization of AI as an extractive industry with countless tentacles reaching across the globe. Nevertheless, we believe that this vision still provides too fragmented and piecemeal an overview of this rapidly consolidating system. Crawford's rich cartography reveals the multiple faces, territories, and manifestations of this complex phenomenon. What it does not do is provide an analytical key, a conceptual framework, a historicization, or a comprehensive theory to explain and understand these different issues as a coherent whole. This descriptive and critical approach does indeed highlight the many relationships of domination, and it offers some illuminating historical insights. But there is still a lack of systematic analysis of the complex intertwining of technology, power, and capital.

The aim of this book is to contribute to collective efforts to construct a critical theory of algorithms and contemporary capitalism. It aims to offer a compass, a theoretical lens, to understand the articulation of these varied processes within a complex historical totality, one that is riddled with contradictions and logics of domination, but also resistance and possibilities for emancipation. It would be both unreasonable and pretentious to claim that we have finally arrived at *the* grand theory of capitalism, subsuming all the reflections, investigations, and theories of those who came before us and those who will critically engage with this book. Critical thinking does not magically emerge from one or two brains that have finally discovered the Holy Grail that will solve all our troubled world's mysteries. We will be happy simply to have made a new

40 Crawford, *Atlas of AI*, p. 8.

and relevant contribution to the efforts of all those who have together created the critical literature on this topic.

Toward a new research program

This section is mainly addressed to readers who wish to learn about the broader foundations of our research project, of which this book is a product. Based on a holistic and interdisciplinary perspective, the reflection in this book aims to shed light on the multiple dimensions and contradictions of algorithmic capitalism as a social totality. We will draw on the tools of philosophy, sociology, political science, history, political economy, cognitive science, and computer science to analyze different aspects of data, algorithms, and other technological innovations that are reconfiguring the social world today.

However, it is also worth understanding that critical theory does not merely describe the social world such as it is; it also seeks to uncover the logic of domination and propose courses of action to overcome it. As Marx points out, critical theory seeks, on the one hand, to shed light on the struggles waged by social movements; but on the other hand, it aims at action: It is "ruthless both in the sense of not being afraid of the results it arrives at and in the sense of being just as little afraid of conflict with the powers that be."[41] A critical theory of algorithmic capital must, therefore, build on various conceptual frameworks and empirical studies to paint a comprehensive picture of the social dynamics of capital in the twenty-first century and the paths to emancipation. Marxian-inspired theories surely have an influence here. Yet we also want to combine insights from multiple perspectives that take seriously the new technologies and their complex effects.

The development of a comprehensive, critical theory of algorithmic capitalism seeks to lay the foundations for a wider research program. What does this mean? First, we need to establish the core theoretical elements that will then be used to develop auxiliary, explanatory, and predictive hypotheses—ones that can be corroborated by empirical studies and our experience of the world. Unlike Karl Popper, for whom any statement of scientific theory must be falsifiable, the philosopher of science Imre Lakatos points out that, in a viable research program, the core hypotheses are often unfalsifiable in nature. This provides a useful

41 Marx to Ruge, September 1843; text from marxists.org.

basis for constructing research hypotheses that will lead to potential explanations, predictions, and discoveries.[42]

In the case that concerns us here, we propose a set of general statements of a sociological, epistemological, methodological, and praxeological nature. This will help to guide the construction of more specific auxiliary and empirical hypotheses, which will, of course, be falsifiable.

A first "added value" of this book—if we can resort to such commercial terminology—is that it offers an explicit and coherent formulation of the theoretical foundations of a research program. Several elements of such a program can already be found in the vast existing critical literature on capitalism and the social impacts of AI. Once we have a clearly defined theoretical core, it then becomes easier to integrate a multitude of works, studies, and explanations drawn from various disciplines of the social sciences. Here are the eleven basic hypotheses that form the core of our research program on algorithmic capitalism:

1. Capitalism is real.
2. Capitalism is an institutionalized social order.
3. A proper analysis of capitalist society requires a holistic approach.
4. Capitalism is an evolutionary process, which thus requires a historical treatment.
5. All technological development takes place within a sociohistorical context.
6. The factors that are driving the development of AI and its effects on the social world cannot be properly analyzed without due consideration of capitalism, and vice versa.
7. Artificial intelligence cannot be reduced to its specific algorithmic technologies; it is a complex infrastructure being deployed around the planet.
8. A critical theory of algorithms and AI must highlight the many relationships of domination that are embedded in social and technological phenomena.
9. A critical theory of algorithms and AI takes the side of those who most suffer its repercussions, and therefore such a theory implies a clear normative commitment to social justice.
10. A critical theory of algorithms and AI must shed light on the possibilities for liberation, prospects for action, and the social alternatives present in today's world.

42 Imre Lakatos, *Proofs and Refutations: The Logic of Mathematical Discovery*, Cambridge: Cambridge University Press, 1976.

11. An emancipatory theory requires the articulation of theory and praxis, and thus intervention aimed at changing the social world, which in turn informs the theory.

Let us now explain each of these statements in order to highlight why they are important to constructing a new research program.

1. It may seem to go without saying that capitalism is something real. Yet that is not quite the case. Several theories uncritically postulate the existence of a "market economy" composed of individuals, companies, and institutions and thus ignore core elements of this system such as the private ownership of the means of production, the exploitation of labor, and the dynamics of accumulation. Although there are several possible interpretations of capitalism (inspired by the works of Marx, Weber, Schumpeter, Veblen, Braudel, Wood, Piketty, Fraser, and others), we believe it is important first to postulate the existence of capitalism as a socioeconomic phenomenon.

2. We believe that in order to study capitalism, it is necessary to analyze not only the capitalist *economy*, but capitalist *society* as a whole, including its relationship with nature, politics, social reproduction, the lived world, and other elements. In short, we see capitalism as an institutionalized social order, a complex totality. Some authors use the term "mode of production," but we prefer to use Nancy Fraser's conceptual framework, which explicitly avoids the dualism of base and superstructure. Still, when we understand "mode of production" in a nonreductionist way—that is, in the sense of the "production" of a global social world—it seems to offer the best starting point for a systematic examination of capitalism. For us, capitalism is not a thing or a structure, but a multidimensional processual social world.

3. In methodological and epistemological terms, a study of capitalism as an institutionalized social order calls for a *holistic* approach. We do not believe it necessary to identify with *ontological* holism, which bases its approach to social reality on the well-known phrase "the whole is greater than the sum of its parts." However, it is necessary to adopt a form of *methodological* holism, in the sense that the explanation of a complex totality such as capitalism does not emerge simply from the isolated analysis of this or that part of this institutionalized social order. A methodological holism maintains that certain explanations of social

phenomena (collective entities, social processes, and emerging social properties) cannot be reduced to or interpreted solely in terms of individual factors linked to actors' mental states, beliefs, or desires.[43] This is not at all to say that detailed analysis of the parts and subregions of the social world is useless. Yet society cannot be studied satisfactorily without analyzing the complex relationship between the whole and its parts.

4. Since capitalism is a social and evolving phenomenon, it is necessary to study not only its historical origins, but also its many metamorphoses and stages of development throughout history. To study the simple logic of capitalism or its fundamental categories in the abstract would leave us with a partial and thus incomplete understanding of this phenomenon. The study of algorithmic capitalism must pay particular attention to recent historical transformations, which require consideration of both the continuities and discontinuities between capitalism's different phases.

5. All technological development is embedded in a sociohistorical context. This in turn means that the study of new (digital, algorithmic, or other) technologies cannot ignore the social world and the historical dynamics that give rise to, select, disseminate, and mobilize the latest technological discoveries for a variety of economic, political, and cultural reasons. The sociology of technology, and indeed STS (science, technology, and society) studies, are thus approaches useful to studying algorithmic capitalism.

6. The thesis that the factors influencing the development of AI are closely linked to the reconfigurations of capitalism, and that capitalism

43 There is a vast critical literature on the limitations of methodological individualism and its claim to explain the entire social world exclusively by way of individual factors. There are several possible positions regarding the legitimate types of explanations in the social sciences, and we support a moderate methodological holism. Julie Zahle distinguishes three positions in this debate: "[1] Strong methodological holism: Holist explanations alone should be offered within the social sciences; they are indispensable. Individualist explanations may, and should, be dispensed with. [2] Moderate methodological holism: In certain cases, holist explanations should be advanced; in other cases individualist explanations should be advanced; both holist and individualist explanations are indispensable within the social sciences. [3] Methodological individualism: Individualist explanations alone should be put forward within the social sciences; they are indispensable. Holist explanations may, and should, be dispensed with." Julie Zahle, "Methodological Holism in the Social Sciences," *Stanford Encyclopedia of Philosophy*, March 21, 2016.

is fundamentally influenced by new algorithmic technologies, lies at the center of our theorizing. It is impossible to study contemporary capitalism in a comprehensive and satisfactory manner without analyzing algorithms, and vice versa. More generally, this statement is based on the thesis of the intertwining of social, technological, and economic phenomena.

7. As Crawford points out, AI consists not only of specific technologies, but also social practices, norms, discourses, institutions, and a gigantic industrial infrastructure with global ramifications. While it is possible to isolate each dimension analytically in order to study a particular component of this system more closely, AI is in reality a highly complex apparatus, distinct from capital and society, yet constantly interacting with them.

8. A simple description of AI and its ramifications would be incomplete without a critical theory to shed light on the relations of domination embedded in this sociotechnological phenomenon. These relations are manifold and cannot be reduced to a few specific instances of discrimination against disadvantaged groups. That is why it is essential to identify the relationships of exploitation, oppression, alienation, reification, control, contempt, invisibilization, and instrumentalization, as well as the social pathologies that are fueled by the intersection of algorithmic technologies and capital. This implies the need for an epistemic mistrust or methodological suspicion. This attitude of suspicion aims to reveal mechanisms, practices, and forms of power that fly under the radar of traditional theory.

9. As well as identifying, describing, and explaining the relations of power and domination, a critical theory of AI also involves a normative assessment. This means a practical commitment in favor of those most severely affected by these technologies, and thus a concern for social suffering and an ethical-political stand for social justice.

10. We have spoken of the descriptive and normative dimensions of theory. Yet a critical perspective on AI cannot settle for an analysis of the problems that algorithms and capitalism cause or stop at an impotent moral condemnation. The theory must also be capable of identifying, within today's social world, the existence of levers for transformation, prospects for action, and potential social alternatives. These are the

"real possibilities" referred to by Herbert Marcuse that make it possible to identify tools that social movements can use in their struggle for liberation.

11. Finally, an emancipatory theory requires the articulation of theory and practice, and thus intervention aimed at changing the social world. As Max Horkheimer points out, if critical theory represents "the intellectual side of the historical process of emancipation," it also enters into a broader process of transformation—a moment of collective action that allows the theory to be given a certain reflexivity. Understanding the social world is not an end in itself, but a means of creating a just and decent world, conducive to human flourishing and the preservation of the conditions for life on Earth. Ultimately, theory cannot just be theory; it must be linked to a practice, which it also feeds in a dynamic way. Too many commentators on AI have hitherto only interpreted the technological world in their various ways; the point is to radically change it, in order to create the social conditions for emancipation and the good life.

Thesis 3

Algorithms accelerate time and keep us glued to our screens

> *What an advantage it will be to … choose some example by which we may guide our lives! This cannot be done without leisure.*
>
> —Seneca

The search for a proper Archimedean point from which to observe the new reality of capitalism poses a real intellectual challenge.[1] As we will see in the theses that follow, the concept of "algorithmic capital" is elaborated at a high level of abstraction. However, the advent of algorithmic capital as a historical reality has already brought major upheavals in our lived experience and everyday existence: in our work, in our free time, in our daily tasks, as well as in our relations with others and the world.

We believe that one compelling starting point for analyzing these new realities is social time—or, more specifically, the inscription of social time in our lived experience. This allows us to conceive the "presence" of algorithmic capital as something intimate and close to us—part of our experience of time itself—before we then elaborate the concept at a more abstract level in the theses that follow. In other words, in order to begin to shed light on the reality of algorithmic capital, we should first ask: What do we do with our time? How do we spend it, and what socio-economic dynamics exert structural pressure on our use of time? In this

1 This thesis is a condensed and revised adaptation of Jonathan Martineau and Jonathan Durand Folco, "Paradoxe de l'accélération des rythmes de vie et capitalisme contemporain: Les catégories sociales de temps à l'ère des technologies algorithmiques," *Politique et Sociétés* 42, no. 1, 2023.

thesis, we shall see that the advent of algorithmic capital has two major effects on time, both of which tend to leave us with less of it.

First, algorithmic capital complicates and blurs the most fundamental time categories inherited from history—sometimes even to the point of rendering them seemingly obsolete. As it does this, it increasingly ties our lived experiences to the market and to capital. Second, this stronger anchoring of our experience to capital heightens the experience of acceleration of time, and particularly the pace of life. Thus, even if we admit that algorithmic capital is surely not the historical origin of acceleration, it is today an aggravating factor. Examining this theme will allow us to lay important bases for a deeper understanding of algorithmic capitalism as a form of society.

A paradox: The acceleration of the pace of life

Until recently, scholarly literature had not devoted much attention to the impact of algorithms on social time.[2] Still, in addressing this issue, we may usefully draw on Hartmut Rosa's thesis of social acceleration.[3] According to Rosa, the structuring principle of modernity is an acceleration of technical time, social change, and the pace of life. This principle has dystopian overtones, engendering forms of temporal alienation. In a way, it can be said that algorithmic capitalism contributes to this acceleration at the three levels identified by Rosa. Algorithms automate production processes and optimize the performance of the technologies to which they are attached, thereby contributing to a process of *technical acceleration*. They also contribute to the *acceleration of social change*, in particular by intensifying and accelerating the means of information and communication. However, our reflection here focuses on the third level, that of the *acceleration of the pace of life*. Algorithmic capitalism deploys new technologies that blur the social categories of time typical of modernity and thereby heightens this acceleration.

Statistical surveys of the use of time, especially in the advanced capitalist countries, have empirically confirmed the experience of an

2 However, recent works have added to the literature on the subject, notably: Rob Kitchin, *Digital Timescapes, Technology, Temporality and Society*, Cambridge: Polity Press, 2023; Ed Finn, "The Black Box of the Present, Time in the Age of Algorithms," *Social Research* 86, no. 2, 2019, pp. 557–79; Ingrid Erickson and Judy Wajcman, "Optimising Temporal Capital: How Big Tech Imagines Time as Auditable," *American Behavioral Scientist* 67, no. 14, 2022.

3 Rosa, *Alienation and Acceleration*.

acceleration in the pace of life across recent decades.[4] For example, the number of Americans who feel that they are "always in a rush" has been growing steadily over the past few decades (with some brief exceptions); a majority of survey respondents today report such an experience. Whether this means a feeling of having no control over one's time, or time pressures, always being in a rush, a lack of time to complete all one's daily tasks, feelings of stress and fatigue, or the sense of never really catching up, most socioeconomic groups share the experience of an accelerated pace of life. Statistical and sociological studies depict such an acceleration in the advanced capitalist societies since at least the 1970s, and it is reaching new heights today.[5] We should add that this acceleration is uneven, in at least two distinct ways. First, although it affects the population in general, it varies according to how time is used, which itself differs according to gender, social class, ethnic origin, type of employment, parenthood, and other factors. Second, there are also various social or individual processes of deceleration: rush-hour congestion, rising rates of depression and burnout, periods of being locked down (think of certain social processes that were "put on hold" during the COVID-19 pandemic) or "islands" of deceleration such as retreats and ecovillages. However, these variations and decelerations do not cancel out the general trend toward acceleration.[6]

This gives rise to an intriguing paradox. While respondents report a qualitative experience—that they lack time—the quantitative data on time use instead points to a fairly stable pace of life. That is, regardless of socioeconomic differences, people have on average as much leisure time as they did before, and on average sleep and work the same number of hours. This is perplexing. Clearly, evidence of a decrease in sleep time, an increase in working hours, a reduction of leisure time, or a combination of such indicators would provide an empirical basis for the experience of

4 Many such surveys are publicly available. See, for example, the data from the American Time-Use Survey published annually by the US Bureau of Labor Statistics, or the general social surveys on time use conducted by Statistics Canada.

5 John Robinson, "Americans Less Rushed But No Happier: 1965–2010 Trends in Subjective Time and Happiness," *Social Indicators Research*, no. 113, 2013, pp. 1091–104; John Robinson and Geoffrey Godbey, *Time for Life: The Surprising Ways Americans Use Their Time*, Philadelphia: Pennsylvania State University Press, 2008; Robinson and Gobdey, "Busyness as Usual," *Social Research* 72, no. 2, 2005, pp. 407–26.

6 Judy Wajcman, *Pressed for Time: The Acceleration of Life in Digital Capitalism*, Chicago: University of Chicago Press, 2015; Michael Bittman and Nancy Folbre, eds., *Family Time: The Social Organisation of Care*, London: Routledge, 2004; Jerry Jacobs and Kathleen Gerson, *The Time Divide: Work, Family, and Gender Inequality*, Cambridge, MA: Harvard University Press, 2004; Rosa, *Alienation and Acceleration.*

acceleration or a lack of time. But this is not what the overall statistical landscape actually shows.

Although some authors defend the theory that sleep time has decreased since the 1970s and 1980s,[7] the unreliability of sleep data from previous eras prevents us from drawing any firm conclusions in this regard. Furthermore, a systematic meta-analysis of sleep duration data from fifteen countries over the last four decades found no significant decrease in hours of sleep.[8] Evidence of an increase in the hours worked would provide a convincing empirical basis for the feeling of acceleration and lack of time. However, the average hours worked have remained fairly stable, although the massive entry of women into the labor market since the 1970s has redistributed these hours—on average, women work more outside the home, while for men these hours remain stable or have even decreased, depending on job type. Overall, then, even though some people work more and others less, the average number of hours worked has remained stable.[9] In this sense, the empirical basis for the feeling of acceleration is more substantiated with regard to women, who really have increased their number of hours worked while continuing to do the majority of household tasks.[10] However, this does not explain the prevalence of this experience among men as well. Finally, while we might expect the feeling of acceleration to be rooted in a reduction of leisure time, in quantitative terms, leisure time has not decreased and may even have increased slightly since 1980, as revealed by sociological studies of time-use diaries. Below, we will explore the question of the *quality* of leisure time, suggesting that it is becoming less and less discretionary. For now, we can simply note that, in purely quantitative terms, there has been no decrease in leisure time across society as a whole. If there are three main suspects imagined to be driving acceleration (increased working hours, decreased sleep hours, the reduction of leisure time) in no case is this directly corroborated by empirical evidence that would allow us to make general claims about the impact of these factors on the overall population.

Why, then, are society and culture in general so heavily colored by the sense that life is accelerating and that we have no time left—when,

---

7 Including Rosa himself (see *Alienation and Acceleration*); see also Jonathan Crary, *24/7: Late Capitalism and the Ends of Sleep*, London: Verso, 2014, p. 11.

8 Yu Sun Bin, Nathaniel Marshall, and Nick Glozier, "Secular Trends in Adult Sleep Duration: A Systematic Review," *Sleep Medicine Reviews* 16, no. 3, June 2012, pp. 223–30.

9 Paul Edwards and Judy Wajcman, *The Politics of Working Life*, Oxford: Oxford University Press, 2005; Wajcman, *Pressed for Time*.

10 Arlie Hochschild, *The Second Shift*, New York: Avon Books, 1990; Bittman and Folbre, eds., *Family Time*.

overall, individuals are not working longer hours, are not sleeping less, and have as much leisure time as before? One possible lead, in finding a solution to this paradox, would be to examine the role that algorithmic capital's new technologies play in the reconfiguration of working time, and especially of leisure time.[11] The new algorithmic technologies are both *intensifying* work time and *disrupting* leisure time, as they intersperse it with new social and commercial demands and pressures. New algorithmic technologies reduce the occurrence of what sociologists of time call "pure leisure," meaning completely free time whose use is at the discretion of the individual.[12] More fundamentally, the rise of a new category of social time—*captive attention time*, or *screen time*, which overlaps with both working and leisure time—accentuates capital's presence in our lives and heightens the experience of acceleration.

Here we will focus on the capturing of attention by various mechanisms that are aimed at maximizing screen time and the extraction of data. These are key in the process of algorithmic capital accumulation, and they distinguish it from previous regimes of capitalist accumulation. We also make the underlying assumption that it is no longer possible to truly "measure" the acceleration of the pace of life based on historical categories of time (work, leisure, sleep), because these are increasingly obsolete in the context of algorithmic capitalism. Let us take a closer look at these points.

Time, technology, and the market

First, it is worth emphasizing the fundamental role of technology in shaping our experience of time. Technique, from flint to tools and from writing to analog and digital media, is an externalization of know-how; technical objects act as a material form for the collective memory. The artefacts, institutions, books, infrastructure, and machines with which we are fundamentally related are repositories of human know-how and memory; they are dynamic "retentional infrastructures," on which our individual and collective experiences of time play out.[13] Algorithmic technological infrastructure is thus a decisive factor in the temporalities experienced in the contemporary era.

11 In Thesis 6, we will examine algorithmic technologies' impact on the time of social reproduction.

12 Michael Bittman and Judy Wajcman, "The Rush Hour: The Character of Leisure Time and Gender Equality," *Social Forces* 79, no. 1, 2000, pp. 165–89.

13 See Bernard Stiegler, *La technique et le temps*, Paris: Fayard, 2016.

Our contention is that algorithmic capital takes control of societies' retentional infrastructures—the technologies that serve as repositories of the collective memory. It also constructs the present and sets the conditions under which the future is anticipated. In one sense, technological infrastructure mediates access to the past. For example, the digitization of documents, writings, images, videos, and other media, and their storage in databases accessible through machine-based computing processes (keyword searches, hashtags), mean that our memory is increasingly dependent on technological media that are ever more governed by algorithms. The data extraction at the basis of Big Data (and we will return to this theme) is, from a temporal point of view, a process of retention and recording, whereby data from past experiences are stored and processed by algorithmic technologies in order to generate behavioral predictions. At the same time, algorithmic technologies are also able to construct the present of experience; just think of the algorithmic curation process that generates the news feeds on different media, or the very production of the text, images, and videos that now populate our online experience. Next, if we consider that our anticipation of the future is based on our *present* engagement with collective and individual memories, and that algorithmic technologies are (as we shall see) *predictive* technologies, we can quickly grasp how important algorithms are to the social construction of our relationship with the future. In other words, the past, present, and future we experience become inseparable from the technological mediations that store memory, curate the present, and predict the future.[14] We must now supplement this philosophical observation with a more sociological analysis: What happens when we examine the impact of algorithmic technologies on traditional categories of time use, such as working and leisure time?

When it comes to working time, it is clear that algorithmic technologies serve to intensify it. The history of capitalism is also a history of putting pressure on labor to make it more intense and productive. Algorithmic technologies have a very complex impact on labor in general, in this regard, and, in Thesis 5, we will return in more detail to the questions concerning algorithmic labor. Algorithms can make labor more productive, that is, by producing more output in a given time. We also note what Judy Wajcman, among others, has amply demonstrated about working time: ICT (information and communication technologies), and today algorithmic technologies, put pressure on working hours and intensify

14 Bernard Stiegler, *La société automatique*, Paris: Fayard, 2015.

them, so that people work more in each hour. In other words, even if the number of hours worked remains stable, workers' experience of this time can accelerate *qualitatively* since these hours are intensified.[15]

However, to dig into the changes of the contemporary era, we must also examine leisure time, beginning with two historically fundamental aspects. First, leisure time is largely so-called "discretionary" time, meaning that it remains under the control of the subject and is organized and populated with activities decided at their discretion. Second, despite certain variations and differences across sociohistorical contexts, there are significant continuities over the long term, notably the fact that leisure time has long been conceived and practiced as a form of "extra-economic" time. From the Greco-Roman world to the medieval or other eras, as a rule, leisure time does not participate in the economic function of life—the production of goods and services that materially reproduce bodies. Leisure time is therefore conceived and practiced as something distinct from productive practices, labor, and the market. Industrial capitalism reproduces this temporal distinction between leisure time and the time spent on economic production; indeed, capitalism radicalizes it in several respects, notably by spatializing it. Work and its time are, in the ideal type of industrial society, confined to a well-defined space-time: the paid work shift. Leisure time, conversely, is spent outside the workplace, either at home or in places where leisure activities are practiced. In this temporal regime, working time and leisure time are clearly distinguished and delimited, and leisure time does not produce value.[16] In this sense, we are the historical heirs of a distinction between working time and leisure time, where the latter is free from productive tasks (often at the expense of the exploitation of others' labor, particularly the domestic labor done by women) and can be devoted to practices of good living and self-fulfillment, away from production and market exchange.

The linking of leisure time to the market, at the level of consumption, clearly predates the era of algorithmic capital; leisure time was populated with mass consumption practices well before the 2000s. However, these processes are today accelerating, particularly due to the growing place—and powerful effectiveness—of the targeted-advertising mechanisms developed by new algorithmic technologies. Moreover, the linking of leisure time to the market has now reached a new stage in the algorithmic

15 Wajcman, *Pressed for Time*.

16 Jonathan Martineau, *Time, Capitalism and Alienation*, Chicago: Haymarket, 2016.

era. Leisure time may be unpaid, nonworking time, but it becomes productive: It produces data which, once extracted and worked up by digital labor, as we will see in Thesis 5, also contributes to the accumulation of capital. In our view, this is such a spectacular development that it forces us to rethink the categories of time and value themselves. After all, it calls into question the historically inherited distinctions between "productive" working time and "unproductive" leisure time.

If we want to fully understand the causes of these upheavals, we have to look at the technological infrastructure of algorithmic capital: the spread of computer screens, tablets, smartphones, laptops, and other devices. These technologies not only help to reorganize working time, but also massively intensify the tendency for leisure time to be reduced to "attention time" spent in front of a screen—"screen time."[17] Online time and screen time have increased dramatically in recent years and now account for a significant portion of waking hours, whether spent working or otherwise. Recent data shows that Internet users aged sixteen and above spend, on average, six hours and thirty-eight minutes connected to the Internet per day. The average American spends seven hours and three minutes looking at a screen each day. Social media apps take up an average of two hours, twenty-one minutes a day. Some of the most popular apps capture important quantities of attention time, decreasing recently for Facebook (still at seventeen hours, seventeen minutes per month), but reaching new heights on TikTok (at a little less than thirty-five hours per month).[18] In Thesis 14, we will see how the COVID-19 pandemic increased the attention time devoted to screens, and the way this colonizes our lived time. Moreover, this attention time is effectively monetizable in advertising markets, and it generates data which, once processed, forms a material, an asset, exploited by the companies that collect it, such as Google and Meta. In short, in the era of algorithmic capitalism, leisure time is increasingly becoming data-producing screen time, complicating the distinction between productive working time and unproductive leisure time.

The net result of these processes, therefore, is an increasing alignment of lived time, whether remunerated or not, with the capitalist market and the accumulation of algorithmic capital. This binding of leisure time to

17 Yves Citton, *Pour une écologie de l'attention*, Paris: Seuil, 2014.

18 Simon Kemp, "Digital 2025 Global Overview Report," February 5, 2025, datareportal.com; Fabio Duarte, "Alarming Average Screen Time Statistics (2025)," *Exploding Topics*, April 24, 2025, explodingtopics.com; Dave Chaffey, "Global Social Media Statistics Research Summary 2023," smartinsights.com, 2023.

productive time undermines its discretionary and ethical potential and exacerbates the experience of an accelerating pace of life. In the following thesis, we will examine in detail the logic of algorithmic-capital accumulation that drives this imperative to capture attention time. For now, let us focus on the mechanisms that generate it.

Captive screen time and behavioral addiction

Generating captive screen time is essential to the accumulation of algorithmic capital. Our time is thus colonized by a whole new range of apps, websites, platforms, interfaces, and games, which are at least partly designed to generate dependence, and even addiction, among users. Facing the screen are individuals who want to be entertained, connect socially, perform tasks, or obtain information. Behind the screen is a vast industry creating applications and interfaces designed to generate maximum attention time and data for commercial purposes. Indeed, tech giants like Google, Facebook, and Netflix, as well as small app and game developers, use psychological knowledge (cognitive behavioral psychology, neuroscience, design, neuromarketing) to make their products as attractive, intrusive, and addictive as possible. The resulting political economy is quite simple: The more popular, intrusive, or addictive a product is, the more attention time and thus data are extracted from the user, the more advertisements can be sold, and the greater the volume of data that can be turned into assets for the product's owner.

The creation of interfaces and applications designed to maximize attention time and the data generated online can draw on several psychological mechanisms. The technological engineering behind the creation of these interfaces makes use of what Adam Alter, a psychologist specializing in Internet addiction, calls the "ingredients of behavioral addiction"—an addiction not to a substance, but to a behavior.[19] Rereading Alter's work and taking some liberty with his categories, we can distinguish between four main mechanisms of addiction. These do not necessarily form distinct categories in practice; rather, they are overlapping mechanisms that use many different "addictive hooks." They are: (1) goal-oriented logic; (2) the temporality of progress and intensification; (3) feedback and social interaction; and (4) the loop of incompleteness. These mechanisms seem obvious at first glance and are not especially

19 Adam Alter, *Irresistible: The Rise of Addictive Technology and the Business of Keeping Us Hooked*, New York: Penguin, 2018.

sophisticated. They are, however, extremely effective. Interfaces, applications, games, and websites are designed and constantly updated to deploy one or more of these mechanisms in order to capture the attention and keep it riveted to the screen for as long as possible.

The first mechanism involves a goal-oriented approach. Obviously, setting oneself goals is perfectly normal and may have considerable ethical value. We may pursue varied goals in life, ranging from work performance to our personal lives—whether that means getting a promotion, eating healthier food, making a list of places to visit, having children, or simply making a to-do list for the day. The logic of goals is a structuring element of the fabric of lived experience and creates powerful mechanisms orienting our behavior.

However, two things exacerbate this goal-oriented logic in our era and steer it in a way that binds our aspirations for self-fulfillment to the imperatives of algorithmic capital accumulation. First, the advent of the Internet and its immense informational power exposes people to an *infinite number* of goals of which they were previously unaware, and that they can now take on board. Second, the "colonization of the lifeworld" by portable devices facilitates the *quantification* of goals and an automatic and effortless measurement in real time of whether they are being achieved. The screens that accompany us throughout our day, seeking to grasp our attention, can both multiply and quantify our goals in "real" life, and present a product (a game, app, social network) that itself offers a whole series of goals to be achieved on screen.[20] These "screen goals" also have significant addictive mechanisms.

In such an era of both infinite and quantified goals, a goal-oriented logic can keep individuals in a constant state of dissatisfaction or unfulfillment. Achieving the next goal or reaching the next target is not only an aim for the future, but one that demands more practice and time and effort in the present. In this lies a significant potential for behavioral addiction. Thus, the technologies that fix our attention on the screen constantly reproduce unfinished quests and goals that are always just beyond the user's reach. This is what has been called the "zone of proximal development" (a concept developed by psychologist Lev Vygotsky), a zone that users can reach with help, instructions, practice, effort, and improvement, but which requires an investment of attention—and screen time.[21] This is a structuring aspect of interfaces, games, and apps: a series of quantifiable goals that engage our time and attention. A characteristic

20 Ibid., p. 109.
21 Ibid., p. 175.

of these screen goals is that they are quantified in such a way that they are constantly created anew. This makes it possible to push these goals ever further, and to transform the achievement of one goal into a new basis for reaching the next level. Finish the difficult level of the video game—to move on to the next one. Lose your last game of *Fortnite*, but only by a narrow margin. You've gathered 1,000 friends on Facebook—but why not 2,000? Why not get to 6,000 followers on Instagram instead of 5,000? Increase your number of retweets on X or TikTok views. Walk 12,000 steps today according to your Fitbit and aim for 13,000 tomorrow. Quantification thus fuels the logic of growth and "always more, always better," anchoring it in a reward system that time and again produces a passing sense of satisfaction, which has to be constantly sought anew. Quantifying goals makes them malleable. Although we are perfectly capable of filling our own zone of proximal development with goals of our own making, this mechanism is prized by developers of addictive technologies. Constantly setting goals just out of immediate reach means spending more time learning the technical aspects of the site, game, or app to achieve the desired results. Most gaming applications and interfaces, whether for entertainment or lifestyle purposes, employ this logic of quantified goals.

The logic of goal orientation often encounters another important logic that massively amplifies it: the logic of progress and intensification. Together, they further enhance the addictive appeal of screens. For instance, the threshold of entry and initial success is set very low, allowing new users to achieve success on the app or in the game right from the start. After this initial, relatively easy success, the degree of difficulty gradually increases, meaning that each new success requires a greater investment of time than did the previous one. For example, the first levels of an addictive game app are usually fairly easy, but it takes much longer to reach each subsequent threshold of success. A similar tactic is to systematically incorporate a mechanism that replicates the experience of "beginner's luck." Offering a large prize or bonus on the first use can guarantee a few subsequent uses—not all of which will be as successful.

It is thus easy to start posting videos on YouTube, but much more time-consuming and difficult to become a YouTuber. Reaching the first level of fame on one of the most downloaded apps of recent years, *Hollywood Story*, is relatively easy, but the higher levels require a huge investment of time and, of course, the repeated payment of fees (in-app purchases) to continue the "game." *Farmville* is relatively easy at first but requires continuous attention to keep your plot of land progressing.

Users must remain ever on guard, since, when they are offline, the game continues with other players in real time—and who may take advantage of their lack of attention. These games, which continue in the player's absence, are highly addictive, deploying a hook beyond the online playing time itself.

Of course, if the goals are always out of reach, the user will eventually lose interest. To avoid this, the setting of goals in the zone of proximal development must work in tandem with the logic of progress. This makes it possible to stimulate the user through serial "micro-rewards" and so-called juice—elements often inspired by the world of gambling and slot machines, which have proven their worth in both real and virtual casinos around the world. This "juice" is one of the feedback mechanisms which, together with social interaction, serve as the third major mechanism for creating behavioral addiction online. Feedback aims to give the user a feeling of control: The screen and interface elements respond to their moves and actions. Social interaction, meanwhile, responds to the fundamental human needs for recognition and community. Online social networks offer a virtual—and sometimes toxic—form of social validation and acknowledgment.

The forms of feedback developed by the technicians of algorithmic capital are largely inspired by these "micro-feedback" mechanisms from the world of gambling, particularly the most addictive slot machines. Tristan Harris, a former designer at Google, referred to smartphones as "the slot machine in your pocket."[22] The success of a slot machine is calculated according to the amount of "time spent at the machine," which generally correlates with the revenue it generates. The engineering of these machines, which seeks to maximize this time, draws on a whole structure of microscopic feedback rules: sounds, light effects, and bright colors that near-instantaneously respond to user inputs.

However, if such feedback has an addictive power, it truly explodes when it is combined with *social* interaction. Doubtless, the best-known form of combining feedback and social interaction is the "Like" button, introduced by Facebook in 2009 and revamped, from 2016, as "reactions" which enable the communication of a wider range of emotions. The "Like" button provides immediate feedback for almost all user interactions with the interface. It also allows for easy and fun social interaction. However, some users develop addictive behaviors toward this form of social validation, which can cause various symptoms such as sharp

22 Cited in Richard Seymour, *The Twittering Machine*, London: Indigo Press, 2019, pp. 51, 76.

spikes in anxiety. A social media post, photo, or comment that receives no reaction is simultaneously seen as a mark of social disapproval. The uncertainty of social feedback guarantees user engagement. A comparison often cited in the literature again concerns games of chance: Sharing something on Facebook, Instagram, or TikTok is equivalent to betting to get reactions. Will the bet pay off? You have to make it to find out, and the incentive to make another one will be there, whether the bet pays off (we will want even more) or not (we will want to redeem ourselves from the social disapproval that came with the last unsuccessful post). Likes and their variants draw on the complex pleasures and vulnerabilities of self-presentation and recognition, sometimes to the point of exhibitionism and voyeurism. They offer a quantitative measure of popularity that fuels social competition; users yearn for a social validation that remains ever uncertain.[23] Many observers have drawn causal inferences between toxic aspects of social media feedback and anxiety issues, notably among young women and girls, who are also the ones spending the most time on social media on average.[24]

The fourth ingredient of behavioral addiction is the loop of incompleteness. One of its best-known variants—familiar to all followers of TV series—is the narrative device of the cliffhanger, which is systematically used to create an addictive relationship with certain products or content. This addictive hook targets individuals' deep desire to resolve a plot, turn the page, conclude a story or experience, or get closure. This is what psychologists call the Zeigarnik effect: Our unresolved or incomplete experiences occupy our thoughts much more than the ones where we have already reached the final page. Creating narrative tension and then leaving it unresolved is, in this sense, a powerful addictive mechanism.

Cliffhangers are a popular tool among technicians working for algorithmic capital, for they increase the likelihood that a user will return to the app or consume the product at some later point. Netflix has innovated in this sense with its auto-play mechanism, which removes the main stop-points in the viewing experience. This puts the viewer in a position of having to act quickly to avoid beginning the next episode. The simple force of inertia will often lead them to keep watching, without always being fully aware of it. Removing the stop mechanisms obviously intensifies the power of addictive hooks: The addition of sequels and

23 Ibid. In Thesis 16, we will return to the question of self-presentation and performance online.

24 See, among others, Jonathan Haidt, *The Anxious Generation*, New York: Penguin, 2024; and Simon Kemp, "Digital 2025 Global Overview Report."

prequels and extensions to series, games, and apps also feeds into this loop of incompleteness. When we combine powerful cliffhangers, the automatic playing of the next episode, the omnipresence of screens, and the inertia generated by passive attention spans, we get "binge-watching." Here, viewers consume several hours of the same TV series, or even an entire show, all in one go, demonstrating the power of these addictive mechanisms.[25]

The addictive hooks of an interface or app that use these four major capture devices are well known to engineers. They will analyze large volumes of user data points to determine which of these hooks work best and then integrate them into their products. After a while, each update to the app, game, or interface becomes a faithful reflection of the mechanisms that have worked best to engage users' attention. In other words, each version is always the most addictive yet. The resulting increased alignment of lived time with market time is thus added to a traditional mechanism employed by capital, through which working time is connected to the market and exchanged for money. But now, *attention* time is captured by *market* time.

The paradox of acceleration and screen time

What can all this teach us about the subjective experience of the pace of life accelerating—and about algorithmic capital itself?

Let us remember the paradox mentioned earlier: While there is a widespread *qualitative* experience of a lack of time, the *quantitative* data on time use shows a fairly stable pace of life. This data tells us that, independently of socioeconomic differences, people have on average as much leisure or discretionary time as before, and sleep and work on average the same number of hours. Do these developments in the means of algorithmic capital's capturing attention time perhaps help to explain in part the feeling that the pace of life is accelerating? As we have seen, this new social category of screen time substantially blurs the distinctions between working time and leisure time. Ultimately, by taking up an increasingly large part of leisure time, screen time degrades this leisure. For it reduces pure leisure time, constantly interrupting it, creating new stimuli, demands, and disruptions. In our view, this is how the feeling of an accelerated pace of life is heightened: by this reduction in

25 Alter, *Irresistible*, pp. 194–5, 211–12.

the discretionary quality of leisure time, through its alignment with the market imperative of data extraction. Here we can draw on Wajcman's similar argument about working time: Information technologies also put pressure on this time to make it more productive and more intensive. Ultimately, both working time and leisure time are caught up in the technological apparatus of algorithmic capital, and captive attention time is becoming the main category of the lived experience of time in our era.

Attention time is a form of time that can easily be aligned with market time, through the mediation of screens and the world of interfaces, platforms, apps, and social media. Under the sway of advertising, marketing, the creation of addiction, and data extraction, it can be subjected to the pressure of the imperatives associated with this form of market time: productivity, speed, efficiency, acceleration, constant solicitation. The growing alignment of leisure time with the market and the capture of attention time by the addictive apparatus of algorithmic capitalism accelerate our time and orient it toward a logic of self-realization. This latter is strongly linked to consumer practices, "screen goals," and the marketing of the self in search of recognition (on social networks), or even a rushed temporality of anxiety-inducing addiction. Furthermore, leisure time tied to the market is subject to various forms of temporal disorganization, which leads to a loss of discretionary control over leisure practices. In other words, as leisure time becomes tied to the market and thus itself becomes economic time, it increasingly escapes the control of the individual. As a result, more market pressures (the incentives to perform, to curate one's digital self, to produce data, to consume) interrupt pure leisure and undermine its discretionary quality.

The increase in screen time is a constant across all population groups in the last decade. The degradation of leisure time is thus a causal factor that can be said to have a generalized impact across society and culture as a whole. This may explain—at least in part—the feeling of lack of time, which reflects the acceleration of the pace of life. In short, algorithmic capital, especially through the development of its technological infrastructure, transforms social categories of time and thus the lived experience of time, thereby exacerbating social acceleration and the resulting temporal alienation. Practices of good living that require long reflection, face-to-face social interactions, the development of skills that take a long time to learn, a space-time free from economic worries, and a correspondence between the means and ends of a given practice are thus increasingly difficult to achieve in a contemporary temporal regime that is so unfavorable to this type of use of time, and in which pure leisure is fading away. It becomes harder to access the temporal quality required by some of the "focal practices" referred to by Albert Borgmann, such as cooking and sharing a meal, running, taking a walk in the forest, or learning a language for its own sake.[26] Filling our schedules with good living and truly free time becomes a Herculean challenge in the age of algorithmic capitalism, screen dependency, and the constant demands and stimuli that undermine discretionary time.

The infernal hamster's wheel of algorithmic acceleration

To conclude this reflection on how algorithmic capitalism affects our experience of time, it is worth revisiting Hartmut Rosa's hypothesis of social acceleration. As we have noted, Rosa distinguishes between three main forms of acceleration: (1) technical acceleration; (2) accelerated socio-cultural change; and (3) an acceleration of the pace of life. He shows that these three moments of acceleration reinforce each other, like a self-perpetuating cycle or a gigantic feedback loop.[27] The acceleration of the means of transport and communication, technological innovations, and economic activity feeds the experience of accelerated change in family, professional, religious, identity, and political and cultural relations as well as romantic relationships. This contributes to the acceleration of the pace of daily life and the feeling that time is always under pressure. This pressure in turn calls for the intensive use of faster and more efficient technologies in order to "save time." Completing the cycle is a growing demand for technological innovations, which makes social acceleration turn ever faster. Rosa does not want to fall into any simplistic or one-dimensional explanation, and he rejects the idea that there is any one structuring cause or "prime mover" behind social acceleration. For him, the capitalist mode of production is not ultimately responsible for this historical process of generalized acceleration. He prefers to focus on the idea of a complex dynamic, in which each specific form of acceleration is also dependent on an overall logic which is not reducible only to capitalism. It should be noted that Rosa is not seeking to develop a critical theory of capitalism, but a general theory of modernity, that is, a social formation that has taken various forms throughout its history, be they capitalist, socialist, fascist, colonial, liberal, democratic, or authoritarian.

26 Albert Borgmann, "Focal Things and Practices," in Craig Hanks, ed., *Technology and Values: Essential Readings*, Hoboken, NJ: Wiley-Blackwell, 2010.

27 Rosa, *Social Acceleration*.

Modernity, examined through the prism of social acceleration, becomes an "iron cage"—to use Max Weber's phrase, who sought to describe the grand logic of "rationalization" that permeates all spheres of modern society, from the state to business and culture.

Rosa contends that, in addition to the feedback loop fueled by the three moments of acceleration, there are two main "drivers" that generate this dynamic: the principle of competition and a conception of the good life that sees acceleration as a promise of personal fulfillment. The principle of competition cannot be reduced to the economic sphere alone, but it does not come from nowhere. This principle is, indeed, one of the central imperatives of capitalism, with competition embedded in the structure of the market. Of course, competition is not omnipresent in the social world, in the sense that many human practices and interactions are based on other logics: gifts, cooperation, care, reciprocity, redistribution, and so on. But we can still support the thesis that capitalism serves to reinforce and spread the imperative of competition throughout society. In fact, this thesis is hardly controversial.

Furthermore, Rosa considers that acceleration cannot function as a simple economic constraint imposed on individuals from the outside, since it also enjoys a broad cultural acceptance among the population. Rosa is clear on this point: The acceleration of means of transport allows us to travel more easily around the world; the acceleration of means of communication allows us to interact more quickly with friends, family, and others close to us; the expanded number of jobs, romantic partners, cultural affiliations, and identities allows us to "live several lives in one," to "live intensely," to pursue a variety of projects as expressions of self-fulfillment. In short, while social acceleration surely has its downsides, it also has its share of immediate benefits and promises of happiness.

Yet there are problems. While Rosa considers the "cultural engine" of social acceleration to be distinct from the "economic engine," this is not the case in our view. The modern ideal of "the life intense,"[28] a freer life with countless options, which makes it possible to broaden "access to the world" in order to "make it [more] available,"[29] is closely linked to the cultural logic of capitalism and its consumerist and productivist ideal. I can do *anything*, achieve *anything*, consume *anything*, choose *anything*, go faster, constantly improve my performance, read new books recommended by my friends on social media, watch the latest series on Netflix,

28 Tristan Garcia, *The Life Intense: A Modern Obsession*, Edinburgh: Edinburgh University Press, 2018.

29 Hartmut Rosa, *The Uncontrollability of the World*, Cambridge: Polity, 2020.

buy new, trendy shoes with a click on my mobile phone—an endless list. In addition to the principle of competition that lies at the heart of capitalism, the ideal of a good life in contemporary society is based on the intensification of consumerism, the ideology of individual performance promoted by neoliberalism, but also a logic of the efficiency of everyday objects that generate a certain level of material comfort.

In other words, we want to show that the two drivers of social acceleration as identified by Rosa—the imperative of competition and the cultural ideal of modernity—are in fact two central dimensions of capitalism as a mode of production and as an institutionalized social order. It is no accident that individuals, companies, cities, and states are set in competition with one another. This is a direct consequence of the functioning of capitalism, which itself contributes to increasing the time pressure on the lived world. The production of commodities, and particularly of digital technologies capable of optimizing a host of practical aspects of life, provides a certain response to the feeling of a lack of time. Here we see the close intertwining of capitalism and the technological acceleration described by Rosa. While he refuses to grant economic factors any primary explanatory power—no doubt to avoid potential criticism that his theory is too economistic, reductionist, or technodeterminist—Rosa does not seem to conceive of capitalism as an institutionalized social order. Rather, he considers it as an economic system, a specific field of the social world, a quite particular sector of modern society. Rosa's social philosophy, centered on the category of modernity, thus prevents him from adequately conceptualizing the specific, yet predominant, role that capitalism plays in the general dynamics of social acceleration.

If we are arguing that capitalism is the main cause of social acceleration, what specific role do algorithms play in this dynamic? We have contended already that algorithmic technologies act as repositories storing up the past and mediating access to it, that they act as curators of our experience of the present, and that they aspire to predict the future. In this, we can see the close intertwining of algorithms and temporalities. Moreover, the logic of capitalist accumulation increasingly relies on data extraction, the capture of attention time, and the proliferation of the addictive technologies discussed in this chapter. Nevertheless, we should also mention the intensive production of algorithmic machines that automate a host of processes, thereby improving the performance and speed of countless phenomena in the social world. This will likely further fuel the overall dynamic of social acceleration that has accompanied capitalist modernity since its inception. In this sense, the advent of AI, social media, and digital technologies does not mark any radical break: They merely prolong and intensify a process of acceleration that has already been underway for centuries.

Thus, the arrival of algorithmic machines in various quarters of the social world will further feed the acceleration of the pace of life. This invalidates the hypothesis that AI could finally make possible the utopia of a "leisure society," as some speculation naïvely claims, when it suggests that the likes of ChatGPT will finally free humanity from work.[30] Instead of freeing up our time, algorithmic capital will, on the contrary, further amplify the experience of "time scarcity," the feeling that everything is always moving too fast, that we are constantly in demand, and that we are perpetually overwhelmed by events and the frenetic technological progress that seems beyond our control.

Capitalism is, indeed, the great "time thief." It literally steals a large part of our lives, not only at work, but also through the devices that drive our addiction to screens, which siphon off our attention time in countless everyday activities. Lack of time, increased burnout, but also cyberaddiction, anxiety, and other psychological disorders associated with digital overconsumption, can be cited as so many forms of social pathologies linked to the development of contemporary capitalism. Fragmented time today heralds a new phase of social acceleration: an *algorithmic acceleration*.

Ultimately, our experience and use of time are increasingly colonized by algorithmic technologies, which work to transform them into a time of producing value and/or data. In so doing, algorithmic capital heightens the acceleration of the pace of life—and demands a rethinking of the categories of time themselves. Yet we must now accept that there is a whole political economy standing behind the deployment of algorithmic technologies in the quest to capture attention time and data. In fact, in the era of algorithmic capital, the capitalist production of economic value is itself reconfigured. This is the object of the next thesis.

30 Gracer Mayer, "AI Will Eventually Free People Up to 'Work When They Want To,' ChatGPT Investor Predicts," *Business Insider*, March 30, 2023.

Thesis 4

Algorithms are machines for extraction, prediction, and manipulation

Search me, God, and know my heart; test me and know my anxious thoughts.

—Psalm 139:23

Next, we need a more detailed analysis of the relationship between algorithms and capitalism. In what sense do algorithms alter some basic processes of capital accumulation? How do capitalist imperatives in turn shape the development of algorithms?

First, to understand the functional role of algorithms within the capitalist system, it is important to look beyond how these computerized calculations perform specific tasks—translating a text, recognizing an image of a cat, playing chess, or driving a car. Rather, we should begin by studying the specific role of the data used to train algorithms developed by companies that want not only to offer services to users but also to make a profit.

In her book *The Age of Surveillance Capitalism*, Shoshana Zuboff reports a major discovery. While her approach has a number of important shortcomings, to which we will return below, her contribution sheds critical light on new mechanisms of capitalist accumulation in the era of Big Data and algorithms. According to her analysis, what she calls "surveillance capitalism" is not primarily based on the exploitation of wage labor, but on human experience in general—for this latter forms the "free raw material for translation into behavioral data."[1] Some of

1 Zuboff, *The Age of Surveillance Capitalism*, p. 14.

this data has a use-value, in the sense that it can be used to meet user needs by improving products and services. Yet beyond this, there is a proprietary *behavioral surplus* that feeds and refines algorithms. These algorithms convert behavioral data into "predictive products": The mass of data drawn from an individual's behavior allows the algorithm to anticipate their future behavior. These predictive products (used for targeted advertising, for example) are a commodity with exchange-value, sold on a behavioral prediction market. Many public and private companies and organizations are scrambling to acquire these products, the better to target potential consumers, optimize and refine their activities, and predict and modulate human behavior toward desired or profitable outcomes.

For Zuboff, therefore, what produces value in contemporary capitalism is not just labor, but also *human experience* itself, insofar as it generates data. Our online activities, conversations, movements, and other elements of daily life are thus digitalized, geolocated, stored, and transformed into behavioral data. Surveillance capital did not immediately emerge with the advent of computers or the Internet, even though in the 1980s the rise of information and communication technologies was already feeding into a transformation of production processes and work organization. To generate value, users' experiences and activities on platforms and in the presence of connected objects must be translated into data (the process known as *datafication*). They are appropriated as "free raw materials" through an extraction process, collected or transformed into predictive products, then monetized and resold to buyers who want to know more about us. It should be noted here that, for Zuboff, the data that interests capital is not data in general, but behavioral data that allows companies to learn about our day-to-day movements and habits. Here, we see the economic imperative to generate attention time, the connected time which produces this raw data. We should note right away as well that in our view, this extracted data does not generate value magically on its own. Zuboff is good at identifying the extraction mechanism, but behavioral data needs to be processed by various forms of digital labor, to generate what we call 'algorithmic capital', to which we will return in Thesis 5.

Zuboff refers to the work of Karl Polanyi, who describes how the self-regulating market economy emerged thanks to the creation of three fictitious commodities: labor, land, and money. Today, the emergence and accumulation of surveillance capital is based on the valorization of a fourth form of fictitious commodity:

> Today's owners of surveillance capital have declared a fourth fictional commodity expropriated from the experiential realities of human beings whose bodies, thoughts, and feelings are as virgin and blameless as nature's once-plentiful meadows and forests before they fell to the market dynamic. In this new logic, *human experience is subjugated to surveillance capitalism's market mechanisms and reborn as "behavior."* These behaviors are rendered into data, ready to take their place in a numberless queue that feeds the machines for fabrication into predictions and eventual exchange in the new behavioral futures markets.[2]

While the production and capture of data are essential, according to Zuboff they generate value only when they lead to the formation of a surplus that can be analyzed (using data analysis techniques and algorithms) and then resold on the market. Indeed, the data mining enabled by various forms of machine learning allows for certain predictions to be made about future behavior (predictive analytics), which constitute the high-value-added "refined product" derived from the raw material of Big Data. The expression "data is the new oil," first coined in 2006 by mathematician Clive Humby, is a metaphor for this extraction-refining process: "Data is the new oil. It's valuable, but if unrefined it cannot really be used … so must data be broken down, analyzed for it to have value."[3] To pursue the metaphor, let us also keep in mind that "refining data" to produce these different "data commodities" is not only a task performed by machines, but one that also necessitates human labor.

The discovery of behavioral surplus

According to Zuboff, this valuable "behavioral surplus" was discovered by Google in the early 2000s. Originally, the powerful Google search engine aimed primarily to draw on user data to improve the system's own performance, that is, the speed, accuracy, and relevance of search results. Even seemingly irrelevant data and traces left by Internet users (cookies, temporary files, geolocation data), also known as "data exhaust," was systematically analyzed, used, and fed back into the search engine to optimize the service. Zuboff's position is that this use of data to improve services represented a certain "balance of power" and a win-win situation between users and Google. Users got better search results, while

2 Ibid., p. 71 (original emphasis).
3 Peter Silva, "Oil vs. Data—Which Is More Valuable?," *F5 Blog*, April 2019.

Google optimized its service and fulfilled its mission: "to organize the world's information and make it universally accessible and useful." Both users and Google contributed to the joint production of use-values: optimized search results.

However, while this model was reciprocal, it did not generate revenue. After all, Google's search results, which, in a certain sense, constitute a service commodity, are not themselves sold to the user.[4] Google is a private company in a capitalist economy. Sooner or later, the imperative to turn a profit would have to make itself felt. Zuboff explains that the bursting of the dot-com bubble in March 2000 tested investors' patience and created an emergency situation at Google. Investors, worrying over their portfolios, wanted to convert the use-values produced by Google into revenue; the company, finding itself mired in a profitability crisis, sought to generate new profits by changing its business model.

The solution, which seems innocuous at first glance, would ultimately produce a revolution. It involved monetizing the huge volume of data that had been accumulated, in the form of targeted advertisements, which were so effective that advertisers soon began to hanker after them. By processing this data, thereby accumulating information about users and building profiles (user profile information, or UPI), Google revolutionized the world of advertising. Instead of linking advertisements to keywords entered in searches, they were now connected to users' profiles. In addition, Google could now accurately calculate the effectiveness of targeted ads by measuring the "click-through rate," the ratio of clicks on a given targeted ad. The more information Google had, the more accurate the UPI was, the more effectively the adverts reached their target audience, and the higher the click-through rate.

Such a system obviously allowed advertisers to precisely measure the return on their advertising spending, making Google's model extremely attractive. It also gave Google the possibility of developing increasingly sophisticated algorithms. Although cofounders Larry Page and Sergey Brin had always refused to subject their search engine to the logic of advertising and commodification, 2001 saw a total about-face under pressure from the imperative of profitability. In the wake of these developments, Google became an extremely profitable company.

Zuboff develops a conceptual framework from her analysis of Google's business model. Drawing inspiration from Marx, for whom *surplus-value* corresponds to the difference in value between what the

4 Evgeny Morozov, "Critique of Techno-Feudal Reason," *New Left Review*, nos. 133/4, January/April 2022, pp. 89–126.

worker produces in a day's work and what she receives in wages, Zuboff defines behavioral surplus as the difference between data reinvested in improving services and the production of surplus data captured by the digital platform and resold in the form of predictive products on the new "future behavior markets" to generate "surveillance revenue." This model is rapidly spreading from beyond Google's walls throughout the business world. It is ever-changing as new players emerge in this new form of capitalism and will soon become dominant.

Zuboff uses the concept of surveillance both metaphorically and literally. Capitalist companies, particularly the Big Five (or GAFAM) and other digital platforms, derive much of their income from "surveilling" our actions, which leave traces that can be transformed into behavioral data, and which in turn can be used and resold to generate profits.[5] But surveillance is more than a metaphor, because we are literally tracked, listened to, and closely monitored by our mobile devices, computers, and smartphones. Through microphones, GPS, cookies, and digital profiles, they operate as outright spies on our activity. The fact that such surveillance is now of only trivial concern for many people—"we know we're all being monitored online, but so what?"—shows how this generalized surveillance has become naturalized, normalized, and internalized, an everyday reality that does not bother us (we will return to this point in Thesis 16). Zuboff's book offers an excellent antidote to this; it is a wake-up call from a "dogmatic slumber" that trivializes the violence of our day-to-day private lives being surveilled. However, while Zuboff's concept of surveillance capital focuses on the relation between users and capitalists, our concept of algorithmic capital, as we will increasingly make clear, focuses as well on the relations between capital and labor, the state, social reproduction, and nature.

The new mechanisms of accumulation

Although Google played a central role in the emergence of surveillance capital, this logic of accumulation is now spreading rapidly across all sectors of the economy. The dynamics of behavioral surplus accumulation take the form of a frantic race to extract ever greater volumes of data about our lives. This *primary imperative of extraction* means that digital platforms and "traditional" companies alike are structurally incentivized

5 GAFAM: Google, Apple, Facebook, Amazon, Microsoft. We use this acronym throughout the book to refer to Big Tech companies.

to redouble their efforts to acquire more behavioral data. They *have to do this* if they want to remain competitive. We think that this massive extraction of personal data is a fundamental mechanism of what we call algorithmic capital—it is an economic necessity that affects virtually all companies in the digital world and even beyond, in the economy in general. Today, companies such as General Electric, John Deere, and Siemens present themselves as data companies. All kinds of manufacturers are equipping their products with sensors, chips, and "smart" mechanisms to extract after-sales data. Those that do not pursue this logic risk being outclassed by their competitors. Thus, the term "digital capitalism" is a somewhat misleading description of what is at stake. For the extractivist logic and the expansion of the scope of algorithms extend far beyond the so-called "online world"; they also reach into and colonize the physical world of connected objects, devices, and cities. We will come back to this point later.

This expansive dynamic is an imperative that drives algorithmic capital to conquer new territories, such as the physical world (homes, workplaces, cars, public transport, cities), our daily lives (smart homes, leisure applications), our bodies (connected glasses, smart toothbrushes, chips under the skin, Fitbits), our social relationships (social media, dating apps), and our mental health (apps for interests, mood management, chatbot therapists). Algorithmic capital runs through the entire commute-work-sleep cycle: reading the news on your feed while crammed into a crowded train in the morning; on Zoom, Teams, or Google Meet calls once at work (in the office or teleworking); or even when going to bed, with smart apps that diligently time and monitor our sleep. For the better-off, this may even mean heading to a "smart bed" that boasts countless features: Bluetooth speakers, USB ports, built-in massage systems that can be controlled with a smartphone, sensors, and mics, and a dedicated space for a laptop so you can idly surf the web from the comfort of your pillow. The imperative to extract is omnipresent in both businesses and a wide range of social organizations, and it is spreading throughout the social world and across our entire lives.

In seeking to describe the introduction of mechanisms of economic rationality and administrative power into the sphere of communication and everyday life, the late philosopher Jürgen Habermas spoke of the "colonization of the lifeworld." Even when he devised this expression, he could not have anticipated that it would so accurately reflect the rapid developments in the logic of surveillance and algorithmic devices as they enter into the most intimate corners of personal life. It would seem that this process has no limits. The race toward generalized interconnectivity, which Zuboff calls the "data hunt," is not the natural result of technological development or even the inevitable evolution of the informational civilization. Rather, it is an *economic imperative* of the highest importance for the accumulation of the behavioral surplus and the production of data commodities. For this accumulation to continue, it must constantly increase not only the quantity of data, but also its *variety* (data on ever-more types of experience), and its *depth* (increasingly intimate data, according to the general idea that the more intimate the data, the more accurate the resulting prediction will be).

The emergence of the control society

This colonization of the lifeworld is not only aimed at collecting data about us, but also at predicting our actions, anticipating our decisions, and prompting our behavior—without us always being aware of it. Zuboff thus identifies a second fundamental mechanism of surveillance capital, namely *the imperative of prediction*, which has a strong behaviorist dimension: "Eventually, surveillance capitalists discovered that the most-predictive behavioral data come from intervening in the state of play in order to nudge, coax, tune, and herd behavior toward profitable outcomes."[6] In other words, a prediction is more likely to come true if they intervene to steer behavior in the direction of the prediction. For example, algorithms may offer a user products at a specific time of the week when they normally spend more, or offer them discounts on tickets to see their favorite band, based on traces left elsewhere on the Internet, collected by smart devices and exchanged between hundreds of third parties without the user's knowledge. Is it a coincidence that an advert for hiking boots appears at the top of your news feed just after you've told a friend on Messenger about your desire to hike the Pacific Crest Trail? The Booking.com site tells you that "there is only one room left at this price" to encourage you to book your hotel stay on the spot. An algorithm from your insurance company that tracks your vehicle remotely rewards you with a discount on your premium because you always obey the speed limit. A police officer stops you pre-emptively because, according to the data points that have been collected, you are in a high-risk area with a high rate of stops, and so on. In addition to these examples, which

6 Zuboff, *The Age of Surveillance Capitalism*, p. 14.

range from nudges by advertisers to more coercive control measures, a growing number of start-ups and apps are using data gathering and the capture of a *surveillance dividend* to try to solve all kinds of social problems (climate change, obesity, poverty, pandemics) through more or less coercive algorithmic control of our behavior.

The imperative to predict does not just impose itself on us surreptitiously. Apps are often presented to us as a means of changing our behavior—as a solution to all our problems, big or small. Thus, the technological solutionism described by Morozov is becoming normalized, in the explicit interest of changing our behavior:

> Consider a smartphone app called BillGuard. It will not only alert you once your intended monthly spending limit has been exceeded but it would also search the Web for coupons that will lower your bills based on your spending patterns. Or take the iBag—an actual bag, equipped with sensors and connectivity, which automatically locks itself—and, presumably, your wallet—when it believes you might be close to overspending. Constant monitoring is what makes such innovations possible.[7]

Like Gilles Deleuze, who spoke of the transition from disciplinary societies to "societies of control,"[8] Zuboff believes that the combination of surveillance capitalism, AI, and the Internet of Things is leading us straight in that direction. The imperative of extraction is combined with the imperative of prediction: This consists not only in anticipating our future behavior, but also in influencing, incentivizing, and controlling it. While we are all free to change our own behavior as we see fit, and to use applications for this purpose, it is important to recognize that the trees of solutionist applications often hide the forest of the imperative of prediction.

This brings to mind the despotism feared by Alexis de Tocqueville, when he studied the possibility of a power capable of intruding into the most secret corners of our lives, behavior, and desires.[9] He imagined this "benign despotism" as the result of a centralization of state power based on the tyranny of the majority—which in his view would impose this herd mentality by directing the pursuit of the good life toward rather

7 Evgeny Morozov, "Like Clueless Guinea Pigs," *Frankfurter Allgemeine Zeitung*, July 30, 2014.

8 Gilles Deleuze, "Postscript on the Societies of Control," available online at the Anarchist Library, theanarchistlibrary.org.

9 Alexis de Tocqueville, *Democracy in America*, Vol. 2, New York: Knopf, 1956 [1840].

mediocre hedonism and entertainment. Yet it must be noted that it is primarily *commercial imperatives*, rather than state or democratic pressures, that today threaten to establish this "benign despotism" that gathers, herds, nudges, coaxes, and manipulates our behavior, making it predictable. As Zuboff notes:

> Competitive pressures produced this shift, in which automated machine processes not only know our behavior but also shape our behavior at scale. With this reorientation from knowledge to power, it is no longer enough to automate information flows about us; the goal now is to automate us. In this phase of surveillance capitalism's evolution, the means of production are subordinated to an increasingly complex and comprehensive "means of behavioral modification."[10]

So, in the era of surveillance capitalism, the "means of production" used to produce commodities are subordinated to the "means of behavioral modification." Our daily actions are literally transformed into commodities to be produced in order to generate more profits. To the question, "What does surveillance capitalism produce?" we might respond with a rather caricatural answer, but one that is not so far from the truth: behaviors.

To illustrate the power of these new means of behavioral modification, Zuboff offers the seemingly innocuous example of the game Pokémon Go. In 2016, Niantic—a Google subsidiary that partnered with Nintendo and the Pokémon company—created an augmented reality game in which players were asked to capture Pokémon in different real-world locations using their smartphones. This viral game, which was downloaded more than 650 million times in seventy countries, and which generated more than $206.5 million in monthly revenue over summer 2016, had a simple but original business model. Its "cost per visit" billing system, analogous to the "cost per click" model, allowed companies to sponsor locations in order to place Pokémon there and attract hundreds or even thousands of people to specific locations, such as in front of or inside their businesses, with a success rate of nearly 98 percent. Zuboff refers to this as "a surveillance capitalist's dream come true," a "living laboratory for telestimulation at scale."[11]

Another more recent example can be found on the social media platform Snapchat, which introduced a chatbot called "My AI" in February

10 Zuboff, *The Age of Surveillance Capitalism*, p. 14.
11 Ibid., p. 198.

2023. The company decided to add this chatbot based on ChatGPT technology in order to increase the interactivity of its platform, interact with users, comment on their photos, and ask for more information about their private lives. Initially reserved for subscribers to the paid version, Snapchat Plus, My AI was then made available to all users in April 2023, with only paid subscribers having the option to disable the bot. Whereas AI was previously seen as an advanced feature, a luxury that had to be paid for, it is now a default and free option, because it allows even more data to be extracted. As the saying goes, "If it's free, then you're the product." This has stoked some disquiet among users put off by the intrusive nature of My AI, and who have suddenly become aware of the surveillance techniques used by this platform, 99 percent of whose revenues are based on targeted advertising.

> While chatting with the chatbot, a plethora of users realized that Snapchat was collecting their personal data, such as their geographical location … Other data collected includes the phone's contact book, information about your smartphone (model, operating system), cookies, IP address and even "the pages you visited before or after accessing our website." … Remarkably, it seems that many users were unaware of the app's practices.[12]

Whereas AI used to be "behind" the screen, like an invisible agent tracking our movements in the shadows, it now appears "in front of us," like a friendly character who wants to interact with us. There is continuity in the business model, based on the mass extraction of personal data. Yet there is a major change in the user experience, which is mediated by an unsolicited conversational agent that we cannot get rid of (unless you pay for it to leave you alone). Snapchat was already monetizing data from our photos, conversations, and geolocation, but My AI now wants to become our "confidant," so that we provide the company with even more intimate information about ourselves. Instead of being a "silent spy," the chatbot becomes an "interactive informant" who talks to us and seeks to establish a relationship of trust.

The widespread use of chatbots, following advances in generative AI such as the rollout of ChatGPT, only intensifies the practices of surveillance identified by Zuboff, which are central to algorithmic capital. It thus generates profits by subtly altering our behavior, whether through

12 Florian Bayard, "Supprimez Snapchat: ChatGPT va-t-il tuer le réseau social?," 01net, April 29, 2023.

recommendations, incentives, targeted advertising, video games, or other features and gadgets. In the age of algorithmic capital, the "benign despotism" foreshadowed by Tocqueville takes the form of a "predictive despotism."

Instrumentarian power and the specter of cybernetic totalitarianism

With algorithmic capital, the production of goods and services operates under a generalized *datafication* of human activities. According to Zuboff, this leads to a new type of domination, "instrumentarian power," which she defines as "the instrumentation and instrumentalization of behavior for the purposes of modification, prediction, monetization, and control." This neologism refers to the specific dynamics by which capital instrumentalizes our lived experience and daily behaviors. In her illustration of this concept, Zuboff draws on the image of the puppet and the puppet master who manipulates it:

> In this formulation, "instrumentation" refers to the puppet: the ubiquitous connected material architecture of sensate computation that renders, interprets, and actuates human experience. "Instrumentalization" denotes the social relations that orient the puppet masters to human experience as surveillance capital wields the machines to transform us into means to others' market ends.[13]

This instrumentarian power is based first on a division of learning within society. The rapid development of the technological capabilities for data collection, control, and analysis allows a small group of hyper powerful companies to accumulate a wealth of knowledge *about* us, but this knowledge is not meant *for* us; rather, it is deliberately concealed and instrumentally used for commercial purposes. Zuboff uses the metaphor of "two texts" to illustrate this idea: The first text is commonly accessible to the public, such as our Facebook page or a web page interface, while the second is hidden and reserved for the exclusive use of surveillance capitalists:

> Much of this public-facing text is composed of what we inscribe on its pages: our posts, blogs, videos, photos, conversations, music, stories,

13 Zuboff, *The Age of Surveillance Capitalism*, pp. 222–3.

> observations, "likes," tweets, and all the great massing hubbub of our lives captured and communicated. Under the regime of surveillance capitalism, however, the first text does not stand alone; it trails a shadow close behind. The first text, full of promise, actually functions as the supply operation for the second text: the *shadow text* ... This one is hidden from our view: "read only" for surveillance capitalists. In this text our experience is dragooned as raw material to be accumulated and analyzed as means to others' market ends.[14]

In a social world dominated by algorithmic capital, all the means necessary for obtaining information, communicating, traveling, socializing, purchasing goods, and accessing many essential services are increasingly in the grip of a handful of digital platforms. They have a "radical monopoly" on various spheres of social interaction. The result is that it is becoming ever more difficult to escape the production of "shadow text" (another name for Big Data or behavioral surplus), which feeds on our most minor daily activities and conversations. To update a powerful phrase from Marx: "Algorithmic capital is dead experience which, vampire-like, lives only by sucking living experience, and lives the more, the more experience it sucks."[15]

Zuboff thus outlines the contours of an "instrumentarian" power that has a strong totalizing tendency. This power is constantly devouring entire areas of social life and of our private existence. Yet it does so not for the totalitarian purposes of imposing ideological conformity, but rather for the sake of the expansion of the behavioral surplus, ever-more accurate predictions, and the automated regulation of our behavior. Zuboff stresses that we should avoid understanding this unique form of power in the familiar language of Big Brother and totalitarianism outlined by George Orwell. Instead of taking the novel *1984* as an example, she suggests rereading *Walden Two*, the utopia published by behavioral psychologist B. F. Skinner in 1948. In this novel, the writer imagines a world in which the latest techniques of behavioral engineering are used to "improve" society. Behaviorism's radical indifference to the subjective lives of individuals, instead settling for an "external" knowledge of others' behavior, allows for a direct influencing of behavior through various conditioning mechanisms. In 1958, Hannah Arendt presciently declared that the "trouble with modern theories of behaviorism is not

---

14 Ibid., p. 124.

15 Marx, *Capital*, Vol. I, p. 342: "Capital is dead labour which, vampire-like, lives only by sucking living labour, and lives the more, the more labour it sucks."

that they are wrong but that they could become true."[16] Today, Zuboff argues that this behaviorist utopia has become not only technically possible, but increasingly desirable and necessary for capital. This is what she attempts to capture with the concept of the Big Other:

> I now name the apparatus *Big Other*: it is the sensate, computational, connected puppet that renders, monitors, computes, and modifies human behavior. Big Other combines these functions of knowing and doing to achieve a pervasive and unprecedented *means of behavioral modification*.[17]

The mysterious concept of the Big Other is in many ways reminiscent of some Marxian themes such as the "vast automaton" or "domination without a subject"—suggesting a kind of logic of unlimited accumulation whose sole purpose is self-expanding domination. The critique of capitalism as the "self-valorization of value," as today advanced for example by the proponents of value-critique (*Wertkritik*), here somehow converges with Zuboff's analysis, which describes how surveillance capital combines with the technological apparatus of AI, the Internet of Things, and ubiquitous computing to produce a new and disturbing form of automated domination.[18] In a way, the Big Other represents a personification of capital, not as an economic logic centered on the extraction of personal data, but as a new form of domination based on predicting and manipulating behavior. Other authors use expressions such as "cybernetic totalitarianism," "soft digital totalitarianism," or the "algorithmic Leviathan" to describe this phenomenon.[19]

However, Zuboff is careful to distinguish this "instrumentarianism" from the more traditional forms of twentieth-century totalitarianism (see Table 4.1). The modalities of power cannot easily be reduced to the traditional categorization of totalitarianism based on terror, ideological uniformity, personality cults, and so on. Rather, we need a new analytical framework in order to fully grasp the novelty of this development, whose

16 Hannah Arendt, *The Human Condition*, Chicago: University of Chicago Press, 1998, p. 322.

17 Zuboff, *The Age of Surveillance Capitalism*, p. 237.

18 See, for example, the works of Robert Kurz, Anselm Jappe, and Roswitha Scholz associated with that current. See also Eric Martin and Maxime Ouellet, eds., *La tyrannie de la valeur: Débats pour le renouvellement de la théorie critique*, Montreal: Écosociété, 2014.

19 Ouellet, *La révolution culturelle du capital*; Sadin, *La silicolonisation du monde*; Sadin, *L'intelligence artificielle ou l'enjeu du siècle.*

contours still remain unclear and fluid, especially today as alliances are being forged between operators of technological power and far-right political forces.

Is the bleak portrait of surveillance capital, as symbolized by Zuboff's disturbing figure of the Big Other, overly pessimistic, alarmist, or exaggerated? Does it point to a future in which her own fears and projections take the form of a digital dystopia? Or does it present us with a form of what Günther Anders calls "exaggeration as a method of knowledge," in the sense that she amplifies processes and trends already at work in the rapid technological developments of the Fourth Industrial Revolution to make it easier to understand? This process is widely used in TV series such as *Black Mirror*, with the added twist that the developments depicted in certain episodes have, in fact, already become reality. Hans Jonas's "heuristics of fear," Jean-Pierre Dupuy's "enlightened catastrophism," and Theodor Adorno and Max Horkheimer's well-known strategy of dramatization in *Dialectic of Enlightenment* represent different ways of exaggerating things in order to help us understand the social world in which we live. We think this is also what Zuboff's method is doing.

Zuboff's blind spots

Zuboff can be credited with monumental findings, and her theoretical framework is valuable for understanding the radical novelty of algorithmic capital. We will make extensive use of Zuboff's concepts, while modifying them, deepening them, and directing them toward areas she has not explored. What she calls surveillance capitalism is certainly a central feature of what we call algorithmic capitalism, although we will need to add a few more layers to her findings to paint a fuller picture of the new institutionalized social order.

However, we also feel it is essential to point out significant blind spots in Zuboff's research, ones that leave us with a great deal of work to do if we are to build on it in contributing to a critical theory of algorithmic capitalism. The greatest weakness is undoubtedly her complete disregard for questions of labor, which we noted earlier. Indeed, she does not address the transformation of labor in contemporary capitalism at all, instead focusing on the relationship between user experience and the accumulation of the behavioral surplus. It is as if extraction is the sole mechanism behind accumulation, while exploitation has left the scene. Reading her work, we might conclude that the value produced

Table 4.1. Zuboff's comparison: Two species of power

Comparative elements	*Totalitarianism*	*Instrumentarianism*
Central metaphor	Big Brother	Big Other
Totalistic vision	Total possession	Total certainty
Transcendent purpose	Perfection of society/species defined by class or race	Automation of market/ society for the certainty of guaranteed outcomes
Locus of power	Control of the means of violence	Control of the division of learning in society
Means of power	Hierarchical administration of terror	Ownership of the means of behavioral modification
Foundational mechanisms	Arbitrary terror; murder	Dispossession of behavioral surplus for computation, control, prediction
Theory and practice	Theory legitimizes practice	Practice conceals theory
Ideological style	Political religion	Radical indifference
Social strategies	Atomization and division; total believers or total enemies	Otherization of predictable organisms
Core social processes	In-group, out-group, for conformity and obedience	Hive mind, social comparison for confluence and predictability
Unit of social production	Mass (political)	Population (statistical)
Vector of social influence	"Re-education" exerts control from inside-out	Behavior modification exerts control from outside-in
Social patterning	Radical isolation	Radical connection
Demands on individuals	Absolute loyalty through subjugation to state/species	Absolute transparency, through subjugation to guaranteed outcomes
Primary source of individual attachment to power	Thwarted identity	Thwarted effectiveness
Primary mode of individual attachment to power	Identification	Dependency

Source: Zuboff, *The Age of Surveillance Capitalism*

in surveillance capitalism is not at all reliant on human labor but rather depends solely on the human experience that generates data. We thus arrive at the rather perplexing idea of a capitalism without labor. In our view, however, it would be misguided to conclude that the link between labor and value has been definitively broken with the advent of algorithmic capital, or that the major players in the new era of capitalism are producing value without any inputs other than datafied human experience. Companies such as Meta and Google certainly do have an extremely high market-value-to-employee ratio, but that does not mean that their wealth is not based on vast operations that can indeed be identified as human labor. Rather, a vast overhaul of the world of work is today in full swing. This is such a profound change that we will need the following two theses to examine it more closely. Data extraction is surely a revolutionary novelty in capitalism, but it is only through its interaction with human labor—that is, it is only in this new configuration of the relationship between extraction and exploitation—that algorithmic capital emerges.

A second blind spot in Zuboff's analysis is her limited consideration of "new means of production," the use of Big Data, algorithms, prediction machines, and AI tools that facilitate data extraction, behavior prediction, and the automation of decision-making processes and the labor process as a whole. This will be the subject of Thesis 7, which will further analyze what might be called "algorithmic production."

Third, as we have pointed out, Zuboff also neglects the role of extractive industries, the enormous consumption of energy, and, more generally, the entire material infrastructure that stands behind AI. In Thesis 15, we will return to the ecological dimension of algorithmic capitalism.

Fourth, the analysis of *instrumentarian power*—a singular form of algorithmic domination outlined by Zuboff—will be examined in more detail, subjected to critique, specified, and adapted by our analysis of algorithmic power in Thesis 9. The metaphor of the Big Other surely lacks some of the subtlety needed to illuminate the various ramifications of algorithmic power. It may be useful as a magnifying mirror of algorithmic capital—but nevertheless, we need to identify more precise mechanisms to understand the complex articulations between algorithms and other forms of power, oppression, and alienation already present in society.

Fifth, Zuboff's framework neglects the question of the state, as well as geopolitical configurations. If our aim is to analyze algorithmic capitalism as an institutionalized social order, we will need to examine these as well, which we will do in Theses 10, 11, and 12.

Finally, Zuboff's analysis falls short of a truly emancipatory scope. Her underlying political perspective seems closer to an advocacy for a return to forms of capitalism more aligned with the protection of privacy and individual liberties. Our aim here is, rather, to contribute to a critical theory that seeks to highlight possible paths *beyond* capitalism itself.

In this sense, although we do make use of some concepts designed by Zuboff to elucidate new mechanisms of accumulation, we need to take a more radical approach in the remaining theses of this book.

All in all, surveillance capital, as conceptualized by Zuboff, serves as a starting point to investigate the various configurations of a new logic of accumulation that is gradually taking hold of various spheres of social life. However, the critical role of surveillance, which emphasizes the tension between the imperative of data extraction and individuals' personal lives, is only one aspect among many. It cannot be said to constitute the *central* aspect of contemporary capitalism. Indeed, capitalism has more than one trick up its sleeve. Algorithmic capital also needs minerals, energy, human labor, business models, cities, states—in short, a complex global institutional apparatus to enable its growth. In Theses 3 and 4, we entered into the realm of experience, via the question of time and Zuboff's threatening silhouette of generalized surveillance by the Big Other. It is now time to head into the secret lair of algorithmic production, starting with today's new and emerging forms of exploitation of human labor.

Thesis 5

AI is not abolishing work, but reorganizing and exploiting it

If every tool, when summoned, or even by intelligent anticipation, could do the work that befits it, just as the creations of Daedalus moved of themselves, or the tripods of Hephaestus went of their own accord to their sacred work, if the weavers' shuttles were to weave of themselves, then there would be no need either of apprentices for the master craftsmen, or of slaves for the lords.

—Aristotle

The myth of the end of work

Political economy has long indulged in myths about the prospect of a society finally freed from labor thanks to the complete mechanization of production.[1] Already in the mid-nineteenth century, John Stuart Mill wondered whether all the new machinery had reduced human toil by even an iota and freed up more time. Marx offered an emphatic response in *Capital*: Capitalist machinery had heightened the exploitation of labor, rather than lightened its burden. Mechanization is a process that arises from the laws of capitalist competition, and the imperative to produce surplus-value. And, as long as the introduction of new machinery takes place within a capitalist context, labor will only be further exploited and subjected to greater upheavals, pressures, and precarity.

1 This thesis and Thesis 6 draw on and adapt the analysis that appears in Jonathan Martineau and Jonathan Durand Folco, "Les quatre moments du travail algorithmique: Vers une synthèse théorique," *Anthropologie et Sociétés* 47, no. 1, 2023.

The concept of automation in vogue today brings with it the latest iteration of this old debate. It is said that the automation of processes producing value and management systems, facilitated to an unprecedented degree by algorithmic systems, robotics, and AI, could well render human labor obsolete, this time around. Yet, just as in the nineteenth century, contemporary automation is primarily driven by an economic logic. It slashes necessary labor time, thereby increasing the production of surplus-value. It reduces the time of circulation, accelerates and automates the movement of financial capital, and establishes new modes of social relations in social reproduction and consumption, which are now mediated by machines. However, this automation process does not simply lead to the end of work, but rather to its complex reorganization into new arrangements, through its either formal or real *subsumption* to algorithmic capital. As Mary Gray and Siddharth Suri aptly summarize, "The great paradox of automation is that the desire to eliminate human labor always generates new tasks for humans."[2] While automation surely is a real and direct threat to many categories of jobs, it is far from leading us into a post-work society. What some people today celebrate or fear as the end of work is, in fact, the beginning of algorithmic labor.

Furthermore, the myth relies on a causal connection between the end of work and technological innovation that can today be put into question. As Aaron Benanav argues, what might appear in statistics as labor shortages are not, in the current context, simply a function of technological change. Rather, they result from longer-term trends of secular stagnation in global capitalism, causing, in turn, a lower demand for labor.[3] This gives rise to *under*employment, which, in our view, acts as a fertile soil for the rise of low-wage and precarious labor, taking forms that we describe below, such as microwork and on-demand digital labor.

We think it is necessary to pay closer analytical attention to the issue of labor, especially given the current state of the critical literature on contemporary capitalism. We can take the example of Zuboff's extremely influential contribution, discussed in the previous thesis, whose analysis disregards the issue of labor and bases the political economy of contemporary capitalism on a relationship of dispossession between *users* and "surveillance capitalists." Strangely, these capitalists do not seem to exploit workers. Moreover, a growing body of critical literature conceptualizes

2 Mary L. Gray and Siddharth Suri, *Ghost Work: How to Stop Silicon Valley from Building a New Global Underclass*, Boston: Houghton Mifflin Harcourt, 2019, p. xxii.

3 Aaron Benanav, *Automation and the Future of Work*, London: Verso, 2022.

the economic power of web giants as a *rentier* power.[4] While rent surely is a crucial aspect of the political economy of platforms—and we will come back to it in Theses 7 and 8—this literature tends sometimes to downplay the role of production, labor, and exploitation in the new economy.

The three moments of algorithmic labor

In a previous study, we proposed a concept of *algorithmic labor* as the unity of four moments;[5] however, in this thesis we will deal with just three of them, and, in the following thesis, we will formulate a complete thesis on the fourth moment, which concerns domestic labor and social reproduction. While most recent contributions focus on the emergence of "digital labor," here we want to adopt a broader perspective. Beyond (and mixed in with) what Antonio Casilli calls digital labor also lies a vast world of human labor that (re)produces the material infrastructure of algorithmic capital.[6] This labor, both prior to and simultaneous with digital labor, takes several forms, which we can break down into two essential moments: the extractive and the industrial/logistical. Such labor ranges from the villager in Kivu extracting the minerals used to manufacture computers and mobile phones to the workers who crowd into a vast industrial complex in Shenzhen to assemble technological devices; to the factory supervisor who uses industrial platforms to automate production processes; and to the order picker at Amazon. The work of all these people coincides with, and underpins and intertwines with, the digital labor of the AI engineer at Google, the freelance web designer, the Uber driver, the Facebook community manager, the YouTuber, the content creator on OnlyFans, and the click worker in Kenya who "assists" and "trains" the ChatGPT algorithm.

Here, we want to create a conceptual map of the reconfiguration of work within and around online platforms—digital labor—but also to head deeper into the "secret lair of production" of the *material infrastructure* of algorithmic capital. Doing this will allow us to see the other moments of algorithmic labor—its extractive and industrial/logistical dimensions—which thus form a single whole together with digital labor.

---

4 Cédric Durand, *How Silicon Valley Unleashed Techno-Feudalism: The Making of the Digital Economy*, London: Verso, 2025; Yanis Varoufakis, *Technofeudalism: What Killed Capitalism*, Brooklyn: Melville House, 2024.

5 Martineau and Folco, "Les quatre moments du travail algorithmique."

6 Antonio Casilli, *Waiting for Robots: The Hired Hands of Automation*, Chicago: University of Chicago Press, 2025.

These moments are often themselves nondigital and include multiple forms of labor, be it forced, salaried, precarious, freelance, service labor, or child labor. All these workers together form a *global worker* who (re)produces the material infrastructure and the social relation of algorithmic capital.

The platform and its ecosystem

The first stage in analyzing algorithmic labor is to examine a new socio-economic entity that is playing an increasingly important role in the organization of work in contemporary capitalism: the *platform*. While they are often presented as a new business model, we can also see them, as in Casilli's account, as *market/company hybrids* that operate according to three principles: coordination, collection, and capture.

First, they coordinate actors on the basis of a system of differential prices. The various actors interact through the platform, in relationships that are calculated and shaped by its algorithms. For example, YouTube coordinates three types of users, each with their own "price": user-advertisers, who pay YouTube to advertise their products; user-viewers, who consume the platform's content at no cost (YouTube "sells" their attention time to advertisers); and user-YouTubers, the popular content creators who are sometimes paid by the platform.

Second, platforms collect massive amounts of data from their users—mainly in order to improve their algorithmic coordination, but also to monetize this data directly. Third, platforms capture the value produced by these connections. They do this using various mechanisms, such as transaction fees, the sale of products and services, selling target audiences to advertisers, and the transfer of nonpaid tasks to users.

The platform model thus breaks out of the closed space of the individual company. It allows value to be produced and captured in commodified relationships that it itself generates, by algorithmically coordinating actors on the basis of the collection and analysis of data. The platform thus creates an ecosystem to which it delegates the production of value—which it then attempts to capture, "enclose," and control. It thus involves dynamics of both centrifugal and centripetal movement that centralizes economic power even as it decentralizes participation in the economy.[7]

7 Martin Kenney and John Zysman, "The Platform Economy: Restructuring the Space of Capitalist Accumulation," *Cambridge Journal of Regions, Economy and Society* 13, no. 1, 2020, p. 18.

Not all platforms have the same functions, as Nick Srnicek has observed.[8] Doubtless the best-known ones are advertising platforms, whose revenues depend largely on the sale of predictive products to advertisers: for instance, Facebook and Google. Cloud platforms, such as Amazon Web Service (accounting for more than half of Amazon's total profits in 2024), Microsoft Azure, and Google Cloud, are concentrations of infrastructure and equipment that are able to store and analyze data, and these digital capabilities are leased to clients. Industrial platforms such as GE's Predix and Siemens's MindSphere focus on transforming industrial production processes into online processes using digital tools and applications that they can lease or sell to companies wanting to automate their production. Lean platforms are truly the archetype of this hybrid form fusing market and company. They stand out, above all, for their outsourcing of costs: They do not own all their means of production and avoid direct wage relations with their "users" as much as possible. Instead, they connect users in order to facilitate the exchange of products or services, from which they extract a fee. Several platforms, and even ones that do produce goods and/or services, can thus generate rents simply by owning the algorithm and the interface that coordinates exchanges (as do, for example, Uber, Airbnb, and Deliveroo).[9]

The digital moment of algorithmic labor

The rise of platforms and algorithmic capital is bringing about a far-reaching reorganization of work and the rise of "digital labor." Casilli defines this key concept as "work broken down into datafied, menial tasks, and which serves to train automatic systems."[10] Casilli focuses solely on work carried out within the platform ecosystem, or using digital tools. However, there is also a broader definition of digital labor, referring to all alienated digital work that produces or uses digital technologies and content. This latter definition, proposed by Christian Fuchs and Marisol Sandoval, includes the production of the material infrastructure of

8 Nick Srnicek, *Platform Capitalism*, Cambridge: Polity, 2016.

9 Jathan Sadowski, "The Internet of Landlords: Digital Platforms and New Mechanisms of Rentier Capitalism," *Antipode* 52, no. 2, March 2020, pp. 562–80; Niels van Doorn, "Platform Labour: On the Gendered and Racialised Exploitation of Low-Income Service Work in the 'On-Demand' Economy," *Information, Communication & Society* 20, no. 6, 2017, pp. 898–914.

10 Antonio Casilli, *En attendant les robots: Enquête sur le travail du clic*, Paris: Seuil, 2019, p. 14 (passage not included in the English edition).

digital technologies, and thus comes closer to what we aim to grasp with our concept of algorithmic labor.[11] Let us first build on Casilli's definition to examine platform-related digital labor before moving on to the extractive and industrial/logistical aspects of algorithmic labor.

"Digital labor" in the strict sense, meaning the labor that takes place within the platform ecosystem, corresponds to an outsourcing and fragmentation of work. It is mainly carried on outside the premises of a company or platform and is often divided into tasks that are even smaller and more specific than those formerly created by the division of industrial labor and by Taylorism. It is also work that requires access to a digital device, and by its nature produces data that provides the platform with an "information service."[12] Casilli identifies three main types of digital labor that underpin the economic architecture of the platform: (1) on-demand digital labor; (2) microwork; and (3) social media labor. Each of these produce—in his terminology—*qualifying*, *monetizing*, and *automating* values.[13] In addition to Casilli's categories, we also add another, (4) "venture work," which we define below.

On-demand digital labor, typical of lean platforms, is performed by providers whom the platform connects with those in demand of some service or product. They may be remunerated in various ways (hourly rates, per piece/task, and so on). For example, an Uber driver will work in response to customer demand on the platform, and handymen on Handy or TaskRabbit visit people who have registered their needs on the app. The platforms and apps take a cut of these transactions, and also collect—indeed, on a massive scale—the data generated, which becomes a valuable asset.[14] In Casilli's terms, on-demand work thus produces: (1) *qualifying* values, where users rate each other, share information, choose certain offers at certain prices, and evaluate and criticize certain locations, products, features, and so forth; (2) *monetizing* values, where the

---

11 Christian Fuchs and Marisol Sandoval, "Digital Workers of the World Unite!: A Framework for Critically Theorising and Analysing Digital Labour," *tripleC: Communication, Capitalism and Critique* 12, no. 2, 2014, p. 11.

12 Niels van Doorn and Adam Badger, "Platform Capitalism's Hidden Abode: Producing Data Assets in the Gig Economy," *Antipode* 52, no. 5, 2020, p. 1476.

13 Digital labor is involved in the production of "value" in the more classical Marxian sense as well. It is important, however, to acknowledge the complexity of the new political economic configurations and consider digital labor as a multiform activity that can produce data as commodities and/or data as raw material, and that can be waged and thus "exploited," but can also in some forms be nonwaged and thus better understood as an activity subjected to a mechanism of extraction or dispossession. We return to this topic in Thesis 7.

14 Van Doorn and Badger, "Platform Capitalism's Hidden Abode," p. 1477.

extracted data is resold to third parties or, in the form of data assets, can attract capital investment; and (3) *automation* value, where the extracted data is used to develop algorithms and optimize systems.[15]

Microwork involves delegating fragmented work actions to platform users, or "click-workers," who perform a series of repetitive, often tedious tasks, carried out as microsteps in a larger process.[16] The rise of algorithmic capital has created a global army of these new click workers, working from their own homes, or sometimes crammed into disused old factories across the globe that have been converted into "click farms." This work is mostly invisible, carried out by online agents who are almost constantly intervening in the interactions between platforms and their users, in what Gray and Suri call "ghost work." Some platforms specifically recruit Internet users to perform these tasks, bringing them together and delegating their own "piecework" to this workforce. Or they may get them to work for clients, who use the microwork platform to outsource various tasks that it would be impractical to perform in-house. These tasks are varied, but one common trait is that they supplement the work of machines or algorithms, or else train them—in other words, they perform calculation or assessment tasks that the algorithms need but are not necessarily able to carry out by themselves. This often means digital "chores," like annotating videos, filtering tweets, transcribing scanned documents, answering questionnaires, correcting database entries, or connecting two similar products in a catalog. When they are paid (and this is not always the case), these "microtaskers" of algorithmic labor are often compensated on a per-task basis, usually a few cents per operation. Remuneration for microwork remains very low, and much microwork remains unpaid, presented as a game, "play-bour"[17]—a hobby or a secondary activity. Here, the worker is not subordinate to a boss but participates in and interacts with the platform, and so is not considered an employee.

15 These three forms of value are a helpful conceptual innovation from Casilli to help understand the exploitation of labor by platforms, though differing from a more classical Marxian conception of value that would attend specifically to the socially necessary labor time that goes into the production of data commodities. On data valuation, see also Jathan Sadowski, "When Data Is Capital: Datafication, Accumulation, and Extraction," *Big Data & Society*, no. 1, January 2019, pp. 5–6, and Amanda Parsons and Salome Viljoen, "Valuing Social Data," *Columbia Law Review* 124, no. 4, 2023. We will return to these questions in Thesis 7.

16 Casilli, *Waiting for Robots*; Ursula Huws, *Labor in the Global Digital Economy: The Cybertariat Comes of Age*, New York: Monthly Review Press, 2014, pp. 166–7.

17 Julian Kücklich, "Precarious Playbour: Modders and the Digital Games Industry," *Fibreculture Journal*, no. 5, 2005.

Amazon and Clickworker's microtask platforms are readily accessible. Other algorithmic capital giants have their own platforms, although they are less easily accessed and sometimes reserved for certain microtaskers who are qualified to train or support a particular algorithm: Google has its RaterHub, Microsoft its UHRS. Meta uses an army of microworkers on a flat-rate basis, mainly in the Philippines, for content moderation and other tasks.[18] The Australian firm Appen offers crowdsourcing and microlabor services to eight of the world's ten largest technology companies. It is emblematic of the growth of this sector.

Thus, behind artificial intelligence and the proliferation of algorithms stands the reality of microwork. For these technologies to function, they require constant inputs from human brains and fingers. We are living in the era of "artificial artificial intelligence" (sic). This was how Jeff Bezos described Amazon Mechanical Turk, whose operation relies on an army of microworkers. The stubborn persistence of human labor in the era of "artificial artificial intelligence" means that the rise of algorithms will demand ever more workers to supervise, optimize, and train them. This brings us to the heart of "the paradox of automation's last mile," where each further development in automation creates new volatile, transient, and unstable microwork markets to support it.[19] Click-workers often share the typical conditions of what Marx called the "stagnant" part of the industrial reserve army: precarity, enforced flexibility, low wages, and the performance of the most thankless tasks.[20]

Leaving aside the case of microwork for Amazon, which remains centered on US soil, this type of labor (and its most arduous tasks) is increasingly outsourced to the Global South.[21] Digital labor is thus building its own digital value-chains according to an outsourcing dynamic that both intermeshes with and updates the international division of labor between Global North and South.[22] To date, most demand is concentrated in the G7 countries, while the majority of microworkers are in the Global South. This type of worker does not enjoy job security, the prospect of unionization, or contracts ensuring them fixed remuneration. While microwork brings little to the microworker, it nevertheless

18 Srnicek, *Platform Capitalism*, p. 49.

19 Gray and Suri, *Ghost Work*, p. 36.

20 Marx, *Capital*, Vol. I, pp. 794–6.

21 Dyer-Witheford, Kjøsen, and Steinhoff, *Inhuman Power*, p. 77; Kenney and Zysman, "The Platform Economy," p. 9.

22 Christian Fuchs, *Digital Labour and Karl Marx*, New York: Routledge, 2014; Phil Jones, *Work Without the Worker: Labour in the Age of Platform Capitalism*, London: Verso, 2021.

produces the three types of value for the platform, as identified by Casilli: qualification value (classification, content evaluation), monetization value (the production of data), and automation value (training, algorithm support).

The third type of digital labor—social media labor—is more controversial, in the sense that current scholarship is divided over whether it can be considered a type of *labor*. On the one hand, interaction between platforms and users is presented as a pleasant form of entertainment, a leisure activity; on the other, it is presented as work.[23] We believe that social networking is a productive activity with economic value and that, in this sense, it can be useful to consider it as work. It may sometimes be a tedious and alienating form of work, and sometimes a fun and enjoyable one, but, ultimately, this is what it remains from the perspective of critical political economy.

Social media labor refers to the work carried out by platform users, often free of charge, to keep these platforms running or to improve them. Social media users who interact with the platform by producing, evaluating, and classifying content are a good example of this form of digital labor, but it takes other forms as well. Casilli distinguishes between four forms of social media workers. First, there are "organic users," who are ordinary users with a profile on Facebook, Instagram, or TikTok, for example; and second, "professional users," such as YouTubers, influencers, and coaches. In addition to these are "moderators," who are content evaluators, often paid a flat fee. This moderation work entails no small amount of drudgery and involves very difficult tasks, such as sorting content on social media, which means repeatedly viewing huge quantities of offensive, violent, and traumatic images, videos, or comments. Developers of chatbots such as ChatGPT, Gemini, and ERNIE employ thousands of microworkers to detoxify the content generated by their high-performance algorithms. Finally, there are the "click farmers." These "manufacture virality," says Casilli, producing "vast volumes of ratings, shares, and 'likes' of all kinds, including from 'click farms' in China or Thailand."[24] This social media labor produces not only *content* (text, images, comments, and videos), but also *data*.

The platform is able to capture different types of value from all these activities. Using Casilli's helpful terminology, these can broken down into the categories of qualification, monetization, and automation. Qualifying value is extracted from the actions of users who produce,

23 Casilli, *Waiting for Robots*, pp. 187ff.
24 Ibid., p. 188.

classify, and evaluate content. This can be as simple as liking something on Facebook, sharing a photo on Instagram, posting a video on TikTok, reviewing a product on Amazon, adding a hashtag on Twitter/X, rating a restaurant on Yelp! or a mechanic on Google Maps. All these actions help the platform to renew its content (to attract and retain user attention), and to classify and evaluate information. This may also mean an evaluation of other users; all reputational mechanisms (rating your Uber driver or following an Instagram page) fall under this category. These mechanisms often lead to digital discrimination in the production and use of algorithms, which reproduce racism, sexism, classism, heterosexism, and other forms of discrimination present in society.[25] We will return to this point later. All these acts also produce monetization value, adding to the platforms' data assets. As in other types of digital labor, the data produced can be sold to advertisers or data brokers (such as Acxiom and Epsilon, for example), or attract capital investment.[26] The automation value of networked labor is obtained when classifying actions (equivalent to applying human judgment to data) are used to feed and train algorithms based on qualified data. It can also be derived from content production: Uploading a photo of a face onto Facebook or Instagram provides material for training facial recognition algorithms.

In addition to these three types of digital labor in algorithmic capital, there are the platforms' direct employees and their core of entrepreneurs and managerial personnel.[27] These highly skilled workers are generally well paid and work in excellent conditions: The wealth of digital labor is concentrated in the hands of this labor elite, known as "venture labor."[28] These workers represent the cream of the crop of highly skilled programmers, AI scientists, senior managers, web engineers, data science experts, and other skilled workers, and their employment relationships stick closer to the traditional salary-based model. Added to this is an army of freelancers and contract workers, who sometimes work at company HQ with the entrepreneurial core, or else in separate spaces, or from home.

---

25 Safiya Umoja Noble, *Algorithms of Oppression: How Search Engines Reinforce Racism*, New York: New York University Press, 2018; Virginia Eubanks, *Automating Inequality: How High-Tech Tools Profile, Police, and Punish the Poor*, New York: Picador, 2019.

26 Chih-Liang Yeh, "Pursuing Consumer Empowerment in the Age of Big Data: A Comprehensive Regulatory Framework for Data Brokers," *Telecommunications Policy* 42, no. 4, May 2018, pp. 282–92; Sadowski, "When Data Is Capital."

27 Kenney and Zysman, "The Platform Economy," pp. 61–4.

28 Gina Neff, *Venture Labor: Work and the Burden of Risk in Innovative Industries*, Cambridge, MA: MIT Press, 2012.

This work, which is often called "independent" but in fact depends on the platform, obeys a similar logic to microwork. It, too, represents a fragmentation and outsourcing of work, albeit with a lower degree of task fragmentation than for the online microworkers and generally higher remuneration.

The development of digital labor clearly illustrates a major trend analyzed by Ursula Huws, namely the general shift in the "digital age" toward activities that are directly productive for individual capitalist companies, but which are not paid,[29] or even recognized as work. Thus, struggles for better conditions and recognition of this work are constantly shaking the world of platforms. From the first class-action lawsuits brought by community leaders against AOL twenty-five years ago, to the complaints brought by contributors to Huffington Post or Wikipedia, or by Vine "produsers," all of them are demanding remuneration for the work they do on and for the platform. These workers see themselves as professionals, or at least aspire to move beyond the stage of so-called hobbyism and amateurism.[30] In this sense, they work in the hope of becoming professionals, and the platforms systematically deny them this possibility. The working conditions of waged employees at some tech giants such as Amazon, whose warehouses are laboratories for the automation of work, are also setting off major struggles.[31] We will return to this subject in Thesis 18.

However, the conflict over digital labor is more explicit and obvious in the case of on-demand work. This labor is low paid, often precarious, and subject to a structural devaluation "rooted in a long history of racialized, gendered, and classed subordination that originated when capitalism first sought to scale globally."[32] Indeed, most cleaners, caretakers, and homecare providers belong to racialized communities, especially in urban settings. This work is often not recognized as work by the platforms, who say that it is performed by "independent contractors." They make this claim despite their enormous control over this work, disciplining it through various hiring and management techniques that tend to maintain a surplus of available labor.[33] Some drivers at Uber, Lyft, and other on-demand transport services have sought recognition as workers or employees of the platforms concerned, in some cases through union

29 Huws, *Labor in the Global Digital Economy*, p. 155.
30 Casilli, *Waiting for Robots*.
31 Kenney and Zysman, "The Platform Economy," pp. 14–15.
32 Van Doorn and Badger, "Platform Capitalism's Hidden Abode," p. 1485.
33 Van Doorn, "Platform Labour," pp. 902, 904, 907.

drives. The on-demand workers generally present their grievances and demands to the platforms, most of which call themselves intermediaries between users, rather than employers. This means washing their hands of responsibility for the working conditions of the "users" who are the ones supplying goods and services.[34] As for microwork, platforms also retreat into a role of supposed neutrality, leaving it up to customers and workers to handle their own conflicts, sometimes through the app itself. Customers may refuse to pay for work they consider poorly done, while workers can sometimes organize to evaluate service users and report abuses. Thus, labor conflicts turn into "disputes between two categories of users."[35]

We will return to the many different struggles in the world of digital labor in Thesis 18, as they reflect the fight for better contracts (even as independent contractors) and, moreover, for these jobs to be duly recognized and treated as salaried employment. Digital workers are developing a collective subjectivity, which is shaped by both their adverse conditions and their struggles to improve them. As historian E. P. Thompson put it, class is not just the product of workers' position relative to the means of production. Rather, class "happens" when people struggle collectively and thus create a shared identity.[36] The so-called "users," "partners," or "contractors" of platforms instead want to be recognized as *workers*. This strategy of seeking recognition and forming a collective subjectivity is an important step in the politicization of the economic relations of algorithmic capital and the formation of a new working class.

Algorithmic labor beyond digital labor: The extractivist moment

The next two sections seek to highlight the global workforce that is producing the devices, goods, and technologies necessary for digital labor and for the accumulation of algorithmic capital. Their labor is needed in other stages in the process, in the phases of extraction, manufacturing, and logistics. Thesis 6 will focus on the production of bodies and on the labor of social reproduction necessary to sustain algorithmic capital.

34 Michael Sainato, "'I Made £3.75 an Hour': Lyft and Uber Drivers Push to Unionise for Better Pay," *The Guardian*, March 22, 2019; Clara Dallaire-Fortier, *Le travail sous le capitalisme de plateforme*, Montreal: IRIS, 2020, p. 4.

35 Casilli, *Waiting for Robots*, p. 151.

36 E. P. Thompson, *The Making of the English Working Class*, London: Penguin, 1963.

Let us first consider the production of the material infrastructure of algorithmic capital in its extractivist phase. By this, we mean the extraction of all the materials, minerals, rare earths, and other natural resources that go into the manufacture of portable devices, computers, fiber optics, communication towers, cables, and other hardware.[37] We do not have room here to review all the different types of extractivist algorithmic labor. But emblematic of this moment is the militarized political economy of labor that has developed around the mining of resources used in the manufacture of computers and smartphones—particularly strategic metals such as coltan and cassiterite. Coltan is especially important because it is used to produce tantalum powder, needed in the manufacture of capacitors found in virtually all electronic devices. More than 80 percent of the world's coltan is found in regions straddling the eastern Democratic Republic of Congo (DRC) and Rwanda, which are the two largest producers worldwide.

Coltan mining takes many forms. It ranges from forced labor at gunpoint organized by paramilitaries—and often by the Congolese army itself—who control certain mines to smaller-scale initiatives or mining by families or groups of Congolese villagers seeking a livelihood. As Michael Nest has shown, armed groups profit from coltan and use it to procure weapons and other military equipment.[38] Mining work is obviously dangerous and often conducted in appalling health and safety conditions. Unlike gold, copper, or diamond mines, for example, the coltan mines in the DRC's Kivu region are often artisanal in nature. Safety rules, where they do exist, are set by those who control them. Individual artisanal mines are often operated by self-employed miners, who try to earn a living through their own labor or even by renting out their mines. Accidents and deaths are a common occurrence. Both waged miners and artisanal mine operators generally earn meager incomes. Intermediaries maintain a monopoly on technical and financial knowledge that often prevents local miners from correctly differentiating between minerals (coltan and cassiterite, for example, look similar but have completely different values; their purity varies greatly from one site to another; and so on), or from being aware of significant fluctuations in market ore prices. The hard toil in the mines is thus somewhat separate from the commercial transactions taking place over the heads of the mineworkers.

37 Peter Dauvergne, *AI in the Wild: Sustainability in the Age of Artificial Intelligence*, Cambridge, MA: MIT Press, 2020, pp. 138–9; Crawford, *Atlas of AI*; Fuchs, *Digital Labour and Karl Marx*, p. 171.

38 Michael Nest, *Coltan*, Hoboken, NJ: Wiley, 2014.

But price volatility can just as easily ruin an intermediary in a matter of hours if they buy coltan in the forest at a high price, only to realize that it is now worth only one-tenth as much when they try to resell it on some urban exchange.

These relations take on many different avatars. There is the Congolese miner who emerges from the ground with a load containing coltan and uses his mobile phone to make a call; the militia leader who sends an instant message to coordinate his attack on a site potentially rich in resources; the coltan buyer who tracks current transaction prices in real time on their phone; or the trader dealing with international clients online. All of them embody complex interconnections between the extractive and digital moments of algorithmic labor.[39]

The industrial/logistical moment

The industrial/logistical moment mediates the extractivist and digital moments of algorithmic labor. In between the extraction of resources and the production of value and data by digital labor, there is both the technical apparatus itself and the labor that produced it. Algorithmic capital relies on technology that must be accessible, individualized, and portable. The smartphone is in this sense archetypal.

If coltan mining in the DRC's Kivu region is emblematic of the extractivist moment of algorithmic labor, the manufacture of iPhones in Shenzhen's Foxconn factories surely represents the industrial moment. Production at Foxconn is not limited to iPhones; the Taiwanese multinational produces a wide range of technological equipment for algorithmic capital. This includes tablets, computers, video game consoles, televisions, and e-readers for several companies (including HP, Sony, Microsoft, Nokia, Huawei, and Dell).[40] Foxconn is China's largest exporter in all categories and the world's largest producer of electronic devices. The company owns gigantic production complexes, the largest of which, in Shenzhen, employs more than 430,000 people. This complex is not just a workplace; the company has also built a "Foxconn campus" that combines production and social reproduction and confines employees to living in a space under the company's panopticon-style control. The

39 James H. Smith, "Tantalus in the Digital Age: Coltan Ore, Temporal Dispossession, and 'Movement' in the Eastern Democratic Republic of Congo," *American Ethnologist* 38, no. 1, February 2011, p. 21.

40 Fuchs, *Digital Labour and Karl Marx*, pp. 182–99.

2.3 km^2 complex includes factories, warehouses, dormitories, a bank, a post office, a fire station, restaurants and supermarkets, sports facilities, and an exclusive TV channel, the Foxconn Network. All of them operate on a schedule that conforms to the rhythms of the production lines. Some reports have described these places and practices as "labor camps," whose workforce also sometimes includes children. As one study puts it, "Such a socio-spatial arrangement strengthens managerial domination, wherein control over labor is extended from the factory shop floor to the sphere of everyday life."[41]

Work at Foxconn facilities is described as monotonous and repetitive. It often involves repeating the same task hundreds or thousands of times over a twelve-hour working day, with working weeks often reaching seventy-two hours. The working environment is highly hierarchical and Taylorized, with carceral-style surveillance, high individual quotas, and up to twenty hours of compulsory overtime per week. At times, some employees have to work up to thirty-four hours straight. Overcrowded dormitories, long working weeks, low wages, and a lack of social and personal life outside of work create conditions likely to producing depression, loneliness, and burnout. A wave of suicides shook the Longhua Town complex in 2010, when eighteen young workers aged between seventeen and twenty-five threw themselves from the windows of their dormitories. At the time, this caused quite a stir in Western media. Foxconn's response was to barricade the windows and install safety nets around the external staircases and dormitories.[42]

Yet it would be wrong to think that the industrial sector contributes to algorithmic capital only because it produces the technological equipment. We do not have enough space here to go into greater detail, but it is worth at least noting three other decisive aspects of the complex and deep-seated interconnection between algorithmic capital and the industrial sector.

First, new industrial platforms are automating production in the industrial and manufacturing sectors. Second, the sector is reorganizing

---

41 Pun Ngai and Jenny Chan, "Global Capital, the State, and Chinese Workers: The Foxconn Experience," *Modern China* 38, no. 4, 2012, pp. 385–7, 390, 392.

42 Andrew Malone, "Revealed: Inside the Chinese Suicide Sweatshop Where Workers Toil in 34-Hour Shifts to Make Your iPod," *Daily Mail*, June 11, 2010; Fuchs, *Digital Labour and Karl Marx*, pp. 186–94; Ngai and Chan, "Global Capital, the State, and Chinese Workers," p. 384; Matt Williams, "Foxconn Audit Finds Illegal Overtime and Unpaid Wages at Apple Factory," *The Guardian*, March 29, 2012; Malcolm Moore, "'Mass Suicide' Protest at Apple Manufacturer Foxconn Factory," *Daily Telegraph*, January 11, 2012.

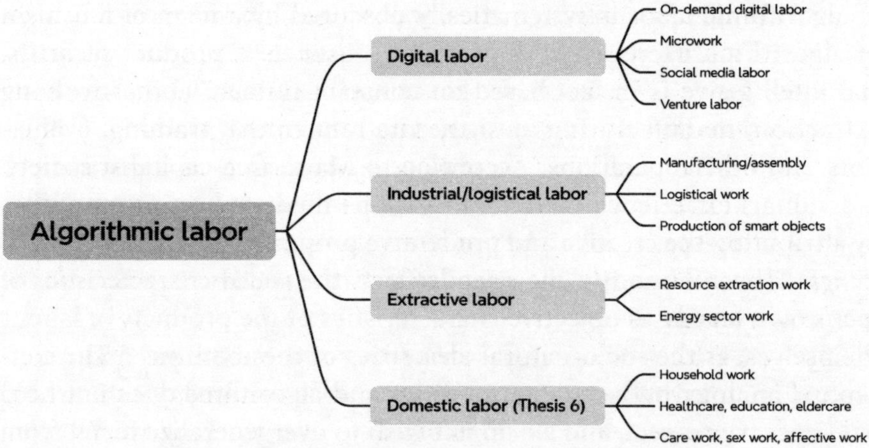

Figure 5.1. Breakdown of the concept of algorithmic labor

itself around the production of smart, connected objects, while the giants of the manufacturing industry are increasingly turning their focus toward algorithmic accumulation. Even such items as beds, cars, refrigerators, and thermostats are now equipped with chips and sensors that can extract after-sales data during use and thus feed into the creation of data assets. Third, logistics work in warehouses, handling, and transport is undergoing a vast process of automation and algorithmization. This is well illustrated by Amazon's warehouses, which are among the most automated and algorithmized working environments in the world.

Ultimately, we can conceptualize the industrial/logistical aspect of algorithmic labor with reference to four general processes. These are: (1) the mass production of digital devices; (2) the transformation of industrial production processes into "smart" production, using algorithmic industrial platforms; (3) designing and producing all kinds of everyday objects as "connected" and "smart" goods; (4) logistical work in warehouses, shipping, and delivery at giants such as Amazon.[43] Figure 5.1 illustrates the breakdown of the concept of algorithmic labor.

The fetishistic nature of the algorithm and its secret

Algorithmic capital is thus a social relation that relies upon and exploits forms of digital, extractive, and industrial labor and involves increasingly sophisticated data extraction mechanisms. However, the existence

43 Srnicek, *Platform Capitalism*, pp. 38–41; Sadowski, "When Data Is Capital"; Sadowski, "The Internet of Landlords."

of algorithmic labor is systematically obscured by a form of fetishism of algorithms. Here, what appears to a user as a product of artificial intelligence is in fact based on constant human labor, involving extraction, manufacturing, design, data refinement, training, evaluation, and other operations. According to Marx, in a capitalist society, individuals experience the social relations mediated by commodities by attributing the creative and productive properties of human labor to *things*. "The commodity," he says, "reflects the social characteristics of men's own labour as objective characteristics of the products of labour themselves, as the socio-natural properties of these things."[44] The conceptual analogy between commodities and algorithms does not hold true in every respect, and we do not wish to overgeneralize it. Yet from this point of view, a form of algorithm fetishism masks the reality of algorithmic labor.

The fetishistic nature of the algorithm obscures the fact that it is a product of digital, extractivist, and industrial algorithmic labor (and domestic labor, as we shall see). Its production is linked to human experience that extends far beyond paid working hours. The algorithm presents itself as the creator of a socioeconomic ecosystem of exchanges and transactions, of "social networks," and even of society itself. To return to Marx's quote, algorithms reflect back to users the social characteristics of labor and human experience as objective characteristics of the algorithms themselves—as qualities they appear to possess by nature. Algorithms seem to be able to play music, drive cars, recommend books that match our tastes, do our shopping, connect us socially, or write our emails, all through their own mechanical movement. They are machines, layers of "artificial neurons," robots, personal assistants that calculate, analyze, and organize, seemingly independently and outside the bounds of human activity; they appear all the more as objects "abounding in metaphysical subtleties and theological niceties."[45] The social relations of algorithmic capitalism appear not so much as "relations between things" but as relations organized and mediated by a thing, a mysterious mechanical and automatic object, an algorithmic system, an artificial intelligence, which establishes social connections, modifies social behavior, and constitutes a new origin of social relations. In this sense, in algorithm fetishism, algorithms—products of human labor and experience—"appear as autonomous figures endowed with a

44 Marx, *Capital*, Vol. I, pp. 164–5.
45 Ibid., p. 163.

life of their own."[46] Ultimately, algorithms, connected objects, and intelligent machines that seem to move on their own are in fact driven by the hidden power of algorithmic "ghost work."[47]

In short, algorithmic systems are presented as "intelligent machines" that are autonomous, rational, and endowed with a life of their own. In reality, they require an entire material infrastructure—and a great deal of labor to train and maintain it. These new mechanisms of labor exploitation, which combine surveillance, control, and the turning of human beings into instruments of capital, have ambivalent and sometimes terrible social consequences. This stems from the fact that algorithmic capital is in a state of constant reconfiguration and subsumption with algorithmic labor and its three digital, extractivist, and industrial/logistical moments.

But there is still one essential form of labor that remains to be analyzed. It is work indispensable to the reproduction of society, the labor force, and human life. This is the labor of social reproduction, which is also greatly affected by the rise of algorithmic capital.

46 Ibid., p. 165.

47 Gray and Suri, *Ghost Work*.

Thesis 6

Algorithmic technologies are colonizing the domestic sphere, housework, care, and emotional labor

Housewifery expands to fill the time available.

—Betty Friedan

Algorithmic capitalism is not confined to the world of work that we explored in Thesis 5. Rather, like the industrial capitalism and the neoliberal capitalism that came before it, this form of capitalism (re) organizes the invisible work that reproduces society, subjectivities, communities, and the labor force. Thesis 6 draws on social reproduction theory and seeks to make its own modest contribution to it, forming the basis for our investigation of the social relations that structure domestic (household, emotional, and affective) labor in the era of algorithmic capitalism.[1]

Note: Elements and sections of this chapter have been published previously in an article in English. See Jonathan Martineau and Jonathan Durand Folco, "The AI Fix? Algorithmic Capital and Social Reproduction," *Spectre*, November 15, 2023.

1 Lise Vogel, *Marxism and the Oppression of Women: Toward a Unitary Theory*, Chicago: Haymarket Books, 2013 [1983]; Sue Ferguson, *Women and Work: Feminism, Labour, and Social Reproduction*, London: Pluto Press, 2020; Tithi Bhattacharya and Lise Vogel, eds., *Social Reproduction Theory: Remapping Class, Recentering Oppression*, London: Pluto Press, 2017.

Algorithmic domestic labor: Toward the automation of housework?

What does the era of algorithmic capitalism mean for social reproduction and domestic labor? The advent of connected objects and the "smart home" is presented to us as an effective response to the time pressures of domestic work. Smart baby bottles, household robots, and connected refrigerators that can automatically order the groceries we need are just a few examples of technological solutions that are supposed to make housework easier and faster and reduce its stresses and strains. The new wave of mechanization and the introduction of AI into homes obviously bring to mind the 1940s and 1950s "golden age," when the advent of dishwashers, kitchen gadgets, and other household appliances promised to free housewives from domestic drudgery. Drawing such a parallel, it might perhaps seem that the automation of homes in the twenty-first century is merely a continuation of this earlier process, now accelerated by algorithms and AI.

However, if we have drawn the proper lesson of this first wave of hybridization of human—in most cases, woman—and machine in domestic work, then we should not expect this new phase of automation to lighten the burden on the women working in the home and free up their time. This, known as the Cowan Paradox, has been recognized by feminist scholarship of different schools: Despite mechanization of domestic work, the time spent on it does not decrease.[2] Betty Friedan aptly noted in the age of washing machines and microwaves: "Housewifery expands to fill the time available."[3] From a Marxist-Feminist perspective, Sylvia Federici argued that the only technology that was truly labor-saving for women was the birth control pill.[4]

Although AI and connected objects are presented as solutions that will lessen the load of domestic work, this is not the main reason why these innovations are being produced. The smart home industry, as well as the fierce rivalry between algorithmic capitalists to introduce their own versions of personalized home management assistance, reflects

---

2 See the illuminating discussion of the Cowan Paradox and its underlying causes (individual isolation of women doing domestic chores, higher standards of cleanliness and hygiene, and the creation of new tasks) in Helen Hester and Nick Srnicek, *After Work: A History of the Home and the Fight for Free Time*, London: Verso, 2023, pp. 22–8.

3 Betty Friedan, *The Feminine Mystique*, New York: W. W. Norton, 2001 [1963], p. 341.

4 Silvia Federici, *Revolution at Point Zero: Housework, Reproduction, and Feminist Struggle*, Oakland: PM Press, 2012, p. 47.

two deeper trends in algorithmic capitalism that directly concern social reproduction. One is a tendency to heighten the privatization and commodification of this sphere by tying it as closely as possible to the market; the other is the quest to extract data in order to fuel the accumulation of algorithmic capital. While the first tendency exacerbates commodification processes that were already widespread under neoliberal capitalism, the second trend is a novel development.

Algorithmic capitalism and the commodification of housework

The neoliberal era saw the emergence of a new reality for domestic work through the establishment of a global market for precarious migrant labor to meet the Global North's growing demand for outsourced domestic work.[5] This labor market is today gradually being subsumed into algorithmic logic through the development of digital platforms that allocate all manner of domestic work. For example, the so-called collaborative economy is reviving the figure of the handyman through digital platforms that facilitate on-demand work. Similarly, the emerging markets for childcare (and pet care), home-cooked meals, housekeeping, and sexual services make it possible to find individuals willing to work for low wages, and with barely a moment's wait. The "gig economy" has already been around for a long time. It is now being promoted by the platform economy and expanded in a context of economic precariousness for young people, people of color, students, single mothers, and members of the lower-middle and working classes. These latter are each looking to make ends meet by occasionally working as drivers, delivery workers, cleaners, babysitters, tutors, home cooks, sex workers, or repair workers of all kinds.

To gain further insight into the commodification of reproductive labor—in the home, in food preparation, in education, and in sex—it's worth taking in a brief overview of some of the platforms and apps driving this process of outsourcing domestic work to markets.[6] For instance, Care.com connects those who are seeking and providing services such as care for children and the elderly, tutoring, pet care, and housekeeping. Platform giants such as TaskRabbit and Handy provide on-demand "handymen" to repair, renovate, maintain, move furniture,

5 Sue Ferguson and David McNally, "Precarious Migrants: Gender, Race and the Social Reproduction of a Global Working Class," *Socialist Register*, no. 51, 2015, pp. 1–23.

6 See also Hester and Srnicek, *After Work*, pp. 36–40.

and do all kinds of odd jobs both inside and outside the home. Some apps are more specialized: Sittercity and UrbanSitter specialize in babysitting, Wizant in tutoring, Justmop and Bark.com in housekeeping, while Zum helps to organize children's journeys to school. All sorts of applications designed especially for children have also been added to the TV in the mechanical arsenal for childcare while caregivers are busy with other tasks—or just in need of a break. Groceries and orders can also be delivered via Postmates, and cooking and meals are also being taken over by algorithmic capitalism. DoorDash, Foodora, Skip the Dishes, and Uber Eats deliver meals from restaurants, while Instacart sends someone to get the groceries you have selected on the app. HelloFresh and Blue Apron deliver weekly meals that need only to be assembled and cooked, while Fresh in your Fridge and Take a Chef take this logic even further, to the point where a cook does the shopping and comes to prepare meals in the home. This platformization of reproductive work heightens the commodification and privatization of these tasks. It does so by creating precarious and complex labor markets, in which low pay and the lack of job protection or stability reproduce the social devaluation of domestic work on a new level.

Algorithmic capitalism is also colonizing the sexual sphere of domestic work. Several forms of sexual services are being platformized and even robotized. For instance, the apps Peppr and Smooci connect sex workers and escorts with clients. Another example of the platformization of sex work is the social network OnlyFans. It grew rapidly at the start of the pandemic, because it offered sex workers a way of continuing to work during the lockdowns and allowed others to generate income after losing their jobs. The sex robot industry is also booming. These machines are becoming increasingly "intelligent," filled with sensors that respond to human touch. Many critics denounce these technologies for reproducing the sexism and machismo present in society—and indeed, sex robots almost exclusively represent women's bodies, thus perpetuating the historical commodification of the female body. Others, however, point to the advantages of potentially ethical applications, designed, for example, to meet the sexual needs of severely disabled people or elderly people living alone.[7] In any case, algorithmic capitalism never misses a chance to generate value by promoting platforms, apps, and robots ready

7 Carlotta Rigotti, "Sex Robots: A Human Rights Discourse?," Open Global Rights, 2019; Ezio Di Nucci, "Sex Robots and the Rights of the Disabled," in *Robot Sex: Social and Ethical Implications*, Cambridge, MA: MIT Press, 2017, pp. 73–88. We will come back to the question of "erobotics" in Thesis 16.

to satisfy sexual desires, and particularly the desires of the heterosexual men who also happen to be largely overrepresented in the field of tech and AI engineering.

Algorithmic capitalism and the imperative of extracting "domestic data"

In the domain of data extraction, algorithmic capitalism is driving an overhaul of the domestic sphere, the better to mold its subsumption. The platforms we've just discussed obviously do collect user data. Here, though, we want to highlight the spectacular entry of algorithmic capitalism into homes via the smart home industry and the Internet of Things. Two developments in particular are important. On the one hand, there are various household objects and appliances equipped with systems capable of capturing data, of being programmed in advance, and of performing certain tasks automatically. These "smart home" accoutrements include fridges, beds, washing machines, vacuum cleaners, irons, coffee machines, burglar alarms, ovens, doorbells, and so on. On the other hand, there are "smart" leisure and entertainment devices: TVs, speakers, game consoles, toys. Today, the manufacture of most of these products involves the integration of various sensors and algorithms that collect data on the task to be performed, and on the given object's online presence, and they may even gather ambient data from the home (such as movements and conversations). They automate certain functions, even when these products may not seem to be primarily electronic devices. This is part of "a conscious, coherent effort to enlist our intimate spaces as a site of continuous technological upgrade, subscription-based services and the perpetual resupply of consumables."[8] Let us look at a few examples.

The Samsung Family Hub smart refrigerator serves this device's primary function of keeping food cool and fresh. But it also includes a smart system that can be operated from a touchscreen on the fridge door or with an app on your phone or tablet. This system allows it to stream music via Bluetooth to speakers throughout the house, and to alert you when the door is opened. Moreover, the integrated cameras allow you to view images of the fridge's contents on your phone, so you will always have an instant answer to that age-old question, "Do we still have any

8 Adam Greenfield, *Radical Technologies: The Design of Everyday Life*, London: Verso, 2018, p. 38.

milk?" The screen displays your schedule (such as Google Calendar), the weather, a whiteboard for leaving notes for the family, an app with daily updates and even inspirational quotes, as well as any other app you might wish to download: podcasts, sports channels, Netflix, or Uber. You can install Doorbell, so that you can hear the bell ring through your fridge's speaker, see the image from the front door camera on the touchscreen, and respond via intercom. Then there's an Internet browser, the integrated Samsung Bixby, to respond to voice commands, and other features. The Deals feature, for example, lets you view supermarket special offers and order groceries directly. If you add the Mastercard Groceries app, payments are made automatically.

In theory, this offloads onto the smart fridge the whole job of keeping track of what food you have and shopping for bargains. In practice, though, grocery shopping now means programming the fridge—and keeping track of the device and all its functions becomes a new household chore in itself. Of course, this smart equipment collects all kinds of data: images, conversations, sounds, family habits and preferences in terms of food, entertainment, visiting hours. At this point in our argument, there is no need to explain further that such a wealth of data is a gold mine for Samsung and for the apps that detect new angles of data extraction, starting from the refrigerator.

Similarly, the iRobot smart vacuum cleaner can be programmed to clean at specific times when no one is home, and/or work in voice-activated mode if it is connected to a smart home system like Google Home. It is armed with cameras and sensors that enable it to create a detailed map of the house, provide cleaning reports—and avoid tumbling down the staircase. This data and these images are then sold to Google, Amazon, and other algorithmic capitalists. A final example is the Sleep Number 360 bed, equipped with SleepIQ smart technology. This bed is a fully fledged data extraction environment. It measures the sleeper's heart rate, movement, and breathing, as well as audio signals in the room, and compiles a detailed biometric report of your night's sleep that you can consult the following morning. The bed can be combined with other health apps (such as ones for diet or exercise) to identify all the elements of your daily routine that have an impact on the quality of your sleep. Its creators believe that it will eventually be able to identify risks of heart attack, sleep apnea, and other potentially dangerous health problems. The amount and depth of data collected by this smart bed reveals a lot about you: your health (insurance companies are fond of this type of data), but also your sexual practices and routines. The same goes for sex robots and other smart sex toys, which are well placed to collect biometric data about your body and the most intimate aspects of your life.

Of course, the whole gamut of home entertainment devices also act as hubs for collecting domestic data—and can be used to push targeted advertising. As such, algorithmic capital is especially plugged into the data extracted from human speech. The extraction of conversations and chatter is part of a frantic race between algorithmic giants to bank terabytes of human speech in order to accumulate what we might call a conversational surplus. Since this is, indeed, human speech, and the conversations in the home are the most spontaneous and natural, these snippets of dialogue are ideal for programmers who train AI and voice recognition algorithms. They also help to anticipate what the speakers want to buy and thus to develop even more sophisticated predictive products.

Children's games and toys are no exception to this trend. In fact, several models are equipped with smart systems that also collect domestic data. These include "surveillance dolls" equipped with cameras and voice recognition software, such as Mattel's Barbie. The doll converses with your children, and her own toy house can be activated by the voice commands of the children whom she is recording. Then there's the winner of the 2019 Toy of the Year award, the Fisher-Price smart home for little ones. It offers replicas of all the smart devices found in grown-ups' homes and is likewise equipped with sensors. Even the saga of Mattel's Aristotle child assistant toy, whose production was finally discontinued in 2017 after a public outcry—Aristotle records and films children—did not kill off algorithmic capital's colonization of this space-time dedicated to children's toys. We could make an endless list of connected household appliances and items on the market: from smart vodka bottles and rectal thermometers to toothbrushes and food processors. Add to this all the various motion-detection, security, and antitheft surveillance systems. These are also powerful machines for extracting behavioral surplus; in many cases they also reproduce control mechanisms through which men can monitor their partners, or even let employers film and monitor domestic workers without their knowledge.

These examples illustrate the scope and depth of the data extraction that "smart" and "connected" domestic work and experience help to enable. Smart homes and the connected objects that populate them play an important role in transforming domestic experiences into data. The Internet of Things in the home is spreading like a vast net, an omnipresent and omniscient sensor that extracts data from the bodies and activities in the home. The economic transaction directed at the purchase of a television, bed, vacuum cleaner, toys, or other items used for household reproduction is no longer limited to the purchase of a product that fulfills its basic use-value: to entertain, help sleep, or clean. These products now pursue a specific economic activity—data extraction—even after being added to the domestic environment. As Hester and Srnicek put it: "The drive to make all household tasks 'smart' is less a response to need than a reflection of the economic and technical capacity for collecting data and producing computer chips."[9] Companies are sharing the spoils of a rapidly growing market, valued close to $175 billion in 2025. These developments are dependent on a commercial process that seeks to predict our behavior: beds, fridges, dildos, toothbrushes, vacuum cleaners, and dolls are all devices that extract behavior, migrating our data to algorithmic capital and helping to hone predictive products and train artificial intelligence. Zuboff is not wrong to call these objects Trojan horses. Although *some* of this data is used to improve the functioning of the objects concerned, the extraction produces a behavioral surplus that—like all other data extracted by the surveillance architecture—is traded on predictive product markets or reinvested in algorithmic capital. Algorithmic capitalism has thus colonized and commodified the domestic sphere in an unprecedented way.

Toward a reconfiguration of affective labor in the era of algorithmic capitalism

Although there are still few studies on this subject, new trends seem to be emerging in the relationship between algorithmic capitalism and affective labor. These especially relate to the household AI market, in the form of smart personal assistants that manage and automate the home.[10] The best-known ones include Amazon's Alexa, Google Assistant, Tencent's Xiaowei, and Apple's Siri. In their AI home-assistant versions, these systems connect all the different smart home management devices and apps; they become, in a way, the smart HQ of household work. Equipped with sophisticated conversation programs, smart personal assistants address us as voices through speakers (Amazon's Echo, Google Home). Highly popular with well-off families and the middle classes in

9 Hester and Srnicek, *After Work*, p. 45.

10 See also Julie Paquette's analysis, "L'assistante personnelle virtuelle, une prédatrice attentionnée," *Politique et Sociétés* 42, no. 3, 2023.

the Global North, they have enjoyed an exponential rise in recent years: More than 600 million Alexas had been sold by September 2025, and ever more people are using these devices in their vehicles as well.[11]

At first glance, these assistants' main function is to automate the activities involved in household management and the reproduction of day-to-day life (cleaning, shopping, food, entertainment, and so on). With a simple voice command, it is possible to access the entire online ecosystem associated with the device's manufacturer. If you are running out of toilet paper, Alexa will automatically place an order with Amazon. The vase fell on the floor and broke: Alexa will order a new one that fits perfectly with the wallpaper. You are low on eggs: Alexa will browse the weekly specials and order a dozen at a discount. It is your son's birthday next week: Alexa will prepare a list of gift ideas. This has the effect that sellers are drastically reorienting their marketing strategies toward personal assistants rather than consumers. Indeed, sales will increasingly depend on the seller's ability to position themselves advantageously on the platform, rather than creating a direct relationship with the consumer. A platform like Amazon has a very real ability to monetize its advantageous position in this regard—indeed, there is a strong tendency for this sector to consolidate around a tiny number of key players.

All this is said to be about lightening the mental load of housework and automating certain chores and errands using AI. The goal here seems to be to ensure that everyday consumer goods flow into the home automatically and continuously, much as water and electricity do in most homes today. Amazon's 2013 patent for an "anticipatory shipping" service is another example of the logic of these developments. The platform ships a series of items to your door that you did not know existed, such as books that might interest you based on your profile; you then choose whether to keep them or return them. In this sense, the assistant comes to predict your purchases before they even cross your mind, a new form of self-fulfilling prophecy if ever there was one.

All kinds of scenarios for the introduction of personal assistants into the domestic sphere are now possible, depending on the household concerned. It is easy to imagine situations where programming, maintaining, notifying, and supervising the personal assistant is itself a complicated household task, adding to the domestic worker's already long list of chores. We can also imagine homeowners programming

11 "It's Amazon's Turn to Show Off New Devices," CNBC, September 30, 2025, cnbc.com.

tasks for the worker via the personal assistant, which becomes a kind of algorithmic supervisor-manager of waged domestic labor. We can also imagine personal assistants that actually do succeed in easing their users' stresses and performing certain tasks. But at what cost?

These systems are powerful extractors of centralized domestic data, and powerful connectors between the domestic sphere and the market. For instance, the household AI Alexa directly connects domestic work to Amazon's globalized supply chains. Alexa and virtual assistants thus exacerbate the commodification of the domestic sphere in several key regards, as already discussed by Amy Schiller and John McMahon:

1. The data, preferences, emotions, and personal stories generated by interaction with Alexa are appropriated by Amazon, which uses them to develop, promote, and sell its products (the value of consumer data thus increases exponentially, since it enables the optimization of the assistant's performance).
2. Alexa automates the purchase of Amazon products.
3. The assistant enables and facilitates the use of other commercial and data-extracting applications ("Alexa, play my Netflix series on the fridge screen").
4. The assistant thus acts as a mechanism for privatizing domestic labor by entrenching individuals and families in a space-time saturated with the capital flows of the manufacturer's ecosystem, free of other humans.

Here, social reproduction can proceed without needing to interact with other people—surely a plus, in times of pandemic.[12]

Social relations in the domestic sphere are in this way being remolded. To project an identity onto their assistant and develop a relationship with it, users also have to put in something of themselves. Amazon is doing all it can to encourage such user efforts, because they are essential to making addictive hooks effective and optimizing the use of Alexa. Alexa thus seems to want to be part of the family. Although it could easily content itself with neutral interactions about the logistics of the home, it can also interact on an emotional and affective level. In 2018, Amazon introduced Alexa Blueprints, a series of extra skills that owners can add to their Alexa. They allow it to memorize information about each family

12 Amy Schiller and John McMahon, "Alexa, Alert Me When the Revolution Comes: Gender, Affect, and Labor in the Age of Home-Based Artificial Intelligence," *New Political Science* 41, no. 2, April 2019, pp. 185–6.

member (birthdays, favorite foods or TV shows, and so forth), motivational quotes, and family stories that Alexa can recite to family or guests to add a touch of conviviality to an evening at home.

Alexa is also a personal assistant that communicates with a female voice—just like Siri and the early versions of Google Assistant, the personal assistant is, in fact, most often female. This is a strategic choice by firms like Amazon and Google. Research into human/machine interaction shows that users imagine the voice having an anthropomorphic source, even though they know that it is only the formal interface of the algorithmic circuits lying beneath. These companies choose female voices because they better reflect social prejudices about secretarial work, personal assistance, and domestic labor. According to research, the projection of a stereotypical identity onto the assistant that is aligned with existing prejudices helps to create an authentic relationship with users. Giving the household AI a female voice thus more easily activates a coherent user identity and maximizes the effect of its addictive hooks.[13] This means that companies producing these virtual assistants must choose between two options: either encourage some of society's most persistent and harmful prejudices and ensure better results for their products, or else combat these prejudices and weaken the product's ability to generate greater user familiarity. Clearly, the first choice will normally win out. This strategy only reproduces and strengthens the gendered psychological dynamics involved in giving orders to a "female assistant."[14]

And even if Alexa supposedly provides relief from housework, this in any case masks the servile labor that sustains the Amazon ecosystem around the world. In a new version of the master–slave dialectic, Alexa presents itself as a servant and creates the illusion that the user is its master. The basic reality is a private company that exploits all the drudgery necessary to transform the user's verbal expression of their desires into the desired product turning up on their doorstep. But instead of that, the user sees a docile servant of which they are master—even though they have no control over the network of algorithms that actually constitutes Alexa. In this sense, Alexa functions as a *double fetish*: a fetish of invisibilization, and a fetish of inversion. The first fetish "contributes to hiding Amazon's material structures of labor and the precarious

13 Simone Natale, "To Believe in Siri: A Critical Analysis of AI Voice Assistants," *Communicative Figurations*, no. 32, 2020, p. 8.

14 Alexa, Siri, and Microsoft's discontinued assistant Cortana were all rolled out in North America with a female voice as default.

workforce that sustains the functionality of the platform."[15] The smart assistant constructs a representation that *invisibilizes exploitation* and guarantees the market domination of algorithmic capital and companies such as Amazon. At the same time, Alexa and its—her—docile voice reverse the master–slave relationship between algorithmic capital and the user; they give the user a sense of control even though they are hemmed in on all sides by Amazon's algorithms.

AI assistance also raises a series of questions related to affective labor and the mental burden in the domestic sphere. We have already mentioned the "housewife" and the migrant domestic worker, who both physically and emotionally clean the homes of the wealthy classes. These women transform the negative emotions of their husbands or employers into positive ones, they absorb their frustration, they manage emotional waste by listening, encouraging, and smiling—they are always available, always in a good mood, but remain as invisible as possible themselves. As Shiloh Whitney theorizes, there is an "affective circulation" in the household, dependent on gendered and racialized power relations, and the transformation of negative emotions into positive ones, the "metabolism of affect," is also part of domestic labor.[16] In this sense, domestic AI also performs emotional and affective work in the home through its interactive presence, its voice, its listening, and its response to human commands. In addition, many users will adopt strongly negative behaviors toward the domestic AI assistant, such as shouting, swearing, or using a harshly authoritarian tone. Amazon has recently patented an innovation that allows Alexa to recognize the emotions or irritation of its "master" and modulate its interactions in order to minimize frustration, offer apologies, or make other adjustments.

On the one hand, as Schiller and McMahon note, AI assistants can serve to relieve the pressure on domestic workers and some of the emotional burden they have to shoulder: "If an elite white family offloads its affective waste onto Alexa rather than an underpaid, overworked, affectively overloaded worker of colour, one could argue that this effects an overall beneficial adjustment in patterns of affective labor and circulation in the home."[17] In this reading, Alexa absorbs some of the household's negative emotions, thus rerouting its emotional circuits. But is not this a

15 Natale, "To Believe in Siri," p. 14.

16 Shiloh Whitney, "Byproductive Labour: A Feminist Theory of Affective Labour Beyond the Productive–Reproductive Distinction," *Philosophy & Social Criticism* 44, no. 6, July 2018, pp. 637–60.

17 Schiller and McMahon, "Alexa, Alert Me When the Revolution Comes," p. 181.

particularly harmful form of enabling or permissiveness—where mood swings and violent, angry, rude, and inappropriate behavior directed at the AI are condoned? Rather than an overall reduction in the negative emotions with which the humans in the household have to deal, perhaps this is a normalization, trivialization, and acceptance of this behavior among the assistant's privileged users. This AI-based normalization does not bode well for the less privileged in wider society. In any case, it reproduces behaviors that treat domestic work as something inferior, done by something, or someone, on whom you can take out your frustrations. In this sense, domestic AI can produce and reconfigure dominant subjectivities built on a classed, gendered, and racialized cultural bedrock. Ultimately, the use of a virtual assistant renders human labor invisible, extends the circuits of overconsumption, and reproduces problematic behaviors.

Health, care, and emotional robots

In both healthcare and education, AI and algorithms are being deployed in increasingly sophisticated ways. The reorganization of these major systems according to an algorithmic logic is already well underway. The global pandemic accelerated many of these dynamics, especially in education, where online courses have exploded, for better or worse. Algorithms that adapt learning processes to the "student profile" and enable distance learning are extremely popular.[18] However, healthcare is the field where AI use is most advanced, far beyond the exercise, fitness, and nutrition apps that more and more people are using on a daily basis. AI in the medical field represents a point where money and important holders of social power converge: university research centers, pharmaceutical companies, tech firms, governments, public administration, and so on. A vast network of private and public interests is involved in the rollout of healthcare management tools and technologies that include algorithmic components or are wholly AI-based. Leading experts in this field estimate that by 2030, no medical diagnosis will be made without AI expertise, and that this is the most promising area of application for so-called "professional" AI. In addition to being able to perform a wide range of tasks such as modeling tumors or monitoring the development

18 This is just the tip of the iceberg. On the development and use of algorithms and AI in the world of education, see Eric Martin and Sébastien Mussi, *Bienvenue dans la machine: Enseigner à l'ère numérique*, Montreal: Écosociété, 2023.

of a disease, AI is creating a new "anticipatory" form of medicine. The use of algorithms in analyzing genetic heritage is a striking example of this. It remains to be seen how the various medical professions will welcome these "innovations," which in some respects risk making their work redundant. General practitioners and specialists are usually well organized, and the adoption of technological solutionism in healthcare will undoubtedly have to go through the mediation of these professionals.

There is also a booming field of "emotional robots" used to provide care, make diagnoses, and perform emotional work with sick, elderly, disabled, or vulnerable people. These are "social and emotional" robots designed to read and interpret, as well as convey emotions. They are equipped with sophisticated sensors that extract a wide range of data related to voice recognition, but also facial expressions (indeed, all nonverbal and paralinguistic language), which assist them in constructing a psychological and emotional profile of the user.[19] There are many applications for these different types of robots: conversational robots that can serve as medical advisors, doctors, nurses, or virtual companions for the elderly or children; robot therapists, which, for example, seem capable of holding the attention of autistic children longer than a human therapist can; "retirement-home robots" that can provide care and offer company to elderly people with reduced autonomy; more general-use robots that can perform certain domestic tasks, but can also remind people when to take their medication or serve as alert mechanisms in case of a problem. The "agetech" market is booming, particularly due to the aging of the global population (by 2050 there will be twice as many people over the age of sixty and three times as many octogenarians, according to the World Health Organization), with an associated explosion in healthcare costs.[20] However, as James Wright has shown,[21] experiences in this field in Japan, a pioneer in care robotics, have produced extremely mixed results. Patients, elderly people, and care workers are more likely to underline the new problems that these machines are creating than to cast them as the solution. Their drawbacks include, among other things, the deskilling of care work and the creation of new robot maintenance tasks; the loss of human contact; the

19 Laurence Devillers, *Les robots émotionnels: Santé, surveillance, sexualité … et l'éthique dans tout ça?*, Paris: Éditions de l'Observatoire, 2020, p. 91.

20 Alain McKenna, "Les technologies à l'assaut de l'âge d'or," *Le Devoir*, January 13, 2023.

21 James Wright, *Robots Won't Save Japan: An Ethnography of Eldercare Automation*, Ithaca, NY: Cornell University Press, 2023.

very high costs; and elderly patients' refusal to comply with a robot's requests if a human is not also present.

Here there is a convergence between household robots and "health-monitoring" robots, which are being developed according to the paradigm of so-called affective computing, based on recognizing human emotions. Indeed, the Holy Grail in this field seems to be the development of a robot capable of reading and interpreting human emotions: one that is able to cheer us up when we are sad, help us when we are lonely or in difficulty, and even develop bonds of attachment, friendship, and love. In Thesis 16, we will look at the robots that are changing our intimate, erotic, and sexual relationships (erobots) in order to explore the changes that algorithmic capital is inducing in both interpersonal relationships and human subjectivity itself.

It seems impossible to accept or reject these developments outright. On the one hand, the assistance of a robot capable of detecting a fall or discomfort in a person with reduced autonomy can indeed be a useful complement to care workers or family carers. That said, even according to a leading proponent of AI such as Kai-Fu Lee, it would be unwise to aim to replace human labor in this field completely.[22] On the other hand, when combined with nudge techniques, algorithmic technologies recognizing human emotions and guiding behavior are potentially dystopian. In each case, of course, we have to take into account the social, cultural, and historical context in which these technologies are being used, inasmuch as they may serve different interests regardless of their immediate use for those most directly concerned.

The core of the problem is that these affective robots and algorithmic technologies designed to facilitate the labor of social reproduction are part of a world where the commodification of care, data extraction, and the pursuit of maximum profit are the norm. Algorithmic capital's many impacts on the process of social reproduction—including the domestic sphere, education, and care work for children and the elderly—need more attention from critical theories of social reproduction. Furthermore, studies aimed at identifying AI's effects on these spheres mostly neglect the role of capitalism in the reconfiguration of these social practices. This subject thus demands far more empirical research to help us understand how AI is impacting the most intimate corners of social reproduction.

22 Kai-Fu Lee, *AI Superpowers*.

Thesis 7

Big Data and machine learning are the main drivers of algorithmic capital

Thought is reified as an autonomous, automatic process, aping the machine it has itself produced, so that it can finally be replaced by the machine.
—Max Horkheimer and Theodor W. Adorno

If we want to grasp algorithmic capital *as capital*, that is, as a social relation but also as dynamic of accumulation, we need to go beyond an analysis focused on social control or the exploitation of labor. More than that, we have to understand how capital is accumulated and reproduces itself through the mediation of the data and algorithms integrated into new means of algorithmic production. One useful way to study this process is to examine the reconfiguration of the ways to produce value within platforms—the new dominant model of capitalist firm.

The accumulation of algorithmic capital can be seen as a process consisting of five moments. First, there is *datafication*, the transformation of human experience into data, married to the triumph of machine-learning techniques that require huge amounts of data to train their algorithms. Next, *commodification* transforms data into predictive products or "data commodities," through the intervention of various forms of digital labor. This involves the production of algorithmic technologies that can be sold on the market; it also aims to transform data into monetizable assets, ones that can be used to attract investments. Third, the process of *algorithmic accumulation* transforms data and algorithms into capital. Fourth, the *automation* of decision-making processes introduces automated decision-making systems, which also draw on human judgments in order to hone the algorithms' autonomous functioning. Fifth, the process of *uneven and combined development* is spreading these algorithms across different spheres of the economy in varying ways.

The era of generalized datafication

The emergence of algorithmic capital is only possible given a certain degree of scientific and technological development. Its functioning requires technological capabilities that enable the production, capture, extraction, and storage of vast quantities of data, or what we now call Big Data. To paraphrase Marx's words in the opening lines of *Capital*: The wealth of those societies in which the algorithmic capitalist mode of production prevails presents itself as "an immense accumulation of data." Our investigation must therefore begin with an analysis of the basic form of this "wealth": data. Big Data does not come from nowhere, but from a process of *datafication*—that is, the transformation of human experience (a complex mixture of attention and behavior) and multiple social, natural, physical, and technical processes into data. This metamorphosis of human experience into data that can be used for value creation, together with various forms of datafication of the world, constitutes the first moment of algorithmic capital.

Datafication differs from simple digitization. Digitization refers to the conversion of information (text, images, sound, video) into digital data that can be read by IT devices, making it easier to transmit, reuse, and manipulate. In the 1980s and 1990s, trillions of pieces of information from businesses, governments, and other institutions were converted into digital formats: floppy disks, CD-ROMs, hard drives, and the like. This Great Digitization represents the first stage of the IT revolution, in which a growing mass of information is converted from analog to digital. In 1986, 2.6 exabytes of information were stored in analog format (paper, VHS, vinyl, and other forms), and only 1 percent of this existed in digital form (0.02 exabytes). The global digitization of information reached 25 percent of the total in 2000, 50 percent in 2002, and 94 percent in 2007, with 280 exabytes of digital data.[1]

Recent developments in technology have opened the way for another, distinct phenomenon: *datafication*. This process can transform any given element into a quantified format, contributing to the advent of

---

1 Martin Hilbert and Priscilla Lopez, "The World's Technological Capacity to Store, Communicate, and Compute Information," *Science* 332, no. 6025, 2011, pp. 60–5.

Big Data.[2] The term "datafication" was first coined in 2013 to describe this vast process of transforming objects, processes, and human activities into data so that they can be manipulated by new analytical techniques, such as data mining, predictive analytics, and machine learning.[3] While digitization uses simple data analysis based on traditional sampling mechanisms, datafication makes it possible to identify nonlinear relationships between huge volumes of data, and it facilitates the creation of predictive tools.[4] Following the rapid spread of social media and smartphones since 2007–8, the massive production of data by billions of connected individuals and organizations exchanging messages, photos, videos, and geolocation data throughout the day has brought global connected society into a new age of accelerated accumulation. In 2013, 4.4 zettabytes (10^{21} bytes) of data were produced, a figure that then reached 44 zettabytes in 2020, or forty times more bytes than there are stars in the universe.[5] It was estimated that by 2025 we would be producing 463 exabytes of data *every day*, equivalent to the total yearly amount of data produced in 2008. The exponential accumulation of data seems to have no limits.

The era of generalized datafication arrives once the technical capabilities for extracting, storing, and processing data have reached a certain level of development. Cloud storage, a means of overcoming the problem of data overload, became available in 1999 in the form of "software as a service" offered by companies such as Salesforce. Companies can now lower their costs by skirting around the need to purchase, maintain, and replace their own IT systems and hard drives: It is easier to rent storage space in data centers, which have been springing up all over the world since the early 2000s. In 2005, Roger Magoulas coined the term "Big Data" to describe the deluge of data that could no longer be managed and processed using businesses' traditional tools. In 2006, Yahoo! created the open-source Hadoop, a digital infrastructure based in part on Google's MapReduce software, which allows huge amounts of unstructured data from any source to be processed simultaneously on different servers.

---

2 Rachel Schutt and Cathy O'Neil, *Doing Data Science*, Sebastopol, CA: O'Reilly Media, 2013.

3 Viktor Mayer-Schönberger and Kenneth Cukier, *Big Data: A Revolution That Will Transform How We Live, Work, and Think*, Boston: Mariner Books/Houghton Mifflin Harcourt, 2014.

4 Gabriele Gattiglia, "Think Big About Data: Archaeology and the Big Data Challenge," *Archäologische Informationen*, no. 38, 2015, pp. 113–24.

5 Jeff Desjardins, "How Much Data Is Generated Each Day?," World Economic Forum, 2019, weforum.org.

In addition to these major computing advances, necessary to store and process an exponentially growing amount of data, another technological revolution has made it possible to exploit this data to the full. Algorithmic capital would not have been possible without a paradigm shift in the field of artificial intelligence: machine learning.

The machine revolution: Learning machines

It's worth briefly reviewing some of the milestones in the history of artificial intelligence in the last seventy years or so.[6]

AI gradually emerged as a discipline in the 1940s and 1950s and fully took off as research program in summer 1956 during a workshop held at Dartmouth College, which brought together researchers from MIT and IBM, including John McCarthy, Herbert Simon, and Marvin Minsky. The advent of the computer, the work of Norbert Wiener and Alan Turing, and certain technical and scientific breakthroughs were then used to teach machines to play chess, generate text, recognize images, and solve various other problems. At this stage, AI was conceived as a computer program that could perform tasks requiring mental processes such as perceptual learning, memory, and logical inferences.

AI pioneers were very optimistic about the possibility of achieving certain technological feats in a relatively short period of time. As early as the 1960s, Herbert Simon, revealing one true incentive behind the development of this technology, claimed that "machines will be capable, within twenty years, of doing any work a man can do," while Marvin Minsky believed that real artificial intelligence would be created within a generation.[7] However, this initial enthusiasm was quickly stifled by the harsh reality of technical constraints—machines at the time had limited memory and computing power. This was followed by the "first winter of AI" (1974–80), the result of a series of scientific, but above all economic, problems. Following a wave of exaggerated and unrealistic expectations, various financing bodies (the UK government, the US Department of Defense, the US National Research Council) cut virtually all their

---

6 See Lee, *AI Superpowers*. For a sober account of AI's limitations, see Erik J. Larson, *The Myth of Artificial Intelligence: Why Computers Can't Think the Way We Do*, Cambridge, MA: Belknap Press, 2021. See also the brilliant historical account of AI, locating its development in labor and social relations, in Matteo Pasquinelli, *The Eye of the Master: A Social History of Artificial Intelligence*, London: Verso, 2023.

7 Quoted in Victor Storchan, "Sam Altman: La loi fondamentale de l'IA," *Le Grand Continent*, April 14, 2023.

funding for AI research. Progress was too slow for their liking, and real-world applications in industry and the military remained limited.

It was not until after the arrival of microprocessors in the late 1970s that research and cash flow resumed and new breakthroughs were made in the world of AI. The dominant approach in the 1980s was that of "expert systems." Also known as "rule-based systems" or "symbolic AI," this approach involves coding rules to teach algorithms to calculate with a high degree of accuracy. In the case of natural language processing, this involves programming rules of syntax, semantics, and vocabulary to teach an algorithm how to recognize, decode, and translate human languages. This approach has the advantage of making the process clearer, more transparent, and more readily explained, and it aids the resolution of problems with well-defined parameters.[8] However, funding for AI again hit the wall in the late 1980s due to the prohibitive cost of these systems. The perceptions of investors and funders have thus played a key role in the phases of boom and stagnation that have punctuated the development of AI. Technological progress is far from a merely endogenous phenomenon; rather, it is closely linked to social history and the moods of capital and the state, which may see these discoveries as more or less promising opportunities for profit and power.

While expert systems lost popularity in the late 1980s, AI enjoyed a renaissance with the advent of machine learning. This approach is based not on programming fixed rules, but on algorithms that build their own inference models from a large data sample. Machine learning has its roots in the invention of formal (or artificial) neural networks developed by Warren McCulloch and Walter Pitts in 1943. While the invention of the "perceptron" by Frank Rosenblatt in 1957 kicked off this technique, it was developed mainly by the work of Geoffrey Hinton, Yann Le Cun, and Yoshua Bengio in the 1980s and 1990s. Neural networks enable automated decision-making: They simply need to be fed huge quantities of examples of a given phenomenon, and the networks then identify patterns without further human intervention.[9] Instead of introducing a priori rules or categories into an algorithmic machine, the machine develops its own categories based on accelerated contact with data. The strength of machine-learning algorithms lies in their ability to learn (whether supervised or not): They are learning machines. Machine-learning algorithms really took off in the early 2000s following the rapid increase in computing power and the massive production of data needed to train

8 Lee, *AI Superpowers*.
9 Ibid.

neural networks. Datafication is a prerequisite for machine learning, which in turn accelerates the datafication process that is necessary for training algorithms.

This has led to another technological revolution: deep learning. This technique uses convolutional neural networks (multilayered networks inspired by the visual cortex of animals) with important increases in power in terms of speech and visual recognition. The three pioneers in this field, Hinton, LeCun, and Bengio, received major funding from the Canadian Institute for Advanced Research (CIFAR), as well as from Microsoft, Google, and IBM over the course of their research. The rapid advances in machine learning and deep learning have enabled the rollout of a host of applications in a multitude of sectors: facial recognition, voice recognition, automated language processing, robotics, bioinformatics, health, security, automated decision-making systems, and so on.

More recently, a major innovation in the field of deep learning has enabled the construction of "large language models" (LLMs), significantly improving the generation of machine-produced text. The invention of the "Transformers" architecture by a Google team in 2017,[10] the wide availability of training data, and advances in the performance of graphics processing units (GPUs), have enabled a new range of extremely powerful algorithmic systems, referred to as "foundational models."

Foundational models provided the basis for a wide array of software that quickly drew public attention in 2022 and 2023: programs like Dall-E, Midjourney, ChatGPT, and Stable Diffusion. These high-performance algorithmic models have two main characteristics: "emergent behavior" and "homogenization." "Emergent behavior" means that these systems come with new properties and capabilities for which they were not trained, while "homogenization" refers to the fact that the same set of algorithms can now be used in a wide variety of situations and generate a broader range of applications. Far from being limited to a single domain, these algorithmic models can now transfer learning from one context to another. This "transfer learning" allows algorithms to mobilize capabilities acquired in a particular task (for example, recognizing objects in an image) to another type of operation (recognizing activities in a video).[11] The arrival of "multimodal AI," that is, the ability of algorithmic systems to interpret diverse types of data and perform various operations

---

10 Ashish Vaswani et al., "Attention Is All You Need: Advances in Neural Information Processing Systems," in I. Guyon et al., eds., *Advances in Neural Information Processing Systems* 30, NIPS 2017, papers.nips.cc.

11 Ibid., p. 4.

by navigating between textual, audio, and visual content, thus represents a new technological breakthrough.[12]

However, "generative AI" is the term for these systems that has become established in everyday use since late 2022.[13] This type of artificial intelligence produces entirely new works based on prompts: texts, images, music, and other forms. Since it runs on existing images, texts, and music clips, it of course gives rise to new forms of plagiarism. Artists, authors, and various agencies are still trying to work through the legal implications of such technology; some organizations such as *The New York Times* and others are even suing the likes of OpenAI for using copyrighted material to train their algorithms.[14] The term "synthetic media" is also used to describe the arrival of new content produced or manipulated by artificial intelligence. Algorithms generate synthetic videos, sounds, images, and texts that do not exist in reality but are nonetheless sometimes very realistic, while on the other hand such synthetic content is sometimes termed "AI slop" due to its low quality or lack of effort or meaning.[15]

Deepfakes, synthetic images and voices, and AI art can be classified as synthetic media and generative AI. Several observers have noted that "synthetic data," which complements "real-world data," is likely to play an increasingly important role in training next-generation algorithms, because they "can be generated to feed digital twins in much less time and money than it takes to collect data from primary sources."[16] In an ideal world, generative AI content could be used to train other algorithms, thereby reducing the need for human labor and data extraction to develop new models. However, this is not the reality: Researchers have shown that feeding the algorithm mainly with AI-generated content

12 Gadi Singer, "Multimodality: A New Frontier in Cognitive AI," *Towards Data Science*, February 2, 2022.

13 The term "large generative models" is also used in the scholarly literature, but the term "generative AI" has the advantage of being simpler and more widely used in the media. See Deep Ganguli et al., "Predictability and Surprise in Large Generative Models," ACM Conference on Fairness, Accountability, and Transparency, 2022, pp. 1747–64; Roberto Gozalo-Brizuela and Eduardo C. Garrido-Merchan, "ChatGPT Is Not All You Need: A State of the Art Review of Large Generative AI Models," 2023, arXiv.org.

14 Nate Lanxon, Dina Bass, and Jackie Davalos, "A Cheat Sheet to AI Buzzwords and Their Meanings," *Bloomberg*, March 10, 2023; Michael M. Grynbaum and Ryan Mac, "The Times Sues OpenAI and Microsoft over AI Use of Copyrighted Work," *New York Times*, December 27, 2023.

15 Aldana Vales, "An Introduction to Synthetic Media and Journalism," *Medium*, October 11, 2019.

16 Rev Lebaredian, "Les données synthétiques, primordiales pour la nouvelle ère de l'IA," *IT for Business*, May 18, 2022.

contributes to the degeneration, or even collapse, of models, with distortion effects that are amplified across training cycles: "As an AI training model is exposed to more AI-generated data, it performs worse over time, producing more errors in the responses and content it generates, and producing far less non-erroneous variety in its responses."[17] The fantasy of algorithmic capital one day freeing itself from human labor and oversight is not about to come true with generative AI.

Another new development linked to the arrival of large language models and generative AI concerns the appropriation and further uses of these technologies by users themselves. Previously, most algorithmic systems were developed by platforms to perform various functions without direct user involvement: Algorithms for ratings or recommendations, or automated decision-making algorithms were not used, perceived, or designed by the average person as "artificial intelligence" tools. In other words, the vast majority of us interacted with these algorithms passively, without realizing that it was possible to use them consciously to perform various tasks, for work, or to create new content. However, the arrival of ChatGPT and other generative AI tools allows much more active and participatory use of these technologies, making human/machine collaboration more conscious, dialogical, and interactive. As philosopher Martin Legros notes, the AI of the past "seemed to operate behind our natural consciousness," whereas robots such as ChatGPT now present themselves as assistants, partners, or copilots: "When someone addresses us, talks to us, answers our questions in a specific and personal way, we cannot help but credit it with intentionality and establish an interactive relationship with it."[18]

We are now witnessing a development akin to the transition from the original Internet of the 1990s to Web 2.0, also known as the "participatory web," in the early 2000s. Whereas most Internet users were once satisfied to consume content by navigating from one static website to another via hyperlinks, the advent of social media and new features has enabled many more individuals to share and create a vast range of content via platforms and sites such as Wikipedia, YouTube, and Facebook. Crowdsourcing, collaboration, and information sharing have thus become much more dynamic. While it is still too early to say whether this is a new stage in the overall development of AI (distinguishing a

17 Carl Frazen, "The AI Feedback Loop: Researchers Warn of Model Collapse as AI Trains on AI-Generated Content," VentureBeat, June 12, 2023.

18 Martin Legros, on the program *Avec Philosophie*, "L'intelligence artificielle, objet philosophique: Épisode 2," *France Culture*, May 16, 2023.

first phase linked to deep learning in the 2010s from a second phase associated with generative AI in the 2020s), the important thing here is that all of these machines ultimately rely on two basic elements: training data and algorithms.

This brief depiction of AI seems to tell the story of a technological miracle, a scientific epic that has overcome various technical and economic obstacles. However, it should be remembered that widespread datafication is based on algorithmic capital's imperatives of data extraction and prediction. Companies must extract ever more data and constantly improve their means of prediction in order to remain competitive on the market. AI research is funded by a complex ecosystem of universities, nonprofits, governments, military agencies, and private companies, but it is mainly the platforms that appropriate innovations, commercialize algorithmic technologies, and reap the lion's share of the profits from the exploitation of data.[19] There is, therefore, no spontaneous abundance of data, and no endogenously developed AI. Rather, we are seeing a hunt for data and algorithmic power driven mainly by capitalists under competitive pressure. Data must always be ever-more voluminous and diverse, and processed by ever-more powerful, well-trained, and experienced machines. It is, then, the *movement of capital* that is pushing Big Data and machine learning to become the main pillars of the twenty-first-century economy.

The commodification of data and algorithms

To produce value, data and algorithms are transformed into commodities —and even into capital. How can we cast this complex process in simple terms? Let us start by considering how data undergoes commodification as it is inserted into the process of capital valorization and accumulation.

In its raw form, data presents itself as material extracted from human experience or socionatural processes, which is then refined and "cleaned" through digital processing and algorithms, thereby acquiring a marketable form that allows it to be monetized. We can thus distinguish between *data as material*, which enters the production process as a resource or "raw material," and *data as commodity*, once it has been processed by

19 For an overview of AI funding in Canada, see Ana Brandusescu, "Politique et financement de l'intelligence artificielle au Canada: Investissements publics, intérêts privés," Centre de recherches interdisciplinaires en études montréalaises, Université McGill, Montréal, March 2021.

productive digital labor. Once commodified, data has two different aspects: a use-value (its usefulness, its qualitative dimension, relative to users of said data product) and an exchange-value (its quantitative relational aspect, allowing the data to be exchanged and monetized). The use-value is realized through consumption (of search engines, apps, and software based on algorithms), while the exchange-value is realized through commercial transactions involving the purchase and sale of data and the products derived from it.

Unlike an employer who purchases labor for a fixed period of time in order to produce commodities and value, data as material is generally not purchased from users. Rather, it is appropriated by platforms mostly free of charge, then commodified and resold to third parties. Zuboff uses the term "accumulation by dispossession" to describe the process by which platforms appropriate ever more data through a cycle of incursion, habituation, adaptation, and redirection of whatever is extracted. Others refer to this process as "accumulation by gift" to illustrate how metaphors of gift-giving, sharing, and reciprocity within social media and the collaborative economy are exploited to normalize the private appropriation of personal data.[20] Still others theorize this data appropriation as "unpaid digital labor."[21] From our perspective, data as material is not in the strict sense the product of "unpaid" or "free labor," but rather the result of a mechanism of *extraction* and dispossession. As such, it enters the productive process as free or cheap raw material, helping to keep down the cost of algorithmic capitalists' means of production.[22] In short, the *value* of data comes from digital labor applied to human datafied experience, extracted during both paid and unpaid productive time, often in the form of attention time. We can thus speak of relations of *exploitation* of digital labor and *dispossession* of human experience by algorithmic capital, which allow for the creation of a new form of monetizable surplus-value (akin to the behavioral surplus described by Zuboff). Likewise, this helps us see digital labor as a multiform activity that is sometimes remunerated, and thus exploited, and sometimes not remunerated, and thus subjected to a mechanism of extraction. When and insofar as it participates in the production of commodities sold on the market, digital labor produces surplus-value. Moreover, the vast

20 Marion Fourcade and Daniel N. Kluttz, "A Maussian Bargain: Accumulation by Gift in the Digital Economy," *Big Data & Society* 7, no. 1, 2020.

21 Fuchs, *Digital Labour and Karl Marx*, pp. 95ff.

22 See the short but illuminating discussion by Beverley Best, *The Automatic Fetish: The Law of Value in Marx's Capital*, London: Verso, 2023, pp. 39–46.

"training" undergone by Internet users in the use of digital technologies develops the "general intellect," and so can also be seen as a way to lower the value of labor power across the economy.[23]

It is important to note that the processes of datafication, appropriation, and monetization of data do not take place in the ethereal realm of cyberspace. To paraphrase Marx in *Capital*: Data cannot itself go to market and perform exchanges in its own right. We must therefore have recourse to its guardians, who are possessors of data. The main "data guardians" of our time are not individuals, but platforms and other capitalist firms that extract and accumulate data. Traditional companies have either started to be supplanted by the arrival of big platforms over the last decade, or are themselves becoming "platformized" by turning to the algorithmic accumulation model as a supplement or replacement for their traditional activities. It is largely within the platform ecosystem that data is extracted, manufactured, and exchanged. Marx once distinguished between the sphere of production (internal to the company) and the sphere of circulation, corresponding to the market. However, the emergence of the platform blurs this division, because this "market/company hybrid," to use Antonio Casilli's term once more, becomes a space for data extraction, algorithmic production, and the circulation of commodities. This form of "market enclosure" or "market cornering," typical of the platform economy, is for economists such as Yanis Varoufakis one sign that we have moved beyond market capitalism and have been propelled "back into the future" of "technofeudalism."[24] We come back to a dialog with this critical literature in Thesis 8, which is right, in our view, in identifying profound transformations in late capitalism and the need for new concepts, but perhaps overstating its case in proposing that these changes are ushering in a postcapitalist technofeudal age.

On platforms, extracted data helps to run services, software, tools, or applications with some practical value. Data is an input into algorithmic production methods that perform various tasks: recognizing images, classifying data, making automated decisions, and generating text or images, as in the case of generative AI such as ChatGPT, Dall-E, or Midjourney. In addition to the data used to feed the algorithms, the platform also captures surplus data, which it turns into predictive products that can be sold to third parties such as advertisers. The sale of such products corresponds to the process of data monetization or commodification described by Zuboff. But we should not forget that

23 Ibid., p. 43.
24 See Varoufakis, *Technofeudalism*.

platforms also produce high-value-added algorithms, companies, governmental agencies, and other organizations. These algorithmic commodities have a unique power: the power to predict. Machine-learning algorithms are first and foremost *prediction machines* that can drastically reduce the costs of prediction. Businesses, as well as public authorities and various social organizations, are making increasing use of these predictive machines to increase productivity, intensify labor, and automate and optimize countless processes.[25] These algorithmic machines go far beyond the behavioral prediction mechanisms studied by Zuboff. And not only personal data is transformed into commodities and sold in the form of targeted advertising. Indeed, a myriad of algorithms are marketed to optimize processes in various spheres of the economy and society, including agriculture, government, industry, logistics, the military, commerce, services, finance, policing, and insurance.

Algorithmic machines can be marketed for the most varied purposes: from workflow optimization to revenue management, sales forecasting, planning, inventory management, dynamic pricing, contract management, predictive assessment, and managing customer satisfaction. In the construction industry, we are also seeing a mini-revolution: building data models, project management information systems, drones, smart sensors, advanced analytics, and 3D printing. In the world of finance, the algorithms used for high-frequency trading in 2014 already accounted for more than 75 percent of global financial transactions. Today, in the age of what Steve Maher and Scott Aquanno term "new finance capital,"[26] the "Big Three" of BlackRock, Vanguard, and State Street, colossal asset management firms that dominate the financial sphere, make heavy use of algorithms, especially for their huge passive index and exchange-traded funds, relying on algorithmic systems like BlackRock's Aladdin platform for trades and data analysis. Added to this is the world of financial technologies (fintechs), which are revolutionizing payment systems, home insurance, risk management, investment strategies using robot advisors, and other areas. The list of sectors that are rapidly becoming algorithmic is growing ever longer: from agriculture and mining to manufacturing and transport. The enthusiasm surrounding algorithmic technologies is fueled by the prospect of automating a large proportion of activities, which often promises to cut costs and raise productivity.

---

25 Ajay Agrawal, Joshua Gans, and Avi Goldfarb, *Prediction Machines: The Simple Economics of Artificial Intelligence*, Boston: Harvard Business Review Press, 2018.

26 Stephen Maher and Scott Aquanno, *The Fall and Rise of American Finance: From J. P. Morgan to Blackrock*, London: Verso, 2024.

This phenomenon of self-growth, where the accelerated accumulation of data multiplies the power of algorithms and vice versa, leads us to the heart of algorithmic capital accumulation.

The general formula for algorithmic capital

Let us now attempt to outline the logic of algorithmic capital using formulas to simplify aspects of what are in reality no doubt much more complex processes. First, human and socionatural experience (E) is transformed into data (D). The process of datafication (E→D) is the first step necessary to set up algorithmic machines (AM), which, combined with algorithmic digital labor (AL), generate a "surplus" of data, that is, data that now has value (D′). In the E→D→AM+AL→D′ circuit, D′ represents the quantitative increase in data, but also the transformation of data into commodities such as predictive products with added value. The predictive products can be targeted ads, but also any commodity with predictive capabilities: These data products possess economic and "predictive value."[27] These are predictive commodities (PC), commodities sold and purchased mainly because of their predictive power.

> experience → data → algorithmic machines + algorithmic labor → surplus data + predictive commodities

For the average user, the use of an algorithmic machine simply helps obtain data or services that are useful for some purpose. For the platform, however, the data of such interactions is captured and then monetized in the form of predictive commodities. The data then enters the circuit of circulation, where the commodity is sold to a buyer for money. From the buyer's point of view, for example, an advertiser on Facebook, predictive commodities (PC) are purchased with money (M) in order to personalize advertising and increase the chances of selling a product in order to realize a profit: M→PC→M′. The money is used to acquire a commodity that will create a surplus of money (M′) in the sense that, for the advertiser, the predictive commodity increases or secures profit realization. From the point of view of a productive enterprise purchasing predictive commodities, they make it possible to optimize production processes,

27 We here borrow a term from Amanda Parsons and Salomé Viljoen's detailed analysis in "Valuing Social Data," *Columbia Law Review* 124, no. 4, July 17, 2023, pp. 993–1080.

for example, through automation. Data commodities and algorithms thus become sought-after commodities, whether through predictive markets or through the use of the various algorithmic machines that optimize business and production processes. The commercial production of data and algorithms is thus linked to the classic imperative of capitalist accumulation (M→C→M′), in that data commodities and algorithms are means to intensify labor, make it more productive, make predictions or realize profits. Platforms accumulate as much data as possible and produce predictive commodities with the aim of making profits and increasing their capital, and various enterprises seek these data commodities to realize their profits or optimize production processes. Big Data is thus the fuel that powers the algorithmic machines that produce predictive commodities which become integrated in capitalist value production, circulation, and realization. This is the general formula for transforming data into profits: D′→AM+AL→PC→M′.

> surplus data → algorithmic machines + algorithmic labor → predictive commodities → surplus money (profits)

Thus, data can be understood as undergoing a process of commodification—indeed, this is how it is described in most of the literature on the subject. However, understanding the process of transforming data into *capital* is extremely important for grasping the logic of accumulation and circulation of data. In other words, data is a means of generating and realizing profit (*data commodities*), but it is also accumulated as an end in itself (*data capital*). The expression "data capital" entered the economic lexicon in 2016. In a report entitled "The Rise of Data Capital," Erik Brynjolfsson points out that "the lasting asset is the data."[28] This means that data is not just a commodity; today, we are seeing accumulation of data as *capital*.[29] On this view, data is not simply commodified and sold on the market; it can also represent a strategic asset for extracting extra profits, and/or attracting venture capital investment through mechanisms of *predation* rather than production per se. In its productive capital functions however, data today can constitute constant capital incorporated into algorithmic means of production, a capital asset, just like various tools and machines used to generate value in any labor process. Furthermore, the profits made from selling data commodities

28 Erik Brynjolfsson et al., "The Rise of Data Capital," *MIT Technology Review*, 2016.

29 Jathan Sadowski, "When Data Is Capital: Datafication, Accumulation, and Extraction," *Big Data & Society* 6, no. 1, January 2019.

and algorithms can be reinvested into more algorithmic machines and mechanisms of data extraction, reproducing algorithmic capital on a larger scale.

Viljoen and Parsons offer a helpful model of their own to summarize ways in which data's "predictive" value can be realized. This can be done directly, by selling data to data brokers or through targeted advertising. It can also be done indirectly, by lowering costs through wage discrimination, or by stabilizing and increasing revenues—for example, when a firm uses data as information that allows it to practice different forms of price discrimination. Data's predictive value can also be converted into economic *power*, whether as a data asset attracting investments with the idea that it will generate revenue eventually, or through various forms of rentiership.[30]

We should note however that while data as capital provides a strategic advantage in the market, simply hoarding data leads to a dead end; massive amounts of accumulated data are valuable only insofar as they can be processed by digital labor, analyzed by algorithmic machines, and reinjected into the cycle of accumulation. This is why an exclusively data-driven capitalism would lead nowhere, except to a form of speculation on data assets that would most likely suffer the same fate as other bubbles in the economy—such as the cryptocurrency market, which crashed dramatically in March 2022, or perhaps even the so-called "AI bubble" threatening to burst as of this writing in early 2026.[31] Massive data extraction makes sense only in light of the potential to transform these data streams into profits through the optimization of algorithmic machines, the production of data commodities, or predation.

In a digital world that sees such fierce competition between platforms that each have their own powerful algorithms, the ability to collect, analyze, use, transform, and monopolize data is becoming a key strategic focus. The imperative to extract, predict, and make profits drives each of the platforms to invest in these algorithmic technologies in order to reduce labor costs, increase productivity, and maximize profits. For the platforms, increasing their means of algorithmic production means building their power. The hunt for data is thus combined with an "algorithmic arms race." Any platform—or country—that does not invest sufficiently in AI research and development risks being outpaced by its competitors. This logic is replicated at the level of cities seeking to attract

30 Parsons and Viljoen, "Valuing Social Data."

31 Jonathan Martineau, "La bulle de l'IA a un prix, et il risque d'être cher à payer," *Le Devoir*, December 10, 2025.

private investment, venture capital, and the best-trained tech workers to their territory. Yet it is also seen at the international level, through the intensified competition between states—particularly China and the US (see Thesis 12).

For an example of this competition, we need only look to the world of chatbots (extremely efficient at extracting data from users), following OpenAI's launch of ChatGPT in November 2022. In response to this move, Microsoft announced the integration of GPT-3 into its Bing search engine in February 2023, and Google created a crisis unit with its cofounders before announcing the launch of Bard (now Gemini) for 2023—fearing that ChatGPT would leave it lagging behind.[32] The race for conversational robots is thus reminiscent of the space race of the last century. ChatGPT represents Sputnik I, and now Google and others are working hard to overtake it in the quest for technological supremacy. Anthropic's Claude, xAI's Grok, Deepseek's chatbot, and others have now entered the race as well. Of course, different companies pursue different "strategies of capture," as Nick Srnicek calls them. For example, as far as AI development is concerned, some aim at monopolizing infrastructure (Amazon), others at building the ultimate AI model (OpenAI), others adopt a more "classic" conglomerate strategy (Google), while still others adopt a more open strategy of AI development (Meta).[33]

How can we make sense of this accumulation process? Algorithmic capital aims to expand by increasing algorithmic productive power. This is its bare logic. The goal of this process is not the satisfaction of human needs, but of the profit imperative. As such, the cycle of production must be constantly renewed on a wider scale, seeking to produce more value through the constant improvement of algorithmic means of production and the exploitation of labor. Ultimately, Big Data and the rapid development of AI are both intertwined with and subordinate to the logic of capital accumulation, which accelerates data extraction and the deployment of algorithmic machines in various spheres of the economy. To put it bluntly: The transformation of money into artificial intelligence and its infrastructure ultimately aims to accumulate capital. The general formula for algorithmic capital is thus M+D-AM+AL-PC-M'+D'. The reconversion of that surplus into labor-power and more algorithmic

---

32 Thomas Urbain, "Google lance la contre-attaque à ChatGPT avec son logiciel Bard," *Le Devoir*, February 6, 2023.

33 See the detailed analysis of business strategies in Nick Srnicek, *Silicon Empires: The Fight for the Future of AI*, Cambridge: Polity, 2026.

machines and their infrastructure, leads to accumulation: "Accumulate, accumulate! That is Moses and the prophets!" This accumulation of algorithmic capital also amounts to general algorithmization of value production in contemporary capitalism.

The horizon of "fully automated decision-making"

At the turn of the twenty-first century, the dominant reading saw the digital revolution as a powerful means of harnessing technological development for the purposes of human progress. The staggering increase in computing power, it was argued, would free individuals from tedious routine tasks, allowing them to enjoy their free time and devote themselves to more uplifting and fulfilling activities. This grand narrative remained prevalent until only recently in the entrepreneurial world, where humans and machines seemed to have forged a kind of alliance or collaboration that gave humans the upper hand, with machines content to automate thankless tasks that eat up time and agency. This particular alliance between the human mind and computers, also known as the "standard partnership," is now being challenged by the rise of algorithmic capital, which is deeply reshaping the terms of collaboration between humans and machines. Algorithmic technologies are already playing an increased role in several key processes, and this can only accelerate with the rapid development of algorithms.

The increased use of algorithmic machines aims to address a seemingly mundane problem: Human beings are fallible. In his bestseller *Thinking, Fast and Slow*, the psychologist and behavioral economics pioneer Daniel Kahneman distinguishes between two systems of thought within human cognition.[34] System 1 operates automatically, involuntarily, intuitively, emotionally—and quickly—requiring less effort and energy. System 2, on the other hand, requires more time, effort, and attention in order to solve complex problems analytically. We use System 1 most of the time—for example, in conversation, or when we recognize anger flashing across someone's face. System 2, on the other hand, is mobilized when the first system fails to solve the problems it faces; we then make a greater effort to accomplish a more complex or difficult task. Unsurprisingly, System 1 is extremely useful in our everyday lives, but it is also full of biases, errors, and imprecision: confirmation bias, stereotypes,

34 Daniel Kahneman, *Thinking, Fast and Slow*, New York: Farrar, Straus & Giroux, 2013.

flawed predictions, overestimation of risks, misleading inferences, and other factors, which can also affect System 2: "Because System 1 operates automatically and cannot be turned off at will, errors of intuitive thought are often difficult to prevent. Biases cannot always be avoided, because System 2 may have no clue to the error."[35] Fortunately for businesses, algorithmic machines are coming to the rescue. They can make reliable and rapid predictions or even make better decisions that are supposedly free of bias. McAfee and Brynjolfsson do not mince their words—and perfectly illustrate the mindset of many champions of AI:

> The most obvious approach is to consider letting the machines make the decisions where and when they can—to let pure digital instances of System 2 … come up with their answers without input from System 1. Over time, this is exactly what more and more companies are doing.[36]

Thus emerges the specter of fully automated decision-making, one main horizon of algorithmic capital.[37]

If until recently these processes sounded like mere science fiction, fully automated decision-making has today become a reality. Amazon's recommendations, ticket prices on Google Flights, hotel room rates on Booking.com, Walmart's inventory management, performance evaluations of managers and employees, and the vast majority of financial transactions are just a few examples of the operations now performed automatically by algorithmic machines.[38] Since March 2018, the South Korean company Midas-IT has been marketing entirely AI-driven hiring software. This recruitment robot can automatically evaluate tens of thousands of applications, saving time, money, and energy. During an hour-long interview, the algorithmic machine analyses the interviewee's emotional reactions and draws their psychological profile in real time by asking complex questions in a slightly robotic female voice: "You have a date at a fancy restaurant and you realize you've forgotten your wallet. What do you say to your partner?"[39]

---

35 Ibid., p. 28.

36 Andrew McAfee and Erik Brynjolfsson, *Machine, Platform, Crowd: Harnessing Our Digital Future*, New York: W. W. Norton, 2017, p. 46.

37 Agrawal, Gans, and Goldfarb, *Prediction Machines*.

38 See Thesis 10 on the dissemination of automated algorithmic decisions in the fields of government, public administration, and policing, and Thesis 12 for their use in the military sphere.

39 Frederic Ojardias, "Intelligence artificielle: En Corée du Sud, les robots deviennent recruteurs," Médiapart, December 2018.

If executives and corporate directors thought that the specter of automation would spare them, they now have to think again. Senior execs' performance is now also evaluated by algorithms. In May 2014, a robot called VITAL (Validating Investment Tool for Advancing Life Sciences) became the first AI to be appointed to the board of Hong Kong–based venture capital fund Deep Knowledge Ventures.[40] Usually, corporate boards' decision-making processes are dominated by the HiPPO (highest-paid person's opinion) principle, which is, of course, vulnerable to the human failings of System 1.[41] As everyone knows, a company CEO is not smarter than the average employee, even if they earn 200 times more. In the era of algorithmic capital, the HiPPO principle is being superseded by fully automated decision-making, which is more effective than ordinary managerial intelligence. Unable to free themselves from capital, workers can at least take comfort in the end of the useful life of executives who have become superfluous due to the full development of the productive forces.

Unfortunately, this somewhat over-optimistic scenario overlooks a crucial detail: Humans will not be entirely replaced by robots, but rather, will be accompanied by them, thus creating a "new mind–machine partnership." Bosses will not give up their privileges so easily. Instead, they will use robot investors, robot managers, and robot advisors to improve their market-forecasting capabilities, enhance work-process productivity, manage and monitor their employees, and maximize returns on investment. The burdensome tasks of management, financial analysis, and investment—so costly in terms of time, money, and energy—will be largely delegated to algorithmic machines, the faithful and obedient servants of capital. We can expect to see a redefinition of the division of labor as algorithms become more effective managers. The owners of the means of algorithmic production will be able to outsource their judgment to machines—in part, at least—freeing up their minds to more fully enjoy the fruits of their capital. Formerly mere servants of capital, capitalists will now be able to shrug off the anxieties of full-blooded competition and retire at forty. For they will be able to offload the complex job of managing and accumulating capital onto algorithmic machines that are perfectly qualified for this purpose and need only minimal supervision. Some authors seem to welcome this scenario, which is already taking shape in the interstices of today's economy. With the rise

40 Rob Wile, "A Venture Capital Firm Just Named an Algorithm to Its Board of Directors: Here's What It Actually Does," *Business Insider*, May 2014.

41 McAfee and Brynjolfsson, *Machine, Platform, Crowd.*

of fully automated decision-making, "this automatic second economy is encroaching into the familiar human-mediated one, with algorithms taking over from experts and HiPPOs."[42]

Of course, whatever techno-optimist narratives such as these might say, it is impossible for the economy to become 100 percent automated. As we saw in Theses 5 and 6, the development of AI and algorithms does not abolish work but reconfigures it. Even so, we are now witnessing a reversal of the standard partnership of mind and machine. Whereas, during the first phase of the digital revolution, computers were used to perform mechanical and routine tasks, leaving the human mind to make more complex decisions, today we see algorithms performing more complex tasks, leaving humans to do thankless jobs (microwork, for example), manual correction of machine errors, and intervention in exceptional situations. For example, in December 2014, a hostage situation in a café in Sydney, Australia, caused hundreds of people to flee the area, with some relying on Uber to escape the scene. The spike in demand caused the surge pricing algorithm to significantly raise prices, which subsequently earned Uber severe criticism: How could it be so insensitive? The tragic evening of terrorist atrocities in Paris on November 13, 2015, could have seen the same mistake repeated. Yet less than thirty minutes after the first attack, Uber employees suspended the surge pricing mechanism and notified all users of the state of alert. Since algorithms cannot think completely for themselves, human intelligence must be added to the algorithms' self-correction mechanism; here, human judgment ultimately serves to hone the machines' autonomous functioning. An expression that well illustrates this phenomenon in the world of machine learning is "human-in-the-loop" (HITL). This turns things on their head: "Instead of having machines provide data as an input to human judgment, they're having judgment serve as an input to an algorithm."[43]

The division of decision-making power between humans and machines reaches a critical point when business performance and capital yields are at stake. This can stir up a feeling of inferiority in us when we're faced with the algorithmic machines' impressive capabilities. Thus, the prowess of AI is inflicting a new "narcissistic wound" on human beings. Sigmund Freud said that "the universal narcissism of men, their self-love, has up to the present suffered three severe blows from the researches of

42 Ibid., p. 47.
43 Ibid., p. 56.

science."[44] What did he mean by this? With the Copernican Revolution, the Earth is no longer the center of the universe; with Darwin, humans are no longer creatures made in God's image but the result of evolution through natural selection; and with the advent of psychoanalysis, the ego is no longer sovereign in its own domain. Today, the complex processes of the human mind (literary analysis, the arts, and economic, administrative, medical, legal, and other decisions) are apparently being surpassed by machines the mind itself has created. The human mind is becoming increasingly obsolete in the economic world, where "intelligent machines" are able to "make better decisions" in less time.

Many will raise doubts about algorithms' real ability to "outdo humans." For example, while ChatGPT has many qualities, it tends to offer approximate or even downright wrong information—the term "hallucinations" is used to refer to the responses that algorithms provide with a confident tone but which turn out to be false, unjustified, or completely fabricated.[45] Far from being programmed to seek the truth, this algorithm works on the basis of predictive inferences, like an autocorrect function based on billions of parameters and a gigantic dataset. As several observers have pointed out, ChatGPT is a large language model that tends to invent "facts," references, and arguments in order to generate *plausible* texts, making it an excellent "smooth-talker,"[46] or "stochastic parrot."[47]

Of course, it could be argued that this technology is still in its infancy, and that the arrival of new filters or plug-ins will gradually correct this problem. For example, in May 2023, OpenAI added an Internet browsing feature that provides more accurate references drawn from a search engine. However, it seems still necessary to confirm the accuracy of the sources and data presented, as we can see from the unfortunate example of the US lawyer Steven A. Schwartz. He used ChatGPT to produce a report—and the document it came up with contained made-up quotes, precedents, and court decisions, causing him humiliation in court during a lawsuit

44 Sigmund Freud, "A Difficulty in the Path of Psycho-Analysis" [1917], available at lutecium.org.

45 Ziwei Ji et al., "Survey of Hallucination in Natural Language Generation," *ACM Computing Surveys* 55, no. 12, 2023, pp. 1–38.

46 Jocelyn Maclure, "ChatGPT, baratineur," *La Presse*, January 18, 2023.

47 Emily M. Bender, Timnit Gebru, Angelina McMillan-Major, and Shmargaret Shmitchell, "On the Dangers of Stochastic Parrots: Can Language Models Be Too Big?," *Proceedings of the 2021 ACM Conference on Fairness, Accountability, and Transparency*, Association for Computing Machinery, New York, 2021, pp. 610–23.

against the airline Avianca.[48] In short, because algorithmic machines will never be completely reliable, it will always be necessary to keep a human in the loop to vouch for the quality and accuracy of the results.

However, the fact that we cannot rely 100 percent on algorithmically produced decisions or texts, and that humans will remain involved in the process, lends credence to the theory of a reversal of the human/machine partnership. That is, human judgment becomes a simple matter of verifying or correcting all manner of automatically generated texts, images, decisions, and products. To take one among many possible examples: Two months after the launch of ChatGPT, more than 200 books available on Amazon's Kindle Store had been "written or co-written" with this software, some taking only a few hours to produce.[49] The main role of algorithmic machines is thus to quicken the pace of production and speed up the completion of countless activities, ranging from writing simple emails to academic papers, scientific articles, and entire books. ChatGPT and other similar algorithms are thus set to become the "sausage machines" of the academic and literary world until this tendency is mitigated by resistance from the artisans, workers, and decision-makers in these areas.

Some have also highlighted the poor quality of the works produced by ChatGPT, or their lack of style and literary sophistication. Yet its true purpose is surely not to excel at writing poems or novels. This algorithm is capable of performing ordinary office tasks in no time at all. It is no coincidence that a growing number of workers are using this tool to complete as many tasks as possible: writing emails, drafting reports, preparing meetings, filling out application forms, and generating questions and ideas to stimulate discussion. A survey indicates that the use of ChatGPT and other algorithmic tools in professional settings in the United States rose from 27 to 43 percent in January 2023 alone, with 70 percent of employees not informing their superiors that they are using these technologies in their work.[50] Another objection to the claim that machines are outmatching humans would be to say that these are in fact good examples of a standard partnership between humans and machines, with workers using algorithmic tools as personal assistants to

48 Benjamin Weiser, "Here's What Happens When Your Lawyer Uses ChatGPT," *New York Times*, May 27, 2023.

49 Mathilde Rochefort, "Oui, ChatGPT est déjà l'auteur (ou le co-auteur) de nombreux livres sur le Kindle Store," *Clubic*, February 22, 2023.

50 "70% of Workers Using ChatGPT at Work Are Not Telling Their Boss; Overall Usage Among Professionals Jumps to 43%," fishbowlapp.com, February 1, 2023.

Figure 7.1 Ways in which newsrooms use generative AI tools

Category	Approx. %
Text Summarization	55%
Simplified Research	40%
Workflow / Efficiency	35%
Text Editing / Correction	35%
Content Creation	30%
Ideation	28%
Translation	28%
More Personalization	15%

Source: Adapted from Teemu Eriksson, "New survey finds half of newsrooms use Generative AI tools; only 20% have guidelines in place," World Association of News Publishers, May 25, 2023

save time. For example, a survey indicates that 49 percent of journalists in newsrooms around the world have started using generative AI tools such as ChatGPT to perform various tasks, as shown in Figure 7.1.

However, while there do seem to be clear advantages to using ChatGPT in terms of individual productivity gains, we may well ask about the problems associated with its use, should it become all-pervasive in the manner of Google searches or the Internet itself. We can identify at least three issues that aggravate existing trends in the world of work.

First, there is the issue of increased dependency on these algorithmic machines. As with computers, the use of ChatGPT and other algorithmic tools at work and in everyday life will lead to the rapid development of dependency, similar to the way Wi-Fi has become a basic need in the Maslow pyramid of digital society. This is evidenced by the integration of GPT-4 into Microsoft Office, Bing, and other office software, as well as the thousands of free or paid algorithmic applications today on the market.[51] This does not seem to pose a problem in itself, since we are already used to using IT and the Internet for a variety of tasks at work or at home. However, the fact that it is difficult, if not actually impossible, to function socially without using these devices shows the enormous power that these machines (and their owners) exert over our lives. The algorithmic world thus translates into a strengthening of the digital realm's influence over our social and economic life.

Second, the mass use of algorithmic tools such as ChatGPT strengthens the dynamics of exploitation and competition. While the time saved

51 For a directory of algorithmic tools that can be used at work or at home, see futurepedia.io.

by using an algorithmic tool may, from the worker's point of view, be helpful in the short term, it will not necessarily allow them to work less. The opposite may indeed be true. Workers who are more adept at using AI will be more productive than ones without AI, and this will raise performance expectations for everyone. Labor will become intensified.[52] There will thus be pressure to use these tools in order to perform better, with employers demanding more from both managers and employees in terms of performance. This is, in a way, the rebound effect of efficiency gains: The harder we work, the higher the expectation on us to work, thus increasing demands on the working class and the community in general. In short, it is not so much that machines will replace humans en masse, as that "algorithm-assisted humans" will outcompete "algorithm-less humans" in the labor market.

Third, this human/machine partnership fuels the process of social acceleration. In Thesis 3, we discussed Hartmut Rosa's claim that one of the main characteristics of modern societies is acceleration: The increased speed of transport and communication, which allows us to move faster (technical acceleration), has the effect of increasing the pace of social change (renewal of ideas, fashions, habits, identities, and so on), which in turn leads to increased pressure on the pace of life. With this time overload and the multiplication of demands to be met on an individual level, this creates pressure to make greater use of more powerful technological tools to save time. This pressure helps to turn the wheel of acceleration, so that we never get to see the long-yearned-for leisure society arrive; rather, we are living in a mad rush where everything is accelerating without us having any significant control over our lives, over the trajectory of society, or over technological development itself.

The effect of this "algorithmic acceleration" is that, rather than algorithmic machines serving human creativity and allowing more free time, it is instead humans who have to serve algorithmic technologies, inasmuch as they are forced to work with and for these technologies. Humans are already expected to constantly adapt to the latest technological innovations, which are producing an ever-growing proportion of formerly material and now immaterial goods, such as texts, images, videos, and concepts of all kinds. Humans will be further reduced to the

---

52 The intensification of work seems to be the main effect of implementing AI into labor processes, according to the most recent studies. See Ray A. Smith, "AI Isn't Lightening Workloads. It's Making Them More Intense," *Wall Street Journal*, March 11, 2026; Aruna Ranganathan and Xingqi Maggie Lee, "AI Doesn't Reduce Work, It Intensifies It," *Harvard Business Review*, February 9, 2026.

role of supervisors of algorithms, correcting manufacturing defects and the imprecise results that predictive machines produce.

Ultimately, the role of humans is to assist the machine's autonomous operation. The theory of a "great replacement" of humans by robots (or immigrants) is absurd and simplistic. We should instead imagine society as a great automated assembly line, in which humans will increasingly be relegated to the role of supervisors, assistants, or appendages of robots and algorithmic processes, or of labor inputs at some point in the algorithmic labor process. This will feed into the increased precarity, outsourcing, and fragmentation of work, as we discussed in Thesis 5. The generalized spread of automation in the economic world thus represents the main trendline and horizon of algorithmic capital.

Toward an uneven and combined automation

This vast process of automation today takes the form of uneven and combined development. This concept, used by Leon Trotsky, was primarily intended to shed light on the complex relationships between "advanced" and "backward" societies from the standpoint of the development of the capitalist mode of production. In Trotsky's reading, this led to the juxtaposition, interference, and mutual entanglement of "modern" and "archaic" relations of production within societies such as turn-of-the-century Russia. Trotsky referred to "unevenness, the most general law of the historic process."[53] Drawing on this idea, we can anticipate that the automation process will not be complete and absolute in every aspect of business, sector of the economy, or sphere of society, nor between societies in the Global North and Global South. Systemic automation represents only a *structural, general, and potential trend.* The material development of algorithmic capital will take various forms as a result of countless small managerial decisions and acts of worker resistance that will have to determine, in each instance, which segment of activity will be entrusted to algorithms or robots.

How will this gradual and uneven dissemination of algorithmic logic take place? Economics, technology, and management textbooks are already beginning to popularize this approach by emphasizing the importance of trade-offs in strategic thinking about AI use in order to improve business performance. For example, there may be a need

---

53 Leon Trotsky, *The History of the Russian Revolution*, London: Victor Gollancz, 1984 [1930], p. 27.

to balance data production and confidentiality, speed and accuracy, or autonomy and control.[54] As we pointed out earlier with regard to the ethics of algorithms, this is essentially a question of reaching the right calibration, in conformity with certain objectives; and in this case, the goal is not moral, but commercial.

One of the trade-offs that every company has to weigh relates to a key mechanism of capitalist valuation: the tension between *absolute surplus-value* and *relative surplus-value*. We may remember that, for Marx, absolute surplus-value represents the surplus-value produced by the extension of the working day. By analogy, in the era of algorithmic capital, one element of the increase in absolute surplus-value corresponds to the increase in the amount of data provided by human experience and digital labor, particularly through time spent in front of screens or connected to data-extracting devices. Here, the imperative of extraction drives things along. As a complement to this, relative surplus-value is the "surplus-value which arises from the curtailment of the necessary labor time, and from the corresponding alteration in the respective lengths of the two components of the working day [necessary and surplus labor]."[55] Capital manages to produce more commodities in less time, or to lower the cost of labor power, and thus creates more surplus-value through increased productivity.

At the level of the firm, increasing relative surplus-value or productivity through algorithms requires breaking down the workflow into tasks, then into microdecisions, in order to (fully or partly) replace some of its elements with algorithmic machines.[56] The productivity gains achieved through algorithmization do not come about through the outright elimination of jobs, but through a significant reconfiguration of tasks and their partial automation within the workflow. New practices related to the use of ChatGPT may usefully illustrate this. In a survey of 1,000 US firms that reported using this software or planning to do so, half had officially implemented its use, and a quarter had begun to replace certain employees or streamline their workforce.[57]

At this stage, it is difficult to study the precise effects of LLMs and other algorithmic systems. For example, a study by investment bank Goldman Sachs predicts that more than 300 million jobs will be

54 Agrawal, Gans, and Goldfarb, *Prediction Machines*, p. 4.

55 Marx, *Capital*, Vol. I, p. 432.

56 Agrawal, Gans, and Goldfarb, *Prediction Machines*, p. 125.

57 Trey Williams, "Some Companies Are Already Replacing Workers with ChatGPT," *Fortune*, February 25, 2023.

eliminated or downgraded in coming years due to the arrival of generative AI.[58] This—alarmist and methodologically unsound—discourse is an example of the "fauxtomation" narrative identified in Thesis 5, which aims to frighten workers and send a positive signal to private investors by dangling the prospect of a new source of growth. In contrast, another, more serious study published in March 2023 considers that this "general purpose technology" can spread widely and more or less rapidly across the labor market, with varying degrees of intensity. Based on a sample of 1,016 jobs in the United States, broken down into detailed work activities and their associated tasks, the authors define the degree of "exposure" as an evaluation of whether access to an LLM would cut by at least 50 percent the time required for a human being to perform a specific task or activity.[59] This study estimates that a total of 80 percent of jobs will be affected to some degree (with a minimum of 10 percent of tasks possible to automate), while 19 percent of jobs will be "highly exposed," with at least 50 percent of their tasks possible to automate.

More soberly, economic historian Aaron Benanav reminds us that studies from the 2010s predicting a massive replacement of workers by robotization across multiple sectors have not proved accurate. Ten years later, we do not have mass joblessness, but low unemployment rates and employers even claiming at times that there is a labor shortage. According to Benanav, we are more likely to see the same process that can be observed throughout the evolution of employment in capitalist societies: a few new types of jobs, some other types that become obsolete, and then a majority of jobs whose tasks change and develop in line with technological changes. In his view, a more accurate estimate would be around 10 percent of jobs threatened by LLMs. To understand what is actually going on, we should not just rely on press releases from the people and companies who are promoting the technologies but wait and observe the real changes on the ground.[60]

Furthermore, some individuals seem to be taking advantage of this situation—for instance, the "overemployed" workers who manage to juggle three or four jobs simultaneously thanks to remote working and

---

58 Jack Kelly, "Goldman Sachs Predicts 300 Million Jobs Will Be Lost or Degraded by Artificial Intelligence," *Forbes*, March 31, 2023.

59 Tyna Eloundou, Sam Manning, Pamela Mishkin, and Daniel Rock, "GPTs Are GPTs: An Early Look at the Labor Market Impact Potential of Large Language Models," 2023, arXiv.org.

60 Aaron Benanav, "The Revolution Will Not Be Brought to You by ChatGPT," *New Statesman*, April 11, 2023.

the arrival of Gen AI.[61] Employers will push to streamline their labor costs, whereas employees will be called upon to increase their work intensity and boost their productivity levels. This intensification of work will sometimes be imposed (with a view to maximizing company profits), but it will sometimes be desired by people seeking to maximize their income through the strategic use of algorithms. For example, one university professor in Britain who runs a digital marketing agency and a start-up on the sly revealed to *Vice* what he had been up to:

> The chatbot has helped him generate business plans, internal system documents, blog posts, and Excel spreadsheets. He estimated that ChatGPT can often pull off 80 percent of the legwork, leaving him to handle the final 20 percent. The tool has even helped him win a grant from the U.K. government (the application was "50-50 me and ChatGPT," he said) and complete coursework for a master's degree necessary to receive a university teaching qualification he was too busy to handle by himself while working three jobs.[62]

The impact of automation will vary depending on the type of work and profession—scientific and critical thinking skills will be less exposed in terms of programming and writing tasks. Thus, instead of having three programmers, two translators, four marketing managers, and one social media manager, an employer well versed in how to use algorithmic machines may decide to keep only one programmer, one translator, two marketing managers, and one social media manager, if it believes that the same amount of work can be performed this way. This is the same old logic of industrialization that has been around since the nineteenth century, but in a new form that now affects "intellectual work," office work, highly skilled jobs, and other independent professions. After blue-collar workers, white-collar workers are now threatened with precarity:

> In the next five years, it is likely that AI will begin to reduce employment for college-educated workers. As the technology continues to advance, it will be able to perform tasks that were previously thought to require a high level of education and skill. This could lead to a displacement of workers in certain industries, as companies look to cut costs

61 Dexter Tilo, "Employees Use AI to Juggle Multiple Jobs," August 3, 2025, hcamag.com.

62 Maxwell Strachan, "Overemployed Hustlers Exploit ChatGPT to Take on Even More Full-Time Jobs," *Vice*, April 12, 2023.

> by automating processes. While it is difficult to predict the exact extent of this trend, it is clear that AI will have a significant impact on the job market for college-educated workers.[63]

Automated tasks that were once performed by humans are now offloaded onto algorithmic machines in order to reduce labor costs. This also curbs the risk of living labor resisting the process of capitalist value production. Algorithms are not about to go on strike to demand wage increases or better working conditions or to assert the "dignity of the machine" violated by this system of accumulation. The labor that supports algorithms, however, might do just that.

In the near future, more and more parts of the company will be replaced by algorithmic machines. To take just one example, the news and entertainment website BuzzFeed, which laid off 12 percent of its workforce in December 2022, saw its share price jump 200 percent in January 2023 after the company announced that it would use ChatGPT to produce articles and quizzes.[64] To understand this transformation, we should turn back to analysis of the "machinism" bound up with the advent of large-scale industry in the nineteenth century. Marx in his day already saw a global reconfiguration of the division of labor, which would take the form of an "automatic system of machinery."[65] As the production apparatus becomes increasingly driven by algorithmic machines, humans will be transformed into assistants to the machines, forming "a lifeless mechanism which is independent of the workers, who are incorporated into it as its living appendages."[66] Digital Taylorism has already become a reality—as illustrated by the wristbands worn by Amazon warehouse employees, guided by an AI machine which uses haptic feedback to detail each movement they must perform. Factory workers, as well as click-workers and employees in the service sector, will have to keep up with the frenetic pace of algorithms that dictate the tasks to be performed.

This system of machinery, which Marx called a "vast automaton" that mobilizes workers as assistants to its autonomous functioning, is now combined with algorithmic machines that are speedily pursuing the

---

63 Annie Lowrey, "How ChatGPT Will Destabilize White-Collar Work," *The Atlantic*, January 20, 2023.

64 Morgan Chittum, "BuzzFeed Stock Surges 200% on Plans to Use OpenAI to Write Stories," *Business Insider*, January 26, 2023.

65 Karl Marx, *Grundrisse: Foundations of the Critique of Political Economy*, London: Penguin, 1993, p. 692.

66 Marx, *Capital*, Vol. I, p. 548.

automation process that began in the nineteenth century. The reconfiguration of human/machine relations is reflected in the predominance of algorithmic processes, which workers increasingly only monitor and assist. While the IT revolution of the 1980s promised to put technical progress at the service of the free flourishing of the human mind, algorithmic capital reverses this relationship. It instead puts the human mind in the service of generalized automation and the increased autonomy of the machines themselves. Marx had glimpsed this "mechanical monster" in his day, emphasizing that it was important to consider machines not in isolation, but in terms of their overall functioning.[67] The mechanical monster of the Fourth Industrial Revolution is already upon us: *It is not humans who produce using algorithms, but algorithms that produce using humans.*

Toward a disembedding of algorithmic technologies?

To summarize, algorithmic capital is driven by a dynamic of data accumulation (the imperative of extraction), increased algorithmic power (the imperative of prediction), close human/machine collaboration in which machines play an increasingly dominant role in decision-making, as well as in the reorganization of the production process (the imperative of productivity). The combination of these mechanisms serves to accelerate the automation of businesses, the economy, and—increasingly—society in general. Yet this vast historical process of algorithmic acceleration does not unfold in any one-dimensional or homogenous way. Indeed, there are many forces opposing and slowing this process, and tensions and resistances that emerge in various sectors of social life. While Big Data and the machine-learning revolution are already having a drastic impact on society, the effects of this new logic of capital are widely differentiated in terms of class, gender, race, age, (dis)ability, and other factors. Advances in algorithmic capital must always be translated by real-world actors within specific organizations, according to particular material, cultural, and institutional constraints. This gives rise to a vast, uneven, and combined development of algorithmic power in various dimensions of society.

But how should we conceptualize the tension between the general dynamics of automation generated by algorithmic capital and the

67 Ibid., p. 503.

opposing forces that prevent this process from being completed in full? How can we explain both the pervasive logic of algorithmic capital, which seems to materialize to varying degrees, and the resistance of society, which constantly hinders our advance to the so-called utopia of fully automated decision-making? Here we can draw on Karl Polanyi's thesis of the disembedding of the market economy in his 1944 work *The Great Transformation*. For Polanyi, every society has an economy that allows its members to meet their needs; markets have existed through different periods, but alongside other forms of economic exchange such as reciprocity and redistribution. However, the nineteenth century saw an unprecedented shift, through which the accelerated development of the market economy and the Industrial Revolution were accompanied by the arrival of a self-regulating market to which it was necessary to adapt in order to ensure prosperity for all.

Whereas, previously, every economy had been "embedded" or integrated into an institutional, cultural, and normative framework proper to each society, which set its limits, the market economy is gradually freeing itself from these constraints, as it instead organizes itself into an autonomous system that feeds back into society. It is no longer the market that is present in society, but society that is "embedded" in the market:

> The control of the economic system by the market is of overwhelming consequence to the whole organization of society: it means no less than the running of society as an adjunct to the market. Instead of economy being embedded in social relations, social relations are embedded in the economic system … For once the economic system is organized in separate institutions, based on specific motives and conferring a special status, society must be shaped in such a manner as to allow that system to function according to its own laws. This is the meaning of the familiar assertion that a market economy can function only in a market society.[68]

However, it is precisely this becoming-autonomous of the market economy, or its disembedding, that poses a problem for societies, which are then exposed to the rending of the social fabric, rising inequalities, and all manner of turmoil and crises. To quote Polanyi once more:

---

68 Karl Polanyi, *The Great Transformation: The Political and Economic Origins of Our Time*, New York: Farrar & Rinehart, 1944, p. 57.

> Our thesis is that the idea of a self-adjusting market implied a stark utopia. Such an institution could not exist for any length of time without annihilating the human and natural substance of society; it would have physically destroyed man and transformed his surroundings into a wilderness. Inevitably, society took measures to protect itself, but whatever measures it took impaired the self-regulation of the market, disorganized industrial life, and thus endangered society in yet another way. It was this dilemma which forced the development of the market system into a definite groove and finally disrupted the social organization based upon it.[69]

Polanyi here posits the existence of a *double movement*: The dynamics of commodification generate a series of disruptions that lead society to protect itself through various measures and mechanisms against the excesses of this commodification. This may involve such things as the adoption of laws and rules, and social mobilizations demanding certain limits on exploitation. This Polanyian analysis can be transposed onto the context of the accelerated development of AI in the twenty-first century. It's true that every society relies on a set of technical tools to function, and computing machines have been around for quite some time already. Yet it is still worth thinking about the possible consequences of the disembedding of technological development from society and its becoming increasingly autonomous. The accelerated arrival of Big Data and algorithmic machines has already led to several economic, social, and political upheavals. These changes are happening faster than societies' ability to regulate them politically and normatively—hence the impression of a constant gap between technological advances and regulators who are constantly lagging behind.

The specific nature of algorithmic development over the past two decades has also been accompanied by new forms of governance and semi- or fully automated decision-making systems, based on the idea of an objectivity, precision, and efficiency that exceeds human capabilities. Ever more financial, legal, economic, administrative, and social decisions are being entrusted to "relatively autonomous" or (partially) automated systems that manage various complex tasks and decisions on our behalf. Instead of technology serving human beings, algorithmic tools are increasingly shaping a world in which humans are becoming assistants or auxiliaries. The invisible hand of the self-regulating market is now giving way to the automated hand of self-regulating algorithms.

---

69 Ibid., pp. 3–4.

Moreover, the capitalist economic system exerts various pressures to rapidly produce ever-higher-performing innovations—for reasons of competitive advantage, productivity, and gaining market share. The overall effect is that this race for technological development takes precedence over any regulation of these machines, determined with reference to well-contemplated social goals. In short, the dynamics of commodification proper to capitalism are now accompanied by a trend toward the "algorithmization" of economic and social processes. This also has its consequences for the social practices, culture, and political life of human communities.

As Polanyi pointed out, there is no such thing as a purely self-regulating market or an economic system that is entirely disembedded from society. If this were the case, society would collapse. The same can be said of technological development: There is no purely autonomous algorithmic system or artificial superintelligence that completely dominates the social world. Nevertheless, there is a tendency for the economy to become increasingly disembedded by means of reliance on the self-regulating market, as well as a tendency for technology to become disembedded by way of algorithms and automated decision-making systems. Just as the market economy tends to fashion a "market society" in which humans become auxiliaries of the market, it could be said that the accelerated development of AI is leading to the advent of an "algorithmic society" in which individuals are increasingly becoming auxiliaries of algorithmic machines.

Such a prospect may seem exaggerated or dystopian. And yet it appears that we are currently seeing a Polanyian double movement, one that pits rising algorithmization against a push for social protection aimed at mitigating the disruptive effects of AI on human societies. When it comes to AI ethics, liberal, reformist, and regulationist approaches constantly speak of regulating or supervising AI. This may proceed through declarations of "responsible development," or else legislative mechanisms to make sure that these innovations benefit humanity rather than a privileged minority of companies and individuals (see Theses 1 and 18). We are seeing the emergence of demands for social protection in the face of a potential loss of control over the algorithmization of social practices, work processes, and other parts of our institutions.

At the same time, we can also see the rise of more radical forms of critique, resistance, and other strong reactions aimed at curbing these machines in various sectors of society—from the sphere of personal life to certain areas of work or education. These criticisms are often dismissed as "Luddite," "backward-looking," "nostalgic," or "reactionary." They are another variant of the push for social protection that seeks to "reembed" technology in society, but with a more radical and confrontational approach. Some (like Zuboff) call for the abolition of personal data markets; others reject technological solutionism and advocate for the socialization of data centers (Morozov); some advocate for the "great refusal" of AI (Sadin), while others call for the reappropriation of algorithmic tools to drive out capitalists and plan the economy (Bastani); and still others advocate for digital sobriety, energy descent, and the use of low tech (see Theses 18 and 19).[70]

There are thus several trends, theoretical perspectives, and resistance movements aimed at countering technological disembedding. This specter haunting a society dominated by algorithmic capital is well illustrated in the 1960 text *The Dream of Machines* by philosopher Günther Anders, for whom the process of technological disembedding is accompanied by the embedding of society in the "world of machines":

> For this revolution consists precisely ... in the fact that machines, since they have taken over our efforts, have also granted themselves the task of setting the goals; once they have joined forces to form a machine park, they have ceased to be a sum of "means" by which we could pursue this or that goal freely or at our will; they have, rather, transformed themselves into our world; a world which, even before we have been able to take any particular action, has already taken action on us; a power which, even before helping us to do anything, has already enslaved us, in other words forged us. Ultimately, this revolutionary process reaches its climax when the world of machines goes so far as to force those to whom it belongs to allow themselves to be possessed by it and to conform to the demands it makes, because of its being, its power and its structure ... And in doing so, even those who are the legal owners of the world of instruments will fall under its diktat and will have to obey it.[71]

Anders's vision may be overly pessimistic, totalizing, and one-sided: There is always an element of human agency, resistance, and countermovement aimed at reembedding technological development in society. However, he does well illustrate the general dynamic of technological

70 Zuboff, *The Age of Surveillance Capitalism*; Evgeny Morozov, "Socialise the Data Centres!," *New Left Review*, no. 91, January/February 2015, pp. 45–66; Éric Sadin, *L'intelligence artificielle ou l'enjeu du siècle*; Bastani, *Fully Automated Luxury Communism.*

71 Günther Anders, *Le rêve des machines*, Paris: Allia, 2022, pp. 44–5.

disembedding as a process or trend that has recurred in different forms in various moments since the Industrial Revolution. This process is particularly intense today with the massive and accelerated spread of applications resulting from the AI revolution, accompanied by a "world of algorithmic machines" from which it is becoming almost impossible to escape. What are we going to do about it? That is the question—and that is where the serious politics begin.

Notwithstanding this resistance of the real, the complex historical process through which technology grows increasingly disembedded is moving toward the horizon of a fully automated capitalism, with the various spheres of human existence gradually subordinated to its pervasive logic. Algorithmic capital is becoming the great "autonomous vehicle" of contemporary society, forcing us to borrow its intelligent machines so that we can exist and move in its world. Its sole concern—regardless of the specific decisions left up to our individual preferences—is to monopolize the means of communication, transport, production, and subsistence so that we increasingly need it in order to perform the simple tasks of everyday life. Having charted the logic of algorithmic capital, we must now place it in its historical context.

Thesis 8

Neoliberalism is giving way to a new stage of capitalism

The unprecedented is necessarily unrecognizable.

—Shoshana Zuboff

Contemporary societies have been transformed over the past two decades in many ways. Neoliberal capitalism has undergone a major crisis, culminating in the "Great Recession" of 2007–8, with the effect that many wonder whether the neoliberal model is still standing or whether it is a thing of the past. After a decade of postcrisis austerity, the "return of the state," and the "emergency Keynesianism" of the COVID-19 pandemic, the rise of various forms of populism and challenges to the free-trade consensus, and indeed the advance of China's authoritarian capitalism, also raise questions about the neoliberal model's enduring dominance.[1] Is neoliberalism still firmly in place, has it been transformed, or has it been superseded by some new form? If so, how can we think about our post-neoliberal era?

1 Alfredo Saad-Filho, "Coronavirus, Crisis, and the End of Neoliberalism," *The Bullet*, April 2020, socialistproject.ca; Felicia Wong, "The Post-Neoliberal World Is Already Here," *Democracy*, May 2020; Claude Vaillancourt, *La fin du néolibéralisme: Regard sur un virage discret*, Montreal: Écosociété, 2023; Gary Gerstle, *The Rise and Fall of the Neoliberal Order*, Oxford: Oxford University Press, 2022; Cédric Durand, "1979 in Reverse," *New Left Review—Sidecar*, March 2021, newleftreview.org/sidecar; William Davies and Nicholas Gane, "Post-Neoliberalism?: An Introduction," *Theory, Culture & Society* 36, no. 6, 2021, pp. 3–28.

The name(s) of our era

While some analysts still use the concept of neoliberalism to characterize our historical period,[2] other highly influential authors now use the terms "technofeudalism" or "neofeudalism" to refer to a post-neoliberal era in which the GAFAM companies are becoming the new rentiers who lord it over the world.[3] Still others argue that our period is pursuing the development of the digital, informational, or cognitive capitalism that emerged with the advent of computers, the Internet, the knowledge economy, and post-Fordism at the turn of the 1980s and 1990s, signaling the advent of postindustrial societies.[4] However, we believe that while this literature makes helpful contributions to our critical conceptual toolkit, its authors fall short of fully grasping today's reality. Above all, we need to come to terms with what has happened to capital since the 2000s, with such developments as Big Data, machine learning, platform hegemony, and a new logic of accumulation.

How can we accurately characterize what is specific about our current era and properly analyze this strange historical shift? Our hypothesis is that, far from simply continuing with digital capitalism or neoliberalism —and still less turning back to a "neo-feudal" future—we are entering into the latest phase of the capitalist mode of production, algorithmic capitalism, which both builds on and goes beyond neoliberal capitalism. In Thesis 2, we listed the various recently coined terms that attempt to capture the essence of contemporary capitalism. We then explained why we consider the concept of "algorithmic capitalism" to be heuristically superior. In the following sections, we will outline what differentiates this new "algorithmic" stage of capitalism from feudalism, or "cognitive," "informational," or "digital" capitalism, as well as neoliberalism.

There are doubtless several similarities between our conception of algorithmic capitalism and the analysis by Nick Dyer-Witheford, Mikkola Kjøsen, and James Steinhoff in *AI Capitalism*.[5] These authors

2 David Harvey, "The Persistence of Neoliberalism Despite the Loss of Its Legitimacy," *Real News Network*, July 2019; Miloš Sumonja, "Neoliberalism Is Not Dead: On Political Implications of Covid-19," *Capital & Class* 45, no. 2, 2020, pp. 215–27.

3 Dean, *Communism or Neo-Feudalism?*; Durand, *How Silicon Valley Unleashed Techno-Feudalism*; Varoufakis, *Technofeudalism*; Don Tapscott, "We're Living in an Era of Digital Feudalism," *Quartz*, September 11, 2019.

4 Negri and Vercellone, "Le rapport capital/travail dans le capitalism cognitif"; Moulier Boutang, *Le capitalism cognitif.*

5 Dyer-Witheford, Kjøsen, and Steinhoff, *Inhuman Power*. See also Dyer-Witheford's new book, coauthored with Alessandra Mularoni, providing a solid argument on the

Table 8.1 Historical development of the general conditions of production

Historical period	***19th century***	***End of 19th cent.–1970***	***1970–2010***	***2010–***
Epoch of production	Industrial capitalism		Digital capitalism	
Period of production	Large-scale industry	Fordism	Post-Fordism, neoliberalism	Algorithmic capitalism
Technologies and organization of production	Industrial machinery, industrial division of labour	Taylorism, mass production	Flexible production, global supply chains	Narrow AI
General conditions of production	Steam engine, railways	Electricity, telephone, motor vehicles	ICT, computer networks, container shipping	Platforms, algorithms, cloud, Big Data, sensors

Source: This table is inspired by, but also adapted and different from the version found in Dyer-Witherford et al., *Inhuman Power*, p. 51

emphasize the decisive role of general conditions of production in the evolution of capitalism: the technologies, institutions, and practices that constitute the environment in which capitalist production takes shape in a given place and time, gradually becoming the essential factors common to all production. Energy and means of transport and communication, such as railways, phones, cars, and electricity, are just some examples of general conditions of production that have had a significant impact on capitalist forms of production, distribution, and consumption. In the account of Dyer-Witheford et al., our era is characterized by the rise of AI, algorithms, and platforms to the rank of general conditions of production. There is thus a shift within digital capitalism itself, with a departure from post-Fordism toward something new that still needs a name. Table 8.1, which is based on the work of Dyer-Witheford and his colleagues, takes into account changes in the general conditions of production.

Dyer-Witheford and his coauthors thus help distinguish between the era of industrial capitalism and the era of digital capitalism, as well as different "periods of production" within digital capitalism itself. This has the advantage of pointing out, as Zuboff also does, that something new has emerged in capitalism over the last twenty years that calls for new concepts. However, the periodization they propose remains perhaps too

function of Big Tech as one of automating circulation: *Cybernetic Circulation Complex: Big Tech and Planetary Crisis*, London: Verso, 2025.

general. If we want to explain the transition between different stages of capitalism, we should take into account not only the technological changes that are altering the general conditions of production, but also their social and institutional dimensions.

One approach useful for integrating these dimensions comes from regulation theory. Somewhat like Nancy Fraser's approach to capitalism as an institutionalized social order, this theory proposes that capitalism manages to stabilize itself and overcome its contradictions by way of specific institutional forms: wage relations, monetary and financial regimes, systems of innovation, state regulations, and international relations.[6] Taking these broader institutional and social elements into account allows us to think in terms of *stages of capitalism* or *regimes of accumulation*. These are "sets of regularities that ensure the general, relatively coherent progression of capital accumulation," a "relatively stabilized institutional matrix" that allows capitalism to reproduce itself in a particular configuration during a given period.[7] The regime of accumulation is itself enveloped in a *mode of regulation*, a "set of individual and collective procedures and behaviors" that reproduces social relations, sustains the regime of accumulation, and makes individual decisions compatible with a dynamic adjustment of the system as a whole.[8] The combination of an accumulation regime and a mode of regulation constitutes a *mode of development*, such as Fordism.[9]

Borrowing from this conceptual framework, we argue that algorithmic capitalism represents a genuinely new regime of accumulation based on the predominance of *algorithmic capital* and enveloped within a mode of regulation supported by *algorithmic governmentality*. We will return later on to the predominant role played by algorithmic governmentality, which makes it possible to regulate social relations and behavior, steer the regime of accumulation, and automatically coordinate various decentralized decisions. The combination of the regime of accumulation and this new model of regulation generates an *algorithmic mode of development*, which takes over from Fordism and post-Fordism, in what is now a post-neoliberal context. Finally, in Theses 9 and 10, we will mount

6 Robert Boyer, *Économie politique des capitalismes: Théorie de la régulation et des crises*, Paris: La Découverte, 2015.

7 Patrick Dieuaide, Bernard Paulre, and Carlo Vercellone, "Le capitalisme cognitif," Working Papers halshs-00226409, HAL, 2003, shs.hal.science; Fraser and Jaeggi, *Capitalism*.

8 Boyer, *Économie politique des capitalismes*.

9 Bernard Billaudot, *L'ordre économique de la société moderne: Un réexamen de la théorie de la régulation*, Paris: L'Harmattan, 1996, p. 15.

a more detailed examination of the algorithmic mechanisms of power and state rationality.

The mirage of cognitive capitalism

The terminology on which we have just drawn, derived from the regulation school, has also been used to advance the hypothesis of a transition to a "cognitive capitalism" superseding industrial capitalism. This transition is said to have brought great possibilities for human emancipation, through the combination of living knowledge and digital technologies. Several authors who adopt this strand of thinking emphasize the central role of information and communication technologies, the valorization of knowledge, and the exploitation of living knowledge in the dynamics of capital accumulation.[10] This analysis converges in part with the theories on the "knowledge economy," the "new economy," or the "knowledge-based society" that were popularized in the 1990s, but it offers a more critical interpretation of the dynamics of value extraction, social relations, and the capital/labor contradiction. Cognitive capitalism is not limited to the advent of new information technologies but relies on knowledge, science, and technology in the creation of value and the accumulation of capital. It operates according to a mode of regulation that is attentive to change, flexibility, innovation, and intellectual property rights. The emphasis in these accounts is not so much on the technologies themselves, but on the central role of the production of knowledge through cooperative labor, which can expand to vast networks via the mediation of IT hardware.[11]

The theorists of cognitive capitalism argue that this new regime of accumulation is a response to the crisis of the Fordist model and the exhaustion of the logic of industrial capitalism. The industrial production of standardized material goods is gradually giving way to a primarily immaterial and intellectual form of productive capital, where "the cognitive and immaterial dimension of labor becomes the key element in value creation."[12] Value capture increasingly takes place through mechanisms

10 Dieuaide, Paulre, and Vercellone, "Le capitalisme cognitif"; Moulier Boutang, *Le capitalisme cognitif*; Negri and Vercellone, "Le rapport capital/travail dans le capitalisme cognitif."

11 Dieuaide, Paulre, and Vercellone, "Le capitalisme cognitif," pp. 4–5; Moulier Boutang, "Le capitalisme cognitif," p. 87.

12 Jean-Marie Monnier and Carlo Vercellone, "Le capitalisme cognitif, nouvelle forme de capitalisme?," *Problèmes économiques: Hors-série*, no. 5, 2014, p. 1.

of rent extraction, intellectual property rights (patents, copyright, trademarks), and the commodification of knowledge.

These authors note that the IT revolution has had ambivalent effects. On the one hand, it enables the emergence of social networks, collaborative practices, free software, and digital commons. André Gorz thus considered that the knowledge economy has "at its core, a negation of the capitalist market economy."[13] For his part, Yochai Benkler in *The Wealth of Networks* emphasized the advent of a new model of "commons-based peer production," able to transcend the mechanisms of both market coordination and centralized planning.[14] Yet theorists of cognitive capitalism also recognize the presence of an opposing dynamic linked to proprietary logics, the commodification of knowledge, and the consolidation of "knowledge oligopolies" that create artificial scarcity in order to capture the value of innovations.[15] The rising share of immaterial capital as a factor of production and the cognitive division of labor contribute to parasitism on the "living labor" of individuals collaborating in networks, who constitute a kind of "collective brain" or "general intellect," as Marx termed it in the "Fragment on Machines" in his *Grundrisse*.[16]

Cognitive capitalism emerges from a restructuring process in which capital seeks to subjugate and exploit knowledge and its "collective conditions of production," while repressing the emancipatory potential of the "general intellect." Thus, in cognitive capitalism "the productive value of intellectual and immaterial labor becomes dominant," and capital valorization takes place through the commodification of knowledge and the expropriation of the commons for rent.[17] We are now seeing, on this reading, the emergence of the "immaterial worker," a "cognitariat," or "multitude" who are increasingly able to create value and organize themselves outside the traditional parameters of the firm and waged work, thus breaking down the logic of industrial capitalism and Fordism and opening up new possibilities for emancipation.

The primacy of "living labor" over the "dead labor" embodied in machines and fixed capital leads to a rather optimistic reading of the evolution of capitalism, said to be immanently creating the conditions for its own transcendence. Instead of imagining that digital technologies

---

13 André Gorz, *L'immatériel: Connaissance, valeur et capital*, Paris: Galilée, 2003, p. 76.

14 Yochai Benkler, *The Wealth of Networks: How Social Production Transforms Markets and Freedom*, New Haven, CT: Yale University Press, 2006.

15 Monnier and Vercellone, "Le capitalisme cognitif, nouvelle forme de capitalisme?"

16 Marx, *Grundrisse*, p. 706.

17 Negri and Vercellone, "Le rapport capital/travail," p. 41.

have the upper hand, Negri and his collaborators emphasize the autonomy of social, cooperative, and living labor. In their account, this labor is able to self-organize outside the grip of capital, which is understood as an external parasite. Hence, the innovation and functioning of information technologies depend more on living knowledge than on capital. The "creative force" of the IT revolution comes from "social networks of labor cooperation" that harbor a potential alternative to enterprise and the market.[18] In this sense, these theorists say there is already a kind of spontaneous production of the "commons" through collective work, giving rise to a form of "communism" in today's world. The multitude just needs to free itself from the capitalist shell that seeks to confine it.

However, this optimistic interpretation of the commons as the fruit of the living knowledge of the multitude, subject to capitalist parasitism, presents various problems. Pierre Dardot and Christian Laval have shown that this conception of the commons as a product of collective self-organization is based on the model of the "spontaneous social force of the commons" inspired by Pierre-Joseph Proudhon, and that the analysis of exploitation used here is based on the logic of *theft*.[19] This perspective conceives of value production as a spontaneous creation of collaborative labor, ignoring the structuring, organizing, and productive role of capital, which is reduced to the status of a mere thief, a rentier mechanism, confiscating the products of labor. In this sense, capital is not productive, but a mere parasite that creates no added value. This is precisely the part of Proudhon that Marx subjected to critique: For the cooperation in capitalism is not spontaneous, but brought about by capital, which organizes labor in such a way as to exploit it and produce surplus-value.

In addition to the fact that this theory has a hard time grasping the mechanisms of domination and control that still permeate the organization of work, it imagines itself breaking free from Taylorist logic, industrial society, and the tutelage of capital, even though capital continues to exploit living labor to its advantage. Dardot and Laval offer a different interpretation of the "Fragment on Machines," according to which Marx envisages a much more pessimistic scenario than the proponents of cognitive capitalism do with regard to capital's ability to incorporate knowledge to its own advantage.[20] It can even be read as

18 Ibid.

19 Pierre Dardot and Christian Laval, *Commun: Essai sur la révolution au XXIe siècle*, Paris: La Découverte, 2014, p. 5.

20 Ibid., p. 52.

an early conceptualization of algorithmic capital, which is essentially an extension of capitalist accumulation incorporating the latest techno-scientific advances.

Another major problem with the theory of cognitive capitalism is that it was largely developed at the dawn of the IT revolution, the beginnings of the Internet, the advent of post-Fordism and the new millennium. It roughly corresponds to the first phase of digital capitalism—the one running from the 1980s to the early 2000s, a period that coincides with the stage of financialized, neoliberal, and globalized capitalism. As a result, theorists of cognitive capitalism were not able to see that the basis of value is not cooperative work and the free association of producers on the Internet, but the extraction of data and the exploitation of digital labor by platforms and other algorithmic capitalists. In this phase, "dead labor" has its revenge over "living labor." The main creative force behind the algorithmic revolution is, indeed, a dynamic of innovation driven by capital (with the support of AI engineers), opening up new possibilities for accumulation in the wake of the exhaustion of neoliberal capitalism and the 2007–8 crisis.

It seems that "cognitive capitalism" was little but a mirage that appeared together with the changes in capitalism that preceded the other decisive developments we analyzed in previous theses: the advent of digital platforms, Big Data, new advances in robotization, and so on. Just as Marx and Engels criticized the utopian socialists of their day, entranced by the first faltering steps in the history of the proletariat and capitalist industry, theorists of cognitive capitalism believed that in "the misty realm of philosophical fantasy" they could see the first seeds of a postcapitalist society. For they were unable to grasp the material conditions of contemporary capitalism and the real possibilities for overcoming it. Thus, the hope placed in the emancipatory potential of "living knowledge" serves to obscure the specific dynamics of "dead knowledge"—that is, the knowledge based on machines and automated cognition, which is now in the process of surpassing various spheres of human intelligence.

Moreover, algorithmic capitalism does not point to the end of industrial capitalism, but rather to the revival of logics of industry and automation, which are increasingly extending their grip over today's societies. When we speak of the Fourth Industrial Revolution, this is not just a metaphor. Rather, the production of mass, connected consumer goods, the standardization of behavior through algorithmic technologies, and the exploitation of algorithmic labor in good old factories are amplifying the dynamics of "large-scale industry" that Marx examined in the nineteenth century. Algorithmic capitalism is not a departure from industrial capitalism, but its renewal, propelled by dead knowledge (data) and automated cognition.

While the theorists of cognitive capitalism celebrate the manifold skills of immaterial workers, their creativity, the importance of self-production, and the new possibilities brought about by this transformation of capitalism, the perspective that algorithmic capitalism offers is in fact far bleaker. It is not the Silicon Valley bourgeoisie that is producing its own gravediggers, but rather the "living labor" producing for digital platforms that is creating *its* own gravediggers: algorithmic machines. This is why it is important not to underestimate the reversal of the balance of power between living labor and dead labor, in which algorithmic capital now holds the upper hand. This does not mean that all resistance is futile, or that we will one day be faced with an apocalyptic scenario drawn straight from *Terminator* or *The Matrix*. But we ought to give up on a techno-optimistic, idealistic, and ultimately utopian view of capitalism, imagined to be bringing about its own demise through peer-to-peer logic, the digital commons, and the zero-marginal-cost society—a Care Bears vision of contemporary capitalism. Rather, we can usefully place the emergence of algorithmic capital within the long history of capitalism and better analyze the tipping points between each historical phase of this mode of production.

A brief overview of the stages of capitalism

Since capitalism's first origins, its history has been marked by several distinct stages or regimes of accumulation: mercantilist and colonial capitalism (from the sixteenth to eighteenth centuries), liberal industrial capitalism (from the nineteenth century up to 1945), state-regulated capitalism (from 1945 to 1980), and neoliberal, financialized, and globalized capitalism (1980–2008).[21] Most theories of capitalism stop at this last stage, even if several point to the fact that we have entered a new era. This is then called the end of globalization, a post-neoliberal society, and so on. In our view, algorithmic capitalism is emerging in the wake of the crisis of neoliberal capitalism that culminated with the great financial crisis of 2007–8. Its rise converges with a series of innovations that are

21 Of course, these categories operate at a very high level of abstraction, encompassing realities that are geographically unequal and historically nonlinear.

changing the general conditions of production: smartphones, Big Data, machine learning, platforms, and others.

One might retort that technological changes are an omnipresent factor in the history of capitalism, that trends toward automation with machines and robots have been at work since at least the nineteenth century, and that the "surveillance" described by Zuboff is not some radically new phenomenon. Some insist that the role of market discipline, so central to neoliberalism, is just as powerful today as ever, despite the last great financial crisis. On such a reading, any diagnosis of the "end of neoliberalism" is bound to be mistaken or at least exaggerated: Such an objection would thus demand more evidence of how far algorithmic capital is becoming predominant in our era (as against financial, industrial, or commercial capital), as well as further arguments to justify the claim of a transition to a new stage of capitalism, specifying the exact timing and causes of this shift.[22] These concerns and warnings are important, and to be taken seriously.

To respond to such skeptical challenges, we will here seek to provide sufficient examples, explanations, facts, and indicators in seeking to demonstrate that a profound transformation of capitalism has indeed taken place. We do so in full acknowledgment that elements of neoliberalism surely do remain present. No historical transition occurs as a clean break; several old and new elements coexist across a certain period. That said, the buildup of transformations of varying importance across several spheres of activity, within a relatively short timeframe, points to a major qualitative change and this not just in "the economy," but on a societal scale. In addition to the technological transformations discussed in the previous chapter, this new phase of capitalism is accompanied by significant changes of various dimensions: forms of enterprise, forms of work, drivers of capital accumulation, the dominant types of capital, the predominant types of markets, class relations, global logic, configuration of states, major societal dynamics, culture, and subjectivities. Table 8.2 presents each accumulation regime as an ideal type to highlight the contrasts between the different phases of capitalism across several different aspects.

Our aim here is not to explain each element of this historical portrait in detail. We have analyzed the first six dimensions in previous theses, while the remaining ones will be addressed from this point onward. In short,

22 We would like to thank our colleague Sébastien Rioux for raising these objections during an online conversation. The thesis that neoliberalism has reached its end remains the subject of debate.

our hypothesis is that the dominance of financial capital is giving way to the dominance of algorithmic capital, and that neoliberal rationality —marked by a certain form of entrepreneurial subjectivity and the celebration of the market as a model of coordination—is being integrated into a tendentially all-encompassing algorithmic logic. Without doubt, financial capital and neoliberal ideology continue to influence economic life and social practices today. And yet they too are now embedded within a mode of algorithmic regulation that centers on the regime of accumulation based on data extraction and increasing algorithmic power we analyzed earlier. Even if key aspects of neoliberalism do persist, such as the imposition of market discipline, the modes and forms of its imposition are thus not limited to indebtedness or labor precarity. Rather, they also include the rise of captive attention time, data extraction, AI, targeted advertising, the automatization of certain behaviors, and other factors. Our arrival in a new phase of capitalism is correlated to the articulation of a new regime of accumulation and a specific mode of regulation that allow for a response to the shortcomings of the previous mode of development. The advent of algorithmic capitalism is thus a product of the crisis of neoliberalism, while also representing the new core of the capitalist system.

Capital's technological fix and the financial crisis of 2007–8

In terms of political economy, the transition from neoliberal to algorithmic capitalism is rooted in a pivotal moment: the 2007–8 financial crisis. As David McNally diagnosed in the aftermath of the crisis, it effectively marked the end of the historic expansion of neoliberal capitalism, which had already been seriously weakened since the late 1990s and was artificially sustained by massive household and bank debt until the moment the crisis hit.[23] McNally presciently foresaw a "decade of austerity" following the huge bank bailouts and the transformation of the financial crisis into a sovereign debt crisis around the world. Yet we should recognize that the austerity policies of the 2010s were more a sign of the decline and historic failure of neoliberalism than a new heyday. The era of globalized neoliberal capitalism had entered its final phase: Although many aspects of this order continue to shape our times (and historical transitions are long and nonlinear processes that never happen all at

23 David McNally, *Global Slump: The Economics and Politics of Crisis and Resistance*, Oakland, CA: PM Press, 2011.

Table 8.2 Schematic table of the stages of capitalism

Stage of Capitalism	*Mercantilist Capitalism*	*Industrial Capitalism*	*State-regulated Capitalism*	*Neoliberal Capitalism*	*Algorithmic Capitalism*
Period	~1500–1800	~1800–1945	~1945–1980	~1980–2007	~2007-...
Main form of enterprise	Charter companies	Industrial firm	Corporation	Multinational firm	Platform
Form(s) of labor	Manufacture & manual labour Slave labor	Mechanized industrial labour	Mechanized industrial labor, service labor	Flexible labor	Algorithmic labor
Dominant capital type	Merchant's capital	Industrial capital	Industrial capital	Financial capital	Algorithmic capital
Typical market	Commodity markets	Industrial goods	Mass consumer goods	Financial markets	Predictive markets
Typical classes and economic actors	Landowners Workers Merchants Masters Slaves	Capitalists Proletarians	Managers Blue collars White collars Consumers	Shareholders Creditors Debtors	Digital workers Users Tech oligarchs
Dominant logic	Colonial logic	Industrial logic	Technocratic logic	Neoliberal logic	Algorithmic logic
Dominant state form	Absolutist state	Nation-state	Welfare state	Neoliberal state	Algorithmic state

once), in the postcrisis period the age of algorithmic capitalism has well and truly arrived.

Obviously, the establishment of the productive infrastructure on which algorithmic capital relies dates back several decades, in the form of post-Fordism and the rise of ICT, followed by the phase of massive investment during the 1990s Internet bubble. The rise of the web giants accelerated in the 2000s: After Amazon's IPO in 1997, Google followed in 2004, then Facebook in 2012. Apple had launched its iPhone in 2007. However, up until the crisis of 2007–8, it had been the financial sector that dominated a highly financialized economy. In the United States in 2003, financial profits accounted for more than 42 percent of total profits.

The global financial collapse did not put an end to financialization but ushered in a period in which investors were less inclined to invest in esoteric financial products and increasingly turned to other sectors, such as algorithms and AI. The quantitative easing policies pursued by several central banks also had the effect of reducing financial returns and contributed to a shift to high-yield investment opportunities in the high-tech sector, including risky assets such as tech companies developing new products and algorithms that were still unprofitable but had the potential to carry high returns in the longer term. As Yanis Varoufakis puts it, the tech giants "helped themselves to the rivers of cash that were being printed by the central banks of developed capitalist states," contributing to the buildup of what he calls "a new type of capital stock," which he terms "cloud capital" and which we have termed "algorithmic capital."[24] The recovery was ultimately in part driven by algorithmic technologies, as investment funds such as the SoftBank Vision Fund, created in 2017, began showering money on tech companies showing potential. Another round of investing some of the vast sums hoarded by corporations further built up the physical infrastructure of algorithmic capital.[25]

Schematically, we can say that neoliberal capitalism has historically used three types of fixes to tip the balance of power in its favor against the global working class.[26] It has resorted to: financial fixes (forms of indebtedness, for the working class and Global South countries); spatial fixes (the offshoring of production to Global South sites where cheap labor is more readily available and the influx of cheap labor from the Global South for social reproduction in the Global North); and technological

24 Varoufakis, *Technofeudalism*, p. 100.

25 Dyer-Witheford, Kjøsen, and Steinhoff, *Inhuman Power*, pp. 73–4; Srnicek, *Platform Capitalism*, pp. 19–23.

26 David Harvey, *The New Imperialism*, Oxford: Oxford University Press, 2003.

fixes, such as mechanization and automation. The economic recovery of the 2010s, although contradictory and uneven in many respects, can be seen as a new technological fix responding to the limitations encountered by the globalized neoliberal model.

The rise of post-2008 struggles, particularly during the Arab Spring, the Indignados, and Occupy movements, but above all in the wave of workers' struggles in China, India, Cambodia, and elsewhere in the Global South, meant that neoliberal capitalism's spatial fix—offshoring—was reaching its limits in the form of a recomposition of the global working class in struggle. Whereas China was once the epicenter of the earthquake of offshoring, it saw its average hourly wages triple between 2005 and 2016, reaching almost the same levels as countries such as Portugal and Greece. Thus, shortly after the crisis, the limits of the use of global cheap labor as a spatial fix were added onto the shortcomings of the financial fix of speculation and credit, which the crisis had exposed.[27] As a result, after 2008 the technological fix was in the best position to give capitalism a new lease of life. Algorithmic machines can discipline and displace an overly demanding working class, open up channels for investment, and even lead to a reindustrialization of the North around smart production.

However, this technological fix does not just put neoliberalism back in the saddle: Rather, it also fuels the formation of a new logic of accumulation. Indeed, it establishes the technology sector as the new engine of global growth and gradually replaces the logic of financialized accumulation with an algorithmic logic based largely on data extraction and analysis. After 2008, the GAFAM companies and their Chinese equivalents became the world's largest firms, ahead of the major extractive, industrial, and even financial companies. Amy Webb coined the term Big Nine, adding IBM, as well as Alibaba, Tencent, and Baidu (roughly the Chinese equivalents of Amazon, Facebook, and Google) to the GAFAM list, since they are now major leaders in AI development. Today, the term "Magnificent Seven" is also in use in economic publications, adding Nvidia and Tesla to the GAFAM five as the dominant forces in today's economy. While Amazon and Microsoft corner the lion's share of the cloud computing and server space market, online advertising is largely dominated by the Google/Meta duopoly. Apple overtook Walmart to become the world's largest music distributor in 2008, via its iTunes platform. In August 2011, it surpassed ExxonMobil to become the world's

27 Dyer-Witheford, Kjøsen, and Steinhoff, *Inhuman Power.*

most valuable company. The adage "data is the new oil" is borne out in global corporate rankings (see Table 8.3).

This ranking of the biggest companies by market capitalization is obviously not an irrefutable proof of the predominance of algorithmic capital. However, it does indicate certain major trends that shape today's economy. Moreover, the growing market value of platforms is in fact an expression of a form of financial capital associated with companies in the tech sector. In this sense, the financialization of the economy, linked to the stage of globalized neoliberal capitalism, is an ongoing dynamic, although it is increasingly clinging to companies that own, control, develop, and accumulate data and algorithmic capital. The steep fall in the digital giants' stock market value in 2022 (representing a total loss of $4.6 trillion) did not invalidate the central role of capitalist platforms, but showed that they remain vulnerable to market corrections, speculation, and other dynamics of the capitalist economy.[28] Just like industrial and commercial capital, algorithmic capital and its financial expressions are still subject to the rules of the capitalist game.

If we now look at the global ranking of large companies by other indicators such as revenues, profits, or assets, the picture is rather different. Big Tech companies are powerful companies that rank very high and accumulate billions in profits, but they still share the rankings with other companies in the automotive, financial, oil, pharmaceutical, retail, and other sectors. Although they have occupied a dominant position in the stock markets since 2016, the total revenues they have generated does not completely surpass the revenues of other companies linked to industrial, commercial, and financial capital (see Table 8.4).

This table thus seems to put in more relative perspective the theory that algorithmic capital is now at the apex of the global economy, or that we have suddenly entered into a new stage of capitalism that has left behind companies linked to industrial, commercial, and fossil capital. Admittedly, such a quantitative approach aimed at identifying the preponderance of algorithmic capital based on the total value produced by the technology sector, stock market value, the volume of AI-related investments, or even the number of jobs linked to these companies (not counting the army of click-workers) provides a wealth of data to explore, essential for any analysis of the transition between two phases of the capitalist system. But it is more interesting to analyze this complex phenomenon from a qualitative perspective, examining the power relations

28 "Ten Big Tech Stocks Lost $4.6 Trillion in Market Cap in 2022," *PYMNTS*, January 2, 2023.

Table 8.3 Top firms by market capitalization, 2006–25

Rank	Firm	Sector	*Value (Million USD)*	Firm	Sector	*Value (Million USD)*
December 2006 (pre-crisis) to December 2011						
1	ExxonMobil	Energy	446,943	ExxonMobil	Energy	406,272
2	General Electric	Conglomerate	383,564	Apple	Technology	376,410
3	Microsoft	Technology	293,537	PetroChina	Energy	276,844
4	Citigroup	Finance	273,691	Royal Dutch Shell	Energy	236,677
5	Gazprom	Energy	271,482	China Ind. Bank	Finance	228,168
6	China Ind. Bank	Finance	254,592	Microsoft	Technology	218,380
7	Toyota	Automobile	241,161	IBM	Technology	216,724
8	Bank of America	Finance	239,758	Chevron	Energy	211,893
9	Royal Dutch Shell	Energy	225,781	Walmart	Retail	204,659
10	BP	Energy	218,643	China Mobile	Technology	196,148
December 2016 to December 2025						
1	Apple	Technology	617,588	Nvidia	Technology	4,713,000
2	Alphabet (Google)	Technology	531,970	Apple	Technology	4,056,000
3	Microsoft	Technology	483,160	Alphabet (Google)	Technology	3,791,000
4	Berkshire Hathaway	Finance	404,390	Microsoft	Technology	3,622,000
5	ExxonMobil	Energy	374,280	Amazon	Technology	2,490,000
6	Amazon	Technology	356,313	Broadcom	Technology	1,672,000
7	Johnson & Johnson	Pharma	313,432	Meta	Technology	1,671,000
8	JPMorgan Chase	Finance	308,768	Saudi Aramco	Energy	1,662,000
9	General Electric	Conglomerate	279,545	Tesla	Technology	1,556,000
10	Wells Fargo	Finance	276,779	TSMC	Technology	1,246,000

Source: Fortune.com

Table 8.4 Ranking of top firms by income, 2025 (Fortune Global 500)

Rank	*Firm*	*Revenues (Million USD)*	*Profits (Million USD)*	*Assets (Million USD)*	*Employees*	*Sector*
1	Walmart (US)	648,125	15,511	252,399	2,100,000	Retail
2	Amazon (US)	574,795	30,425	527,854	1,525,000	Technology
3	State Grid (China)	545,948	9,204	781,126	1,361,423	Utilities
4	Saudi Aramco (Saudi Arabia)	494,890	120,699	660,819	73,311	Energy
5	Sinopec Group (China)	429,699	9,393	382,688	513,434	Energy
6	China Petroleum (China)	421,713	21,294	630,562	1,026,301	Energy
7	Apple (US)	383,285	96,995	352,583	161,000	Technology
8	United Health Group (US)	371,622	22,381	273,720	440,000	Healthcare
9	Berkshire Hathaway (US)	364,482	96,223	1,069,978	396,500	Finance
10	CVS Health (US)	357,776	8,344	249,728	259,500	Healthcare
17	Alphabet (Google) (US)	307,394	73,795	402,392	182,502	Technology
26	Microsoft (US)	211,915	72,361	411,376	221,000	Technology
66	Meta (US)	134,902	39,098	229,623	67,317	Technology
110	Tesla (US)	96,773	14,997	106,608	140,473	Technology
222	Nvidia	60,922	29,760	65,728	29,600	Technology

Source: Full List of Fortune Global 500 Companies (2026), us500.com

between different economic sectors and the growing role of platforms and algorithmic machines in a growing number of spheres of activity. Michael Hardt and Antonio Negri make a similar argument. They take the example of the transition to industrial production, where agriculture still occupies an important place from a quantitative point of view, even as it is transformed and subordinated by the arrival of industrial capital.

> Quantitative indicators cannot grasp either the *qualitative* transformation in the progression from one paradigm to another or the *hierarchy* among the economic sectors in the context of each paradigm. In the process of modernization and the passage toward the paradigm of industrial dominance, not only did agricultural production decline quantitatively (both in percentage of workers employed and in proportion of the total value produced), but also, more important, agriculture

> itself was transformed. When agriculture came under the domination of industry, even when agriculture was still predominant in quantitative terms, it became subject to the social and financial pressures of industry, and moreover agricultural production itself was industrialized … Historical change, in other words, has to be recognized in terms of the power relationships throughout the economic sphere.[29]

In the previous theses, we sought to demonstrate that the advent of algorithmic capital has already transformed many areas of social and economic life, particularly in terms of the means of production, businesses, labor, and other factors. Thus, the argument that tech companies are *not* the most profitable, do *not* create the most jobs, and do *not* generate the majority of revenues in the contemporary economy cannot lead us to the conclusion that algorithmic capital does not currently play a central role in the capitalist system. We should bear in mind that algorithmic technologies extend to the industrial, agricultural, and mining sectors, among others, and that the vast majority of economic activities are increasingly mediated by platforms, data extraction, and the development of AI—key elements of the tech giants' power.[30] We can thus say that in a world where constant technological improvement plays a pivotal role, the vast majority of economic sectors and organizations (whether private or public) are becoming increasingly dependent on the companies that control these innovations. Here we see the monopolistic tendency of algorithmic capital, which is sometimes confused with the specter of technofeudalism.

The specter of technofeudalism

A specter is haunting the world: the specter of technofeudalism. A growing body of literature draws on images of ruthless lords, digital fiefs, guarded fortresses, and fragmented sovereignty—the imaginary of a bygone era, suddenly resurfacing in new form. Digital platforms are said to be virtual landowners extracting rents like medieval lords:

29 Michael Hardt and Antonio Negri, *Empire*, Cambridge, MA: Harvard University Press, 2000, pp. 281–2.

30 The as-yet-unpublished thesis by Lucie Violland on the transformation of the business model of John Deere provides a telling illustration of the subsumption of the agricultural sector by the logic of algorithmic capital. See Lucie Violland, "Agriculture par abonnement: Analyse critique de la valorisation des données agricoles chez John Deere entre 2012 et 2023," Université du Québec à Montréal, 2025.

> If all of the above sounds like a frightfully medieval problem for society, that is probably because it really is an uncanny echo of that era in history. The only real difference now is the digital form of the landscape. The nature of the lords taking tribute payments remains the same.[31]

The theory of a return to a technofeudal future conceives of cyberspace as a vast territory whose borders must be protected in order to collect rent. Control of intangible assets and data thus takes the place of the territorial fiefs of yesteryear. This is a novel interpretation of the institution of *dominium*, where domination is exercised over both people and land. In addition to *dominium*, a cornerstone of feudalism, economist Cédric Durand also notes that we are seeing the emergence of a new form of serfdom. Serfs, who are neither slaves nor "free workers" under capitalism, remain tied to the "fief," just as users are now largely dependent on digital platforms from which it is difficult to escape. Serfdom is also considered a form of predation, of appropriation of the labor of others through violence and coercion, similar to the accumulation by dispossession of data described earlier. According to Durand, the spread of increasingly invasive apps is forging close links between our lives and the "cyber-territories" of "digital fiefs." Individuals and organizations alike are now dependent on "structures that exercise monopolistic control over data and algorithms."[32]

But are these merely metaphors, or a conceptual description of a new mode of production that is taking shape before our eyes? According to authors such as Durand, Yanis Varoufakis, and Jodi Dean, this "technofeudal hypothesis" represents an analytical category aimed at demonstrating that the changes in contemporary capitalism are leading us toward a form of outright neofeudalism. For Durand, the concept seems to be used to point in a general direction, a tendency toward a feudalization of the economy. Varoufakis's position is more blunt: We have indeed transitioned to technofeudalism, as "capitalism is now dead, in the sense that its dynamics no longer govern our economies."[33] On this reading, GAFAM and other platforms are real "digital lords," exercising a form of domination that is inseparably both economic and political, leading to an enormous concentration of wealth that favors the emergence of an "an all-powerful and spendthrift aristocracy," with

---

31 Indy Johar, "The Sharing Economy Will Go Medieval on You," *Financial Times*, May 21, 2015.

32 Durand, *How Silicon Valley Unleashed Techno-Feudalism*, quote from p. 181.

33 Varoufakis, *Technofeudalism*, p. x.

this accumulation of wealth taking place mainly through predation and control of "digital territory."[34]

As in the hypothesis of cognitive capitalism, this new take on the contemporary mode of production is said to be based on a rentier and parasitic dynamic, external to living labor. The rent mechanism is in fact a form of appropriation "disconnected from engagement in production."[35] The feudal lord appropriates his serfs' surplus labor without himself organizing production or imposing productivity standards at work. Neither labor nor land are commodities. Feudal rents are thus extracted in a nonmarket fashion, through extra-economic appropriation based on the ownership of land and supported by force, law, and customs.[36]

But is it true that value is now mainly captured by way of predation and extracted outside of the markets? Are platforms simply plundering data, or do they also have mechanisms linked to the exploitation of "free labor" and productive investments that grow their capital and promote the emergence of new (predictive) markets? Has not surplus-value in capitalism always taken forms other than profits alone—and was not this what Marx noted, when he theorized that rent, interest, commercial profits, and the profits of industry are all forms surplus-value takes *in capitalism*? What is more, several authors, such as David Harvey and Nancy Fraser, have shown that the mechanisms of predation, expropriation, and accumulation by dispossession are not characteristics of so-called "primitive accumulation" at the dawn of capitalist modernity; they are not "prior to" or "outside of" capitalism, but rather continuous processes parallel to exploitation that have allowed capital to generate value in different phases of capitalism up to the present day.[37]

It should be acknowledged that the mechanism of rent extraction, and especially differential rent, is indeed at the heart of platforms' income streams.[38] However, the revenues derived from owning the technological infrastructure used by others are not strictly speaking "rent," but rather forms of "surplus profits," if we are to stick to Marxian

34 Durand, *How Silicon Valley Unleashed Techno-Feudalism*, p. 106.

35 Ibid., p. 183.

36 Ellen Meiksins Wood, *The Origin of Capitalism: A Longer View*, London: Verso, 2002.

37 Harvey, *The New Imperialism*; Nancy Fraser, "Behind Marx's Hidden Abode: For an Expanded Conception of Capitalism," *New Left Review*, no. 86, March/April 2014, pp. 55–72.

38 Sadowski, "The Internet of Landlords"; Doorn and Badger, "Platform Capitalism's Hidden Abode."

terminology.[39] More importantly, the political economy of platforms is not only, or ultimately, based on rent extraction. As market/enterprise hybrids, platforms use commodification processes and market mechanisms to commodify data assets, commodify the relationships between their users, and—as we saw in Thesis 5—exploit the digital labor of people who are *workers* and not *serfs*. Platforms, unlike feudal lords, actively organize production by coordinating different types of users through their algorithms. Furthermore, the existence of rentier mechanisms is itself entirely compatible with capitalism and does not necessarily mean going back to the feudal future. Rather, it would seem that algorithmic capitalism is simply a new variant of the *capitalist* mode of production. The existence of "parasitic" relations and an "all-powerful aristocracy" or "tiny oligarchy" are not characteristics peculiar to feudalism alone.

The rise of intangible assets, the dynamics of monopolization by means of intellectual property rights, and the rent-seeking mechanisms that we see today are inscribed in a capitalist logic. Ugo Pagano uses the term "intellectual monopoly capitalism" to describe this phase of tightening property rights at the end of the last century.[40] Durand himself describes a series of rent-seeking mechanisms (licensing of intellectual property, monopoly rent, differential rent from intangibles, dynamic innovation rent) used by capitalist companies and platforms seeking to capture value.[41] Durand maintains that these rent-seeking mechanisms are disconnected from engagement in production. In his account, value extraction today is essentially parasitic, based on network effects, predation, and coercion. However, in previous theses, we analyzed the algorithmic production process, showing how capital seizes data and AI innovations to produce predictive products, services, algorithmic machines, and other commodities that contain surplus-value. This is why we should be careful not to see tech giants as mere rentiers who monopolize our data for free and derive their riches solely from their monopoly position by collecting huge rents without creating added value, in the manner of the landed aristocracy of the feudal era. This perspective obscures the specifically *productive* nature of algorithmic capital, which is driven by a hyperdynamic bourgeoisie, as well as the

39 See Karl Marx, *Capital*, Volume 3, London: Penguin, 1981, part 6.

40 Ugo Pagano, "The Crisis of Intellectual Monopoly Capitalism," *Cambridge Journal of Economics* 38, no. 6, 2014, pp. 1409–29.

41 For a detailed overview of rentier mechanisms, see Durand, *How Silicon Valley Unleashed Techno-Feudalism*, pp. 183ff.

forms of exploitation of labor that persist under this new version of capitalism. It also runs the risk of Eurocentrism, in placing the European feudal past as the epistemological standard to understand the economic reality of much of the globe today.[42]

When we take a closer look, we find that the web giants do not behave like feudal lords. If they did—if they reproduced their social power like rentiers—they would not need to make productive investments. Indeed, feudal lords reproduced their position as lords and their social power according to rules of reproduction that did not require productive investment. Not being capitalist enterprises subjected to market competition, the lords reproduced their social position through legal-political and military mechanisms (titles of nobility, tributes, war, and so on).[43] However, the web giants and platforms behave like true capitalists, as Morozov points out:

> Alphabet's R&D spending in 2017, 2018, 2019 and 2020 was $16.6 billion, $21.4 billion, $26 billion and $27.5 billion respectively … Amazon, too, spent $42.7 billion in 2020 alone on research and development, while employing over a million people across the globe. In the US, the company employs more people than the entire residential construction industry: one out of every 153 employed Americans … More telling still, a close analysis of the balance sheets of Google, Amazon and Facebook shows that they have fewer intangible assets than other big corporations.[44]

What is more, GAFAM and the high-tech sector, as news on the state of the economy periodically reminds us, are subject to the vagaries, and the contradictions, of capitalist economic cycles. They will sometimes suffer considerable losses of income and stock market valuation and have to make significant layoffs. It is hard to imagine feudal lords firing serfs, but algorithmic capitalists, like all capitalist companies, can and must adjust their variable capital to market pressures and fluctuations.

---

42 See Giuliana Facciolli and Jonathan Martineau, "L'intelligence artificielle et les algorithmes: Au cœur d'une reconfiguration des relations internationales capitalistes," *Nouveaux Cahiers du Socialisme*, no. 31, Winter 2024, pp. 68–78.

43 Ellen Meiksins Wood, "The Separation of the Economic and the Political in Capitalism," *New Left Review*, no. 127, May/June 1981, pp. 66–95; Robert Brenner, "Agrarian Class Structure and Economic Development in Pre-Industrial Europe," *Past & Present*, no. 70, 1976, pp. 30–75.

44 Evgeny Morozov, "Critique of Techno-Feudal Reason," *New Left Review*, nos. 133/134, January/April 2022, pp. 89–126.

Ultimately, it seems that the notion of technofeudalism is more useful as a metaphor than a plausible historical theory or a concept with full heuristic value. Like Zuboff's Big Other or Marx's "automaton," it succeeds in capturing the imagination by conjuring up a specter, a disturbing figure seeking to alert us to an imminent threat. The authors of the technofeudal hypothesis are right, in our view, to seek new concepts as they, like us, recognize the novelty of our current moment. Their contributions are also highly valuable in alerting us to the fact that postcapitalist developments might take dystopian rather than socialist or democratic paths. However, platforms are not new "digital lords," but capitalist enterprises in their own right, market/enterprise hybrids as we defined them in Thesis 5, benefiting from the key mechanisms of algorithmic capital. And these Big Tech platforms now offer countless technological solutions to today's problems.

Technological solutionism as a response to the crisis

With the financial crisis of 2007–8, algorithmic capital arrived at just the right moment to provide various solutions to the problems of a discredited model. The destabilization of neoliberal capitalism prompted strong reactions, both from protest movements and from elites seeking to "moralize capitalism" through a doctrine of greater openness, collaboration, and sharing. In this transitional moment, a dialectic gradually emerged between the pursuit of neoliberal logic, which stepped up its austerity measures, and technological promises that sought to solve the problems created by this same system. Morozov highlights the intimate relationship between technological solutionism and the problems generated by neoliberalism. The scarcity and austerity advocated by neoliberal Wall Street cross paths with the abundance and innovation of Silicon Valley: "The two complement each other perfectly in the good cop/bad cop routine."[45]

With the rapid rise of the so-called collaborative economy in the 2010s, platforms such as Uber, TaskRabbit, Foodora, and Airbnb made it possible to find gig work and to monetize one's assets, in a context of labor market insecurity and economic drudgery fueled by neoliberal austerity. Supermarket deliveries, mobility services, odd jobs, mentoring,

45 Evgeny Morozov, "Résister à l'ubérisation du monde," *Le Monde diplomatique*, September 2015.

babysitting and pet sitting, house rentals, and selling items, photos, and videos: Hundreds of apps now allow individuals to earn money and become "self-employed"—or as the French term has it, "autoentrepreneurs." As Morozov points out: "The sharing economy amplifies the worst excesses of the dominant economic system: it is neoliberalism on steroids." Although the ravages of neoliberalism and the "disruptive innovations" of the collaborative economy can be seen as independent developments, Silicon Valley presents itself as the great remedy for the ills of our time. Health apps are becoming increasingly common, in a context of an aging population and problems accessing healthcare. "As the welfare state crumbles," Morozov points out, "as it proves incapable of keeping its promises to the population, Silicon Valley is offering us a new form of protection ... If the magic of Silicon Valley can provide all these services, could we do without the state?"[46]

With the lofty rhetoric of TED talks, and start-ups that promise to solve the problems of public health, poverty, and climate change using revolutionary new apps, the algorithmic capitalist imaginary is filling the vacuum left by the decline of the neoliberal utopia. Where this latter advocated competition at all costs and the quest for individual performance, we are now seeing the emergence of a "therapy of connectivity." Champions of this therapeutic approach, such as Mark Zuckerberg, interpret the world's greatest ills, such as violent conflict in the Middle East, with astonishing naïveté, taking them as the result of a simple lack of "connection" and "empathy."[47]

We can see, then, that the advent of algorithmic capitalism is accompanied by an ideology strongly influenced by technological solutionism, even if this logic is not entirely new. We could look back to the myth of Prometheus in ancient Greece, Francis Bacon's utopia of New Atlantis, René Descartes's idea of man as "master and possessor of nature," the naturalism of the Enlightenment, or even the nineteenth-century "age of productivism," with its pervasive idea that technical progress, the quest for efficiency, and the full development of productive forces would bring collective riches.[48] While a Promethean, technocentric, and productivist worldview is not new to the 2000s, the term "technological solutionism," popularized by Morozov, is representative of

46 Evgeny Morozov, *Le mirage numérique: Pour une politique des big data*, Paris: Les Prairies ordinaires, 2015, p. 14.

47 Sadin, *La silicolonisation du monde*, p. 91.

48 Serge Audier, *L'âge du productivisme: Hégémonie prométhéenne, brèches et alternatives écologiques*, Paris: La Découverte, 2019.

what is today a widely circulated ideology that largely emanates from Silicon Valley.[49]

What is the main thrust of this ideology? First, it is a reductive logic that frames any complex social, economic, political, or ecological issue as a technical problem that can be solved simply with an application, an algorithm, or other similar solution. In this sense, statements such as "technology can solve the climate crisis" or "AI will allow us to fight crime" are examples of technological solutionism, as is the assertion by former Google CEO Eric Schmidt, speaking of technological "connectivity," that "if we get this right, then we can fix all the other problems, too."[50] As István Mészáros already pointed out with regard to technocratic optimism: "To say that 'science and technology can solve all our problems in the long run' is much worse than to believe in witchcraft."[51]

Technological solutionism is problematic on several levels. It is a blinkered and simplistic understanding of the world that creates an addiction to digital technologies. These latter are presented as the last word in resolving the great problems of the century, in a way that marginalizes other possible solutions. As Yann Pezzini notes:

> It is a strategic, technical and managerial rationality: we seek the "best" solutions in a given context; solutions that are supposedly effective, rational and proactive, aimed at "solving" social or environmental "problems." These are the solutions that will be valued and applauded and rally support. However, ... this reasoning allows dominant actors to *erase the conflictual and political dimension* of social and environmental issues and to *limit the range of possibilities* to solutions they deem pragmatic/effective (while denigrating others in the process).[52]

Solutionism can cause collateral damage such as misallocation of resources—for instance, investing in predictive policing rather than community organizations or creating AI-based "responsible consumption" applications instead of imposing tighter regulations on the oil industry.

---

49 Evgeny Morozov, *To Save Everything, Click Here: The Folly of Technological Solutionism*, New York: Public Affairs/Perseus, 2013.

50 Quoted in Françoise Laugée, "Solutionnisme," *Revue européenne des médias et du numérique*, no. 33, Winter 2015, English trans. at news.berkeley.edu.

51 István Mészáros, *The Necessity of Social Control*, New York: Monthly Review Press, 2014 [1974], p. 29.

52 Yann Pezzini, "Critique du solutionnisme dans la 'résolution' des enjeux sociaux et environnementaux," *Carnets du Centre de recherche sur les innovations et transformations sociales* (CRITS), March 2023 (original emphasis).

Solutionism can also focus media attention on innovations that help cope with the effects of social problems without addressing their root causes. Doubtless, the term is sometimes used as a label to shut down discussion, to point the finger at one's interlocutor, or to denigrate certain solutions that might otherwise be quite legitimate. Yet, when used properly, this expression refers to an ideological discourse that is indeed widespread in the public sphere, and which seeks to confer legitimacy on the increased use of algorithmic technologies for humanitarian purposes.

In addition to its solutionism, Silicon Valley's techno-optimism is accompanied by other discourses and imaginaries, such as the "Californian ideology," which we will return to in more detail in Thesis 13. According to Richard Barbrook and Andy Cameron:

> This new faith has emerged from a bizarre fusion of the cultural bohemianism of San Francisco with the hi-tech industries of Silicon Valley … the Californian Ideology promiscuously combines the freewheeling spirit of the hippies and the entrepreneurial zeal of the yuppies.[53]

Stewart Brand and the launch of the *Whole Earth Catalog* in 1968 embodies this optimistic worldview: "Individual autonomy ('Do It Yourself'), constant innovation, technological prowess, all characteristics that, in marketing, would later be associated with the personal computer and the new economy."[54] However, this worldview, which emerged in the 1970s and was widely promoted in the 1980s and 1990s, seems more linked to the first phase of digital capitalism, with its mixture of neoliberalism and cyberlibertarianism.[55] In other words, what we have here is a mixture of a neoliberal vision, centered on the entrepreneur's role as savior, and the utopia of a free and horizontal world emerging from digital networks. This Californian ideology thus predates the rise of algorithmic capital, and although it is flourishing with the emergence of this new stage of capitalism, it is an exoteric ideological formation that accompanies another, more esoteric core.[56]

---

53 Richard Barbrook and Andy Cameron, "The Californian Ideology," *Science as Culture* 6, no. 1, January 1996, pp. 44–72, citation from pp. 44–5.

54 Monique Dagneau, "Derrière Internet, l'imaginaire hippie," *Slate*, June 25, 2013. See also Fred Turner, *From Counterculture to Cyberculture: Stewart Brand, the Whole Earth Network, and the Rise of Digital Utopianism*, Chicago: University of Chicago Press, 2008.

55 Hubert Guillaud, "Le problème de la gauche avec Internet," *Le Monde* blog, May 16, 2014.

56 We come back to the Californian Ideology in Thesis 13.

The strange world of TESCREAL

What then is algorithmic capitalism's specific ideology? One answer can be found in a strange philosophical galaxy called TESCREAL. This acronym encapsulates several distinct but connected ideas: transhumanism, extropianism, singularitarianism, cosmism, rationalism, effective altruism, and longtermism.[57] The term TESCREAL was coined by Émile P. Torres and Timnit Gebru to describe a bundle of beliefs, values, and ideologies that are more or less intertwined.[58] They together form an all-encompassing worldview held by several intellectuals, entrepreneurs, and influential personalities linked to AI, space exploration, philanthropy, investment funds in technological innovation, and other associated research centers.[59] Figures such as Nick Bostrom, Elon Musk, Peter Thiel, Sam Bankman-Fried, Toby Ord, and William MacAskill belong to these circles. The rationale for conceptualizing these different ideologies in combination lies in the fact that these ideas overlap in time and space: Indeed, most individuals who adhere to one of these schools of thought also adhere to the views indicated by the other letters in the acronym, so it makes sense to speak of a TESCREAL community.[60] But what exactly is TESCREAL? Let us briefly review each of its components.

First, *transhumanism* is an intellectual movement that advocates the use of technology to improve the human condition, whether by increasing humans' physical or cognitive abilities or by maximizing life expectancy. Although the quest for immortality can be traced back all the way to the Mesopotamian myth of Gilgamesh, and the idea of "perfectibility" has its origins in the Renaissance philosophy of Giovanni Pico della Mirandola, the term "transhumanist" was coined by Julian Huxley in the 1950s. That was before it gained a following among AI researchers such as Marvin Minsky, Raymond Kurzweil, and Nick Bostrom. Although there are left-wing and right-wing strands within this vein of

57 Dave Troy, "The Wide Angle: Understanding TESCREAL: The Weird Ideologies Behind Silicon Valley's Right-Wing Turn," *Washington Spectator*, May 1, 2023.

58 Timnit Gebru and Émile P. Torres, "The TESCREAL Bundle: Eugenics and the Promise of Utopia Through Artificial General Intelligence," *First Monday* 29, no. 4, April 2024.

59 We explore other connected aspects of this ideology such as eugenics and escapism in Jonathan Martineau, Jonathan Durand Folco, and Myriam Lavoie-Moore, "Longtermisme, eugénisme et évasionisme: Le rapport au futur de la nébuleuse TESCREAL en tension entre techno-utopie et catastrophisme," in *Industrie et futurs imaginés*, E. Caccamo and M. Catoir-Brisson, dirs. Paris: MKF éditions, 2026.

60 Émile P. Torres, "The Acronym Behind Our Wildest AI Dreams and Nightmares," *Truthdig*, June 15, 2023.

thinking, it should be noted that transhumanism has had an ambiguous relationship with eugenics since its inception, and also has forged various links with a Promethean, productivist, neoliberal, and capitalist imaginary aimed at making human beings more efficient and "adapted" to the sociotechnological conditions of today's world.[61] The company Neuralink, created by Elon Musk in 2016, which manufactures "brainchips" for implant, is part of this movement.

Next, *extropianism* is a branch of transhumanism that aims to combat the process of entropy (the thermodynamic law according to which all energy tends to dissipate) through the indefinite extension of human life. To this end it draws on various techniques such as cryogenic freezing of the body or "mind uploading." This school of thought, developed by Max More and Tom W. Bell from 1988 onward, is based on five pillars: unlimited expansion, personal transformation, dynamic optimism, artificial intelligence, and spontaneous order.[62] Several companies have been created in this spirit, including the 2045 Initiative funded by Russian billionaire Dmitry Itskov, the Alphabet subsidiary Calico hosted by Google X Lab (Larry Page invested $750 million in Calico in 2013), and the Alcor Life Extension Foundation, which had already cryogenically frozen more than 149 bodies by 2017.[63] This philosophy represents an extension of historical humanism, while radicalizing its claim that reason can master human destiny:

> Extropianism is a *transhumanist* philosophy ... We challenge the inevitability of aging and death, and we seek continuing enhancements to our intellectual abilities, our physical capacities, and our emotional development. We see humanity as a transitory stage in the evolutionary development of intelligence. We advocate using science to accelerate our move from human to a transhuman or posthuman condition.[64]

Then comes *singularitarianism*, an ideology based on the belief in the advent of technological singularity (also known as "superintelligence"), expected to occur sometime around 2045 according to futurologist Ray Kurzweil.[65] Writer Eliezer S. Yudkowsky founded the Machine Intelli-

61 See Nicolas Le Dévédec, *Le mythe de l'humain augmenté: Une critique politique et écologique du transhumanisme*, Montreal: Écosociété, 2021.

62 Ed Regis, "Meet the Extropians," *Wired*, October 1, 1994.

63 Xavier Ridel, "Les 'extropiens,' ces chasseurs de corps," *Slate*, February 18, 2017.

64 Max More, "The Extropian Principles," mrob.com.

65 Raymond Kurzeil, *The Singularity Is Near: When Humans Transcend Biology*, New York: Penguin, 2005.

gence Research Institute in 2000, and he describes the followers of this philosophy as follows: "Singularitarian believes that the Singularity is possible and desirable. A Singularitarian actually works to bring about the Singularity."[66] Interestingly, Yudkowsky has since radically changed his mind about the positive nature of singularity, now believing that superhuman intelligence poses a major existential risk to humanity.[67] Far from being satisfied with a six-month pause on AI, Yudkowsky is instead calling for a halt to the training cycles of overly powerful algorithmic systems, strict control of research laboratories, and the adoption of international agreements similar to those in the nuclear industry. He also advocates radical measures in the event of such a moratorium being violated: We should "be willing to destroy a rogue datacenter by airstrike."[68]

The letter *C* refers to *cosmism*, the ideology promoting space exploration and the colonization of new territories beyond our own planet. Although the idea of space conquest dates back at least to the writings of Jules Verne and beyond, cosmism was promoted by Russian scientists such as Konstantin Tsiolkovsky and Vladimir Vernadsky and popularized in the United States by figures such as Walt Disney, before becoming a focus of the Cold War in the 1950s. It has been taken up over the past decade by private companies such as Blue Origin (Bezos) and SpaceX (Musk). According to Mary-Jane Rubenstein, this ideology extends the pattern of European colonialism, the humanitarian imperialism of the Church, and the American myth of the "final frontier." She explains how it also takes advantage of a certain privatization of the space race after legislative changes during Barack Obama's presidency that would allow private corporations to claim ownership of the subsoil of other planets.[69]

For its part, *rationalism* is a well-established philosophical movement that stipulates that reason should be the source of knowledge and morality. Dating back to the origins of Western philosophy, an ideological version of the rationalist ideal has now materialized in certain communities such as lesswrong.com, Leverage Research, and other hyper-rationalist currents. They have set the aim of reducing cognitive biases and perfecting human reason through advances in cognitive science, AI, and other technologies.[70] While rationalism takes various

66 Eliezer S. Yudkowsky, "The Singularitarian Principles," April 5, 2001.

67 Eliezer S. Yudkowsky, "Pausing AI Developments Isn't Enough: We Need to Shut It All Down," *Time*, March 29, 2023.

68 Ibid.

69 Mary-Jane Rubenstein, *Astrotopia: The Dangerous Religion of the Corporate Space Race*, Chicago: University of Chicago Press, 2022.

70 Troy, "The Wide Angle: Understanding TESCREAL."

forms, a particular variant of this worldview has now emerged in the world of cryptocurrencies, AI, and philanthropy, with an abstract, mathematical, and disembodied vision of rationality.

In the same vein, *effective altruism* represents an original and updated version of utilitarian philosophy. It seeks to determine the best ways to maximize one's positive impact on the world through evidence-based data and a rigorously "scientific" approach. While philosopher Peter Singer provided the initial inspiration for this attitude,[71] Toby Ord and William MacAskill are the real creators and promoters of this ideology. In 2009, they jointly launched the international organization Giving What We Can (GWWC), which seeks to bring together people who agree to donate at least 10 percent of their income to the NGOs considered to be most "effective." The term "effective altruism" was first used in 2011, at the same time as the creation of the organization 80,000 Hours (cofounded by MacAskill), which promotes "high-impact" professional careers based on the principle of "earn to give."

The ideas promoted by these organizations aim to rationalize the allocation of philanthropic resources according to various presumptions: (1) That social problems and causes should be prioritized according to their scale, how far they have been neglected, and the possibility of solving them; and (2) that it is preferable to focus on high-income careers (in finance, technology, and so on) and then redistribute, rather than investing directly in social causes (the "earn to give" principle). Making one's career on Wall Street may turn out to be more effective than working for some community group or humanitarian organization. In this we can see the logic of "philanthrocapitalism," the introduction of methods and practices from private business into the world of philanthropy to make it more "effective."[72] The effective altruism community has grown significantly since its inception. For example, GWWC had 64 donors and a total fund of $20 million in 2009, reaching 7,400 members and $46 billion in 2021.[73] One of the best-known members and contributors to the effective altruism community, Sam Bankman-Fried (founder of

---

71 Peter Singer, "Famine, Affluence and Morality," *Philosophy and Public Affairs* 1, no. 3, 1972, pp. 229–43; Peter Singer, *The Life You Can Save: Acting Now to End World Poverty*, New York: Random House, 2009; Peter Singer, *The Most Good You Can Do: How Effective Altruism Is Changing Ideas About Living Ethically*, New Haven, CT: Yale University Press, 2015.

72 Matthew Bishop and Michael Green, *Philanthrocapitalism: How the Rich Can Save the World*, London: Bloomsbury, 2008.

73 Benjamin Todd, "Is Effective Altruism Growing?: An Update on the Stock of Funding People," 80,000 Hours, July 28, 2021.

cryptocurrency firm FTX), followed MacAskill's advice in 2013 to choose a career in finance and then served briefly as a member of the board of the Centre for Effective Altruism until 2019, before he came under suspicion of illegal activities prior to his firm's collapse in 2022. He is now serving a twenty-five-year sentence for fraud and conspiracy after a notorious trial in 2024.[74]

Finally, all the philosophies represented by the acronym TESCREAL converge in "longtermism," the philosophy that has been advanced by MacAskill, Toby Ord, and Nick Bostrom since the 2010s. In his book *The Precipice*, Ord writes that longtermism "is especially concerned with the impacts of our actions upon the longterm future. It takes seriously the fact that our own generation is but one page in a much longer story," which should not be measured in terms of a few future generations, but in terms of hundreds, thousands, or even millions of years.[75] This means that classical utilitarian calculation should not be limited to considering those of us who are alive today or those who will be alive in the near future, but potentially as many as 10^{58} future people (including a hypothetical number of simulated consciousnesses in a future where humanity has colonized the galaxy and computer-simulated worlds are also growing in number).[76]

The unique character of longtermist philosophy lies in the fact that transhumanist, extropian, singularitarian, cosmist, rationalist, and effective-altruist visions converge in an overall calculation of pleasures and pains over a very long period of time, making a short-term humanitarian catastrophe seem like a drop in the ocean. And there lies the problem with longtermist ideology. For Bostrom, disasters and atrocities such as the two world wars, the Holocaust, AIDS, and the Chernobyl nuclear accident—"tragic as those events are to the people immediately affected"—"have not determined the long-term fate of our species. Even the worst of those catastrophes were mere ripples on the surface of the great sea of life."[77] Longtermist logic is utilitarianism on steroids. The principle of maximizing the happiness of the greatest number, developed by the philosopher Jeremy Bentham, is here combined with a futuristic vision with scientific pretensions, using probabilistic calculations to

74 Charlotte Alter, "Exclusive: Effective Altruist Leaders Were Repeatedly Warned About Sam Bankman-Fried Years Before FTX Collapsed," *Time*, March 15, 2023.

75 Toby Ord, *The Precipice: Existential Risk and the Future of Humanity*, London: Bloomsbury, 2020.

76 Bostrom, *Superintelligence*.

77 Nick Bostrom, "Dinosaurs, Dodos, Humans?," Oxford: Oxford Future of Humanity Institute, 2006.

quantify the total value of a vast number of consciousnesses downloaded onto computers and future humans who have colonized the galaxy. As Émile P. Torres notes:

> The underlying reasoning here is based on the idea that people—you and I—are nothing more than means to an end. We don't matter in ourselves; we have no inherent value of our own. Instead, people are understood as the "containers" of value, and hence we matter only insofar as we "contain" value, and therefore contribute to the overall net amount of value in the Universe between the Big Bang and the heat death. Since utilitarianism tells us to maximise value, it follows that the more people (value containers) who exist with net-positive amounts of value (pleasure), the better the Universe will become, morally speaking. In a phrase: people exist for the sake of maximising value, rather than value existing for the sake of benefitting people.[78]

This may seem harmless at first glance, but the practical consequences of this (very) long-term ethic are sometimes surprising. For example, longtermist Nicholas Beckstead argues that "saving a life in a rich country is substantially more important than saving a life in a poor country, other things being equal," on the grounds that "richer countries have substantially more innovation, and their workers are much more economically productive."[79] It should be noted that Beckstead was a research fellow at the Future of Humanity Institute founded by Bostrom and was CEO of the FTX Foundation created by Sam Bankman-Fried, who also identifies with longtermism—although the term perhaps resonates differently for him now that he is in prison. This Western-centric outlook, with colonial hues, can be found in the work of another former researcher at the Future of Humanity Institute, Robin Hansen, who argues that should industrial civilization collapse, the priority would be to "retrace the growth path of our human ancestors" by going through the various stages of civilization (hunter-gatherers, Neolithic, industrial society).[80] To do this, "it might make sense to stock a refuge with real hunter-gatherers and subsistence farmers, together with the tools they find useful. Of course, such people

78 Émile P. Torres, "Against Longtermism," *Aeon*, October 19, 2021.

79 Nicholas Beckstead, "On the Overwhelming Importance of Shaping the Far Future," PhD thesis, Rutgers, State University of New Jersey, 2013.

80 Robin Hansen, "Catastrophe, Social Collapse and Human Extinction," in Nick Bostrom and Milan Ćirković, eds., *Global Catastrophic Risks*, Oxford: Oxford University Press, 2008, pp. 363–7.

would need to be disciplined enough to wait peacefully in the refuge until the time to emerge was right."[81]

There are harder and softer versions of longtermist philosophy. In 2021, William MacAskill made the stronger claim that considering the long term is the priority that should guide our actions in the present, rather than one priority among others (the softer version).[82] From a mathematical point of view, this implies investing more in increasing human capabilities, space exploration, creating "digital lives," and reducing the existential risks associated with the advent of superintelligence, rather than worrying about more prosaic issues such as global poverty and climate change, which affect only a few billion people. In his latest bestseller, *What We Owe the Future*, MacAskill argues that:

> Advances in biotechnology could provide another pathway to rebooting growth. If scientists with Einstein-level research abilities were cloned and trained from an early age, or if human beings were genetically engineered to have greater research abilities, this could compensate for having fewer people overall and thereby sustain technological progress.[83]

MacAskill, like almost all longtermists, is more concerned with technological progress than with the present-day well-being of humans or of living species in general. For example, he argues that while it is surely important to care about the climate crisis and reduce our consumption of fossil fuels, the reason we should limit the exploitation of fossil fuels is that we may need them in the future if we have to rebuild industrial civilization after its collapse.[84] In other words, protecting human lives today matters less than protecting the technological conditions for the eventual reconstruction of industrial society. Ultimately, the priority is to maintain the "light of consciousness" in the long term and maximize the chances of creating up to 10^{58} containers of positive value in the future, rather than finding solutions to social problems affecting the most vulnerable of us in the here and now. Utilitarian morality here boils down to a mathematical calculation.

---

81 Robin Hansen, quoted in Émile P. Torres, "Understanding 'Longtermism': Why This Suddenly Influential Philosophy Is So Toxic," *Salon*, August 20, 2022.

82 Hilary Greaves and William MacAskill, "The Case for Strong Longtermism," Global Priorities Institute, University of Oxford, Working Paper No. 5-2021, 2021.

83 William MacAskill, *What We Owe the Future*, New York: Basic Books, 2022, p. 166.

84 Ibid., p. 147.

Even philosopher Peter Singer, once associated with effective altruism, now believes that this extreme version of utilitarianism leads to absurd conclusions—and that perhaps we should prioritize current and future generations rather than placing trillions of human lives spread over millions of years on an equal footing. This particular long-term perspective has a distorting effect that can shrink the problems of the present "to almost nothing, while justifying almost anything that increases our odds of surviving long enough to spread beyond Earth."[85]

Why this long detour via the world of TESCREAL—and what is its real connection to algorithmic capitalism? Here we will submit five hypotheses. First, the TESCREAL vision represents the ideological bearings of the ruling class at the head of large digital companies, the development of AI, and contemporary capitalist philanthropy focused on long-term global risks. To take one of many possible examples, Elon Musk retweeted MacAskill's latest book on its release, commenting: "Worth reading. This is a close match for my philosophy."[86] Philosophers such as Bostrom, Ord, and MacAskill represent the "organic intellectuals" of this wing of the bourgeoisie focused on the development of AI, space exploration, and biotechnology.[87] As Torres points out, longtermism is one of the most influential ideologies among the elites of Silicon Valley and beyond:

> Longtermists have directly influenced reports from the secretary-general of the United Nations; a longtermist is currently running the RAND Corporation; they have the ears of billionaires like Musk; and the so-called Effective Altruism community, which gave rise to the longtermist ideology, has a mind-boggling $46.1 billion in committed funding. Longtermism is everywhere behind the scenes—it has a huge following in the tech sector—and champions of this view are increasingly pulling the strings of both major world governments and the business elite.[88]

AI and longtermism share institutional and social resources and connections, but also project similar narrative forms, ultimately feeding off each other, as Fenwick McKelvey and Jonathan Roberge note:

---

85 Peter Singer, "The Hinge of History," *Project Syndicate*, October 8, 2021.

86 Elon Musk, Twitter, August 2, 2022.

87 According to Antonio Gramsci, "organic intellectuals" produce ideas, discourses, values, and worldviews linked to a particular social class. Organic intellectuals may be linked to subaltern classes, but also to ruling classes.

88 Torres, "Understanding 'Longtermism.'"

> The fact of the matter is that longtermism calls for more longtermism and that AI calls for more AI. This is what we mean by the recursive nature of their power. Through it, longtermism and AI become the causes and consequences of one another.[89]

Second, the TESCREAL ideology espouses a logic according to which human beings are destined to be surpassed by technological progress, whether physically (transhumanism, extropianism), cognitively (singularitarianism, rationalism), spatially (cosmism), morally (effective altruism), or temporally (longtermism). This worldview represents a form of global rationality in which data, machines, and algorithmic optimization are central to decision-making and, by extension, guide our behavior, and determine which social issues are a priority and how we should solve our problems. It prescribes a certain vision of the good life focused on infinite growth and continuously improving our well-being through the full development of technological capabilities.

Third, although the foundations of TESCREAL ideology date back to the 1950s with the emergence of transhumanism and the advent of cybernetics,[90] it was not until the early 2000s that it crystallized into a coherent worldview. Indeed, it was not until the 2010s that it began to spread more widely through various research centers, scientific articles, institutes, podcasts, international conferences, communities of practice, popular books, and lobbying of governments and public institutions. This means that it is an emerging philosophy, contemporary with the shift toward a new stage of capitalism beginning around 2008.

Fourth, neoliberal ideology did not disappear overnight. Indeed, it persisted despite the crisis of 2007–8. Yet it was taken up anew and reconfigured, reintegrated, and superseded with the arrival of algorithmic capitalism. This is evident in the continued presence of neoliberal ideology via the "collaborative economy" (generalized Uberization), or in the "assetization of everything" via data monetization, blockchains, NFTs, and other methods.[91] Nevertheless, neoliberal ideology and its constituent elements, such as its emphasis on individual freedom, the

89 Fenwick McKelvey and Jonathan Roberge, "Recursive Power, AI Governmentality and Technofutures," *EspacesINRS*, 2023, p. 10.

90 This is a thesis advanced by Maxime Ouellet in *La révolution culturelle du capital: Le capitalisme cybernétique dans la société globale de l'information*, Montreal: Écosociété, 2014.

91 Kean Birch and Fabian Muniesa, *Assetization: Turning Things into Assets in Technoscientific Capitalism*, Cambridge, MA: MIT Press, 2020.

entrepreneurial ethos, and the omnipotence of the market, are no longer at the forefront of public discourse. Rather, the algorithmic regulation of behavior, the platform model, all-powerful AI, optimization based on Big Data, and the unprecedented return of utilitarianism in the longtermist vision all dovetail in a comprehensive philosophy that retrieves elements of the previous ideology. In Hegelian terms, we could hypothesize an *Aufhebung*, that is, that neoliberal ideology is simultaneously negated, preserved, and superseded by longtermist ideology and the world of TESCREAL.

Fifth, we hypothesize that the specific ideology of algorithmic capitalism plays out on two levels of discourse, each addressing different audiences. On the one hand, there is the mainstream "exoteric" discourse, that is, the kind of discourse intended for the layperson, the masses, mere mortals. This is the solutionist, techno-optimistic discourse that claims to be able to "make the world a better place" through the arrival of new technological products, digital technology, and AI. This solutionist discourse is widely promoted by the technology sector, Silicon Valley, and the Californian ideology, which promises happiness, individual freedom, personal development, greater empathy, and the resolution of major social and environmental problems. This is the "progressive," "sunny," "green" face of algorithmic capitalism.

On the other hand, there is the "esoteric" discourse (from the Greek *esôteros*, meaning "internal"), which is reserved for insiders, elites, and people who are closely interested in the latest advances in AI and the possibility of technological "singularity." This is the TESCREAL vision, shared mainly by leading entrepreneurs, scientists concerned about the advent of superintelligence, longtermist philosophers, transhumanists and libertarians funding cryogenics projects, and survivalist billionaires building luxury bunkers in New Zealand or planning to flee to outer space to escape climate crisis.[92] As journalist Douglas Rushkoff notes:

> These people once showered the world with madly optimistic business plans for how technology might benefit human society. Now they've reduced technological progress to a video game that one of them wins by finding the escape hatch. Will it be Jeff Bezos migrating to space, Thiel to his New Zealand compound, or Mark Zuckerberg to his virtual

92 Douglas Rushkoff, *Survival of the Richest: Escape Fantasies of the Tech Billionaires*, New York: W. W. Norton, 2022.

> metaverse? And these catastrophising billionaires are the presumptive winners of the digital economy—the supposed champions of the survival-of-the-fittest business landscape that's fuelling most of this speculation to begin with.[93]

The TESCREAL vision described here is not widely shared by the population, but in a way it represents the ideological core of algorithmic capital. It is a vision which aims for infinite growth in value; the celebration of data as a model of knowledge; increased use of algorithms to augment human capabilities; and the pursuit of colonization projects to cross the "final frontier" of intergalactic space. All this comes along with a more troubling dimension linked to the advent of superintelligence and the imminence of the climate crisis, which the elites are seeking to escape through various technological means. Even more worrying is the presence in this ideological mix of certain elitist, natalist, colonialist, eugenicist, and even downright racist ideas.[94] For example, Bostrom has stated in the past that black people are less intelligent than white people, while describing "dysgenic pressures" as an existential risk, worrying that people with low IQs could one day reproduce faster than people with high IQs.[95] This obsession with "intelligence genes" is widespread among transhumanists, but especially within a pronatalist strand among certain Silicon Valley billionaires, such as Elon Musk with his fourteen children. Or take the example of Simone and Malcolm Collins, the "eugenics hipsters" who believe in the superiority of their genes and use various biotechnologies to select the best embryos for their offspring, with the aim of having between seven and thirteen children.[96] As Lucie Ronfaut puts it:

> For the followers [of this philosophy], it is necessary to have many children and, above all, to invest in new technologies to "*reinvent reproduction*": that is, fertility assistance, designing artificial wombs, genetic testing on embryos, etc. The movement is, logically enough, accused of promoting eugenics, in start-up style. This is not only about reproducing

93 Douglas Rushkoff, "The Super-Rich 'Preppers' Planning to Save Themselves from the Apocalypse," *The Guardian*, September 4, 2022.

94 Samara Linton, "Tech Elites' AI Ideologies Have Racist Foundations, Say AI Ethicists," *People of Color in Tech*, May 24, 2023.

95 Émile P. Torres, "Eugenics in the Twenty-First Century: New Names, Old Ideas," *Truthdig*, January 23, 2023.

96 Julia Black, "Ces hipsters de l'eugénisme qui veulent changer l'avenir de l'humanité," *Le Courrier international*, May 21, 2023.

> "better," but above all encouraging the "right people" to have children in order to save the world, i.e. those who are the most intelligent (according to their definition) and wealthy enough to afford all these technologies, as well as amenities like private schools and even a strange project for a "start-up city" reserved for entrepreneurs and their offspring.[97]

Far from the radiant, progressive, and democratic guise of solutionist discourse and the Californian ideology, here we discover the dark, elitist, and expertocratic face of the TESCREAL vision shared by the ruling classes associated with algorithmic capital. We can thus speak of "dark AI," which is very different from the cheerful AI presented to the general public.

This is, of course, a conceptual distinction, while in reality the same person may subscribe to either one or both of these discourses depending on their audience or their mood at the time. A good example is Sam Altman, CEO of OpenAI, a billionaire entrepreneur who has invested in various start-ups in the fields of social media, biotechnology, nuclear fusion, cryptocurrencies, and AI. He "says he is optimistic, but he is also a survivalist, according to the *New Yorker*: he stockpiles weapons, gold, water and antibiotics on his property in Big Sur, on the Californian coast."[98] Altman is both hyperoptimistic about AI's ability to advance humanity and pessimistic about the existential risks posed by a potential superintelligence that could destroy our species. In other words, his longtermist vision is buttressed with a layer of solutionism. He sums up his view with this declaration:

> We can build AGI. We can colonize space. We can get fusion to work and solar to mass scale. We can cure all human disease. We can build new realities. We are only a few breakthroughs away from abundance at a scale that is difficult to imagine.[99]

Like most ideologies, there is, of course, a certain amount of internal diversity and a multitude of attitudes within the world of TESCREAL:

---

97 Lucie Ronfaut, "Pourquoi les dirigeants de la tech veulent-ils faire tellement d'enfants?," *Numérama*, December 4, 2022.

98 Julie Jammot, "Sam Altman, l'entrepreneur qui construit l'IA pour en protéger les humains," *La Presse*, May 16, 2023.

99 Sam Altman, Twitter, February 4, 2022.

> Consequently, a range of positions have emerged within the TESCREAL community, from "AI doomers" to "AI accelerationists" … In between are various "moderate" positions that see the dangers as real but not insuperable, a position exemplified by the Future of Life Institute, which merely calls for a six-month "pause" on AGI research.[100]

In short, the gradual replacement of neoliberalism's atomized vision by Silicon Valley's hyperconnected and empathetic imaginary is not all there is to the transition to a new stage of capitalism. The shift to a new regime of accumulation is accompanied by the arrival of a strange new ideology, TESCREAL, and a new form of power that structures behavior, social relations, and subjectivities. Neoliberalism represents both a rationality that influences the functioning of institutions (according to market principles) and a form of subjectivation (which guides behavior according to the model of the "entrepreneur of the self"). We now need to explain the emergence of a mode of regulation based on *algorithmic governmentality*, which makes it possible to regulate social relations and behaviors while ensuring the automatic coordination of decentralized decisions. This is the subject of Thesis 9.

---

100 Torres, "The Acronym Behind Our Wildest AI Dreams and Nightmares."

Thesis 9

A new form of power is emerging: Algorithmic governmentality

You were more inward to me than my most inward part and higher than my highest.

—Saint Augustine

If we want to properly analyze the new form of power taking form in algorithmic capital, we first need to make a number of conceptual distinctions. First, we should take a step back and dissect the concept of power itself rather than taking it for granted. This "godfather of political concepts" may indeed take various forms, and as the basis of the social and political order it thus warrants close attention.[1]

The concept of power

The scholarly literature generally distinguishes between two main types of power: "power to," which refers to the ability to take action and have an impact on the world (for instance: the power to perform a certain task), and "power over," which involves one actor's ability to influence the actions of another. This corresponds to the classic definition provided by Robert Dahl: "A has power over B to the extent that he can get B to do something that B would not otherwise do."[2] Here we see that "power to" represents the most general category, while "power over" is more specific

1 Jamie Susskind, *Future Politics: Living Together in a World Transformed by Tech*, Oxford: Oxford University Press, 2018, p. 89.

2 Robert A. Dahl, "The Concept of Power," *Behavioral Science* 2, no. 3, 1957, pp. 202–3.

and eminently relational, opening the way to various forms of influence, and also to abuses.

Next, we can distinguish between various subcategories of "power over," which may involve force, coercion, influence, authority, or manipulation. Indeed, the power that an individual, group, or institution can exercise over others involves many different variants: from direct and physical actions affecting others (force) to the use of threats, intimidation, and constraint (coercion); the use of persuasion, or the presentation of select information (influence); the existence of charismatic power, or of a person with great legitimacy (authority); or strategies based on concealing the true motives for action (manipulation). At this stage, the important thing is to determine how algorithms, and the individuals and organizations using them, are able to exert different forms of power. For instance, force can be applied through the use of robots or military drones; coercion can be exercised by predictive policing, automated decision-making systems, or even automatic messages demanding payment under threat of penalty; the use of algorithms can influence other people's perceptions through targeted advertising and predictive analytics; tech brands or execs may gain a certain authority and leverage charismatic power (Google, Steve Jobs, Elon Musk); manipulation can be used to spread misinformation or even to rig elections. Algorithms can thus be instruments of "power over." Their spread throughout society raises the tangled issue of *algorithmic power*.

We need to push the analysis still further and recognize that all power relations are inherently part of social relations. No serious analysis of power can stop at a methodological-individualist perspective, according to which power is a kind of interaction between atomized and isolated individuals. It is vital to analyze the distribution of power in the social world, for a series of processes of power concentration, stratification, and classification of actors according to their social position (income, gender, ethnicity, property, and so on) generate more or less permanent inequalities and asymmetries between different individuals, organizations, and social groups. In this sense, Danièle Kergoat makes a useful distinction between *social relationships*, which are "immanent to concrete individuals," and *social relations*, which are "abstract and pit groups into opposition around some issue." We must, therefore, take into account the social relations that precede individuals and structure their relationships in society, which are generally more fluid. A social relation of power is thus one that connects groups in a lasting asymmetry. Its three "canonical forms" are exploitation (economic appropriation), domination (legal-political subordination), and oppression (discrimination, exclusion, violence, social marginalization). These three canonical forms combine in complex assemblages that reproduce social relations, for instance, of gender, class, and race. We can thus refine the analysis of power, for example, with regard to the relation between men and women. As Kergoat notes: "The situation has indeed changed in terms of social relationships between the sexes and within couples," but this does not mean the end of gendered power relations, since "the social relation continues to express itself in its three canonical forms: exploitation, domination, and oppression (which could be illustrated by wage differentials, glass ceilings, and violence)."[3]

Social relations thus take consolidated form over time, in complex and changing combinations, as systems of exploitation, domination, or oppression. They involve enduring economic, political and/or social asymmetries between two or more groups, and a series of processes of material and symbolic power. Economic exploitation, political domination, and social oppression are undesirable social phenomena because of the negative consequences they entail for the individuals and groups subjected to them.[4] This is why it is important to distinguish between a more neutral use of the term "power to" (which may be beneficial and feed collective action, cooperation, and the realization of various goals) and negative forms of "power over": exploitation, domination, and oppression. The key problem in the quest for social justice or democracy is not, therefore, a matter of destroying power as such. Rather, it is about eliminating the negative relations of "power over" and achieving a more equal distribution of power among the different members of a community.

In this book, we discuss the three canonical forms of systems of power. Algorithms calculate and predict, but algorithmic capital is an active element in power structures that exploit, dominate, and oppress. We have seen that algorithmic capital exploits digital labor, attention time, and human experience. We will also discuss in greater depth the domination exercised by the state and algorithmic policing over society (Thesis 10). We also examine one of the major issues of our time, the automation of social inequalities. This is connected to forms of "algorithmic oppression"—a term we use to describe the way in which algorithms amplify and reorganize oppressive relations in society that existed prior to the advent of algorithmic capital.

3 Danièle Kergoat, "Comprendre les rapports sociaux," *Raison présente*, no. 178, 2011, pp. 12–13.

4 Steven Lukes, *Power: A Radical View*, New York: Palgrave Macmillan, 2004.

However, it seems that algorithmic capital brings about more than just a reconfiguration or amplification of oppressions that had already existed. Rather, it engenders a new and distinctive form of power over individuals, communities, and institutions. A new social relation seems to be emerging that pits two antagonistic groups into opposition around a single issue: control of information, predictive capacity, and the resulting orientation of human behavior. The individuals or organizations that currently control the means of algorithmic production (or prediction) can exploit, dominate, and oppress those who do not have access to, or control of, these machines.[5] What is thus emerging is a new relation of subordination, one that cannot be reduced to social relations of class, gender, or race, although algorithmic power rearticulates these as well. A new sphere or dynamic of power is at work in today's society, pitting those who produce, control, and manipulate algorithmic power against those who are subjected to it. This power is consolidated through algorithmic capital, which is constantly honing the latest social control technologies based on surveillance, nudges, ratings systems, feedback loops, and other means.

A final particularity of power, which is worth underlining, is that it does not only circulate between individuals and groups, but it also *accumulates*. Power may be accumulated in various forms, whether in terms of resources, opportunity hoarding, cultural and symbolic capital, property and inheritance, or the ability to adopt and amend laws. Alain Savard defines capitalism not as a mode of production or a mode of exploitation, but as a *process* or *structure of power appropriation*.[6] We may also distinguish between various forms of power accumulation linked to different institutions, such as the state, capital, the family, and civil society organizations. Table 9.1 provides an overview of the various sites that allow for the accumulation of power, in different ways.

Table 9.1 highlights the fact that the two main centers of power accumulation in modern societies are within the sphere of the market (controlled by capital and large private companies) and the state, whereas the domestic sphere and civil society associations retain minor or negligible forms of power. But if algorithmic capital can exploit, dominate, and oppress, what role does it play in this process of power accumulation? How did we get from a world where neoliberal rationality governed

5 Maximilian Kasy, *The Means of Prediction: How AI Really Works (and Who Benefits)*, Chicago: University of Chicago Press, 2025.

6 Alain Savard, "Power from Below: Power, Democracy and Socialist Theory," PhD thesis, York University, 2020, p. 121.

Table 9.1 Centers of the production and accumulation of power

Sphere of production	*Form of power accumulation*	*Process of power accumulation*	*Relative power per unit*
Mediated by the market	Capital	Exploitation and monopolization	Major
State-organized	State	Taxation, colonialism, imperialism	Major
Personal/domestic	Patriarchal family	Patriarchal marriage	Negligible
Community-based	Churches, associations, political parties	Confidence and influence	Minor

Source: Alain Savard, "Power from Below," p. 183

behavior to one where the social relation of algorithmic power is leading toward a new way of managing organizations, regulating institutions, and guiding individual behavior, furthering the control over our lives by capital and the state?

From neoliberal rationality to algorithmic regulation

In Thesis 8, we explored aspects of the transition from neoliberal to algorithmic capitalism. This shift obviously implies that certain characteristics of neoliberalism themselves served as a foundation for the rise of algorithmic capitalism. Algorithmic power has not appeared out of nowhere; it is a product of history—and it is important to remember that neoliberalism is likewise a product of history. The core ideas of neoliberal philosophy (promotion of the market and free trade, privatization, deregulation, criticism of state intervention and welfarism, the conception of the individual as a *homo oeconomicus*, the fight against inflation) were formulated by a group of economists, philosophers, and intellectuals (including Friedrich Hayek, Ludwig von Mises, Milton Friedman, Gary Becker, and James Buchanan) who came together in the Mont Pèlerin Society, founded in 1947. However, these ideas would not be applied on a large scale until almost three decades later in the neoliberal "laboratory" of Augusto Pinochet's dictatorship in Chile. There, the "Chicago boys"—Latin American neoliberal economists who were disciples of Milton Friedman and Arnold Harberger of the University of Chicago—implemented the first neoliberal policies under a military regime. Then the rise to power of Margaret Thatcher in Britain in 1979 and Ronald Reagan in the United States in 1981, along with the structural adjustment plans imposed by the World Bank and the International Monetary Fund on Global South countries, marked the transition from the Fordist, state-regulated capitalism of the postwar decades (known in French as the *Trente Glorieuses*) to the neoliberal, financialized, and globalized capitalism that took off in the early 1980s.

While neoliberalism is generally associated with a cocktail of economic policies advocated by the Washington Consensus (austerity, free trade, privatization, deregulation), it would not be right to reduce it to the idea of cutting the state back and promoting laissez-faire. As Pierre Dardot and Christian Laval point out, drawing on Michel Foucault's work, neoliberalism is, first and foremost, a political rationality aimed at transforming the role of institutions and a form of subjectivation that guides behavior according to a certain norm for life itself. "At stake," Dardot and Laval emphasize, "is nothing more, nor less, than the *form of our existence*." The neoliberal norm generalizes competition and the economic struggle of all against all. It imposes market discipline and interferes even in individuals' relationships with themselves, encouraging everyone "to think of themselves … as the holder of capital to be valorized." The neoliberal individual is efficient and competitive, "seek[ing] to maximize his human capital in all areas," and is always "work[ing] on himself so as constantly to transform himself, improve himself, and make himself ever more efficient."[7]

Seen in this light, it is clear that neoliberal rationality did not suddenly vanish with the financial crisis of 2007–8. As Philip Mirowski, one of the leading experts in this field, points out, neoliberalism survived the near-collapse of the financial system because it represents, above all, an all-encompassing worldview that continues to influence the behavior of states, organizations, and individuals around the world.[8] Although it is true that the Silicon Valley imaginary less explicitly champions the virtues of competition, this multifaceted ideology is still shaped by the glorification of entrepreneurship, the understanding that innovation drives social change, and the promotion of a flexibilized model of diffuse entrepreneurship in the collaborative economy. In this spirit, some authors consider that cybernetics, Big Data, and neoliberalism

---

7 Pierre Dardot and Christian Laval, *The New Way of the World: On Neoliberal Society*, London: Verso, 2013, pp. 8, 294.

8 Philip Mirowski, *Never Let a Serious Crisis Go to Waste: How Neoliberalism Survived the Financial Meltdown*, London: Verso, 2013.

feed into each other to form one common ideology,[9] also known as the "Californian ideology."[10]

Furthermore, neoliberalism has provided an especially favorable context for processes of commodification (of data), deregulation (of the technology sector), and self-regulation of (digital) companies, a context that provides fertile ground for the emergence of algorithmic capital. We can thus speak of a *continuation* of neoliberal logic by other (technological) means, which fits comfortably into the new version of algorithmic capital. This thesis of continuity highlights the fact that neoliberalism has profoundly shaped the particular configuration that the Internet has taken, stimulated the rise of deregulated digital markets, and paved the way for the monopolistic dynamics that engorge the web giants.[11]

Nevertheless, it should be noted that the arrival of algorithmic capital is also accompanied by a specific form of power that is not strictly speaking a continuation of neoliberalism, one that has been called "algocracy,"[12] or "algorithmic regulation."[13] Data science is combined with a set of predictive analysis techniques, systems of recommendation and choice-guidance, automated decision-making processes, and incentives that allow human behavior to be guided, modified, and controlled. Algorithmic regulation comprises a set of knowledges and devices that make it possible to *represent* social reality through the collection and analysis of Big Data, to *direct* the spaces of interaction through the creation of rules, standards, and classification systems operated by algorithms, and to *intervene* more or less directly to alter individuals' behavior.[14]

If we want to get a full sense of this sea change, it is not enough to assert that Big Tech enjoys untrammeled domination over the world or that it is monitoring us without our knowledge. Rather, we must look at how we are moving from "government by laws" to a model of

---

9 Ouellet, *La révolution culturelle du capital*; Henrichon, *Big Data*.

10 Durand, *How Silicon Valley Unleashed Techno-Feudalism*, pp. 7–70.

11 Paul Starr, "How Neoliberal Policy Shaped the Internet—and What to Do About It Now," *American Prospect*, April 2019.

12 Aneesh Aneesh, "Global Labour: Algocratic Modes of Organisation," *Sociological Theory* 27, no. 4, 2009, pp. 347–70.

13 Morozov, *To Save Everything, Click Here*; Karen Yeung and Martin Lodge, eds., *Algorithmic Regulation*, Oxford: Oxford University Press, 2019; Lena Ulbricht and Karen Yeung, "Algorithmic Regulation: A Maturing Concept for Investigating Regulation of and Through Algorithms," *Regulation & Governance* 16, no. 1, 2022, pp. 3–22.

14 Florian Eyert, Florian Irgmaier, and Lena Ulbricht, "Extending the Framework of Algorithmic Regulation: The Uber Case," *Regulation & Governance* 16, no. 1, 2022, pp. 23–44.

"governance by numbers,"[15] which is transforming before our eyes into algorithmic regulation. Platforms such as Uber are already using these modes of algorithmic power and control over their own drivers and users,[16] but algorithmic regulation is spreading in the social sphere beyond the purpose of labor-discipline.

Researcher Karen Yeung offers a detailed analysis of the multiple systems of algorithmic regulation. These are cybernetic processes based on three components: (1) setting of standards, objectives, and targets; (2) means of collecting information; (3) enforcement and behavior modification to correct anything that might deviate from these standards.[17] Two configurations are possible for each element. First, standards can be defined in a *fixed*, simple, and preprogrammed manner (such as password protection systems); or else *adaptively*, in the manner of systems that modulate flows of road traffic in real time through processes based on machine learning.[18]

Second, information can be collected and monitored *reactively* by analyzing data in real time to detect violations of certain rules (such as automated speed detection systems on motorways, or credit card fraud detection software based on individuals' spending patterns). This can also be done *preemptively* by predicting future behavior. We see this in the case of automatic text entry, or AI-powered price discrimination systems that dynamically set prices based on detailed assessments of users' ability to pay.[19]

Third, when it comes to enforcement, an algorithm can automatically apply a defined penalty without any human intervention, either after the infraction has been committed or preemptively in the case of algorithmic predictions for loan applications, insurance coverage, automated hiring systems, and so on.[20] "Although a human operator might be authorized

15 Alain Supiot, *La gouvernance par les nombres: Cours au Collège de France, 2012–2014*, Paris: Fayard, 2015.

16 Rabih Jamil, "On the Road with Uber Drivers: Navigating Through Algorithmic Management of Labour, Panoptic Surveillance and the Generalized State of Precarity," PhD diss., Université de Montréal, 2024.

17 Karen Yeung, "Algorithmic Regulation: A Critical Interrogation," *Regulation & Governance* 12, no. 4, 2018, pp. 505–23.

18 Yisheng Lv et al., "Traffic Flow Prediction with Big Data: A Deep Learning Approach," *IEEE Transactions on Intelligent Transportation Systems*, no. 16, 2015, pp. 865–73.

19 Akiva A. Miller, "What Do We Worry About When We Worry About Price Discrimination?: The Law and Ethics of Using Personal Information for Pricing," *Journal of Technology Law & Policy*, no. 19, 2014, pp. 41–95.

20 Many examples of this type appear in Cathy O'Neil's *Weapons of Math Destruction: How Big Data Increases Inequality and Threatens Democracy*, New York: Crown, 2016.

Table 9.2 A taxonomy of algorithmic regulation systems

Standard setting	*Monitoring*	*Enforcement*	*Description*
Fixed	Real-time reactive violation detection	Automated	Simple real-time sanction administration systems for administering penalties in real time
Fixed	Real-time reactive violation detection	Recommender system	Simple real-time warning systems
Fixed	Preemptive violation prediction	Automated	Simple preemptive sanction administration systems
Fixed	Preemptive violation prediction	Recommender system	Simple predictive recommender system
Adaptive	Real-time reactive violation detection	Automated	Complex sanction administration systems
Adaptive	Real-time reactive violation detection	Recommender system	Complex real-time prioritization systems
Adaptive	Preemptive violation violations	Automated	Complex predictive sanctioning systems
Adaptive	Preemptive violation prediction	Recommender system	Complex predictive recommender systems

Source: Karen Yeung, "Algorithmic Regulation," p. 508

to review and override the automated decision at a later stage, automation is relied upon to make and implement a decision that has real, consequential effects for the individual."[21] Finally, algorithms can also act as *recommender systems* designed to assist, direct, or guide human decision-making, with humans making the final decision based on these suggestions, as in the case of ratings on Amazon or judicial decisions on prisoners' parole applications led by automated recidivism-risk calculations. The combination of these different variables makes it possible to distinguish between at least eight algorithmic regulation systems (see Table 9.2).

Table 9.2 shows the diversity of algorithmic systems that can exert some form of power or control over human behavior. It also allows us to provide a more precise and granular analysis of algorithmic power, going beyond Zuboff's metaphor of the Big Other, which may seem rudimentary or overly general. Yeung notes that these multiple variants of algorithmic regulation can be used by both state and nonstate actors, and by both private and public organizations. This explains their widespread use in society. Algorithmic regulation can be used in a variety of contexts

21 Karen Yeung, "Algorithmic Regulation," p. 508.

and linked to other models of governance, such as outcome-based regulation, data-driven performance management, risk assessment, and actuarial justice.

These various regulatory systems, considered in isolation, do not in themselves constitute a new form of social relation or structure of domination. Yet several contributions to the literature converge with Yeung's analysis in arguing that algorithmic regulation or "algocracy" represents a new form of coordination of social activities that is distinct from both the market and bureaucracy, although it can be combined with them.[22] Furthermore, the proliferation of algorithmic regulation mechanisms in various social spheres and the control of these systems by a handful of actors capable of producing and perfecting them promotes an accumulation of algorithmic power: a concentration of the ability to influence human behavior through technologies that did not exist as recently as twenty years ago. As Maximilian Kasy remarks, AI systems reflect the priorities of those who own the "means of prediction."[23] Yeung refers to this as a "hypernudge," meaning a new, expanded mode of regulation based on Big Data and algorithms.[24] Cumulatively and globally, algorithmic power is extending its tentacles and gradually forming a new social relation that is consolidating over time. This gives rise to a "new way of the world," an all-encompassing rationality guiding behavior in the era of algorithmic capital. Here we enter into the realm of the "means of behavioral modification" described by Zuboff, which take the form of a "smartification of everyday life" through personal assistants and connected objects. "Algorithmic regulation" is that mode of governance in which "so much of our everyday behaviour is already captured, analysed and nudged."[25] The "invisible hand of the market" is giving way to the "automated invisible hand" and "automated behaviour management." A series of devices in our phones, homes, cars, televisions, toothbrushes, and mattresses already make it possible to encourage, suggest, prescribe, and even compel certain behaviors. In Thesis 10, we will also look at the ramifications of algorithmic regulation as applied to state power and public administration.

---

22 Shiv Issar and Aneesh Aneesh, "What Is Algorithmic Governance?," *Sociology Compass* 16, no. 1, 2022; Arthur Grimonpont, *Algocratie: Vivre libre à l'heure des algorithmes*, Arles: Actes Sud, 2022.

23 Maximilian Kasy, *The Means of Prediction, How AI Really Works (and Who Benefits)*, Chicago: University of Chicago Press, 2025.

24 Karen Yeung, "'Hypernudge': Big Data as a Mode of Regulation by Design," *Information, Communication & Society* 20, no. 1, 2017, pp. 118–36.

25 Evgeny Morozov, "The Rise of Data and the Death of Politics," *The Guardian*, July 20, 2014.

A new form of governmentality

In their Foucauldian analysis of this new power, Antoinette Rouvroy and Thomas Berns conceptualize it as algorithmic governmentality. Algorithms are used as a new form of regulating or governing populations, aiming to "make use of men and things" by managing individuals' behavior and relationships. Algorithmic governmentality is distinct from neoliberal rationality in two ways. First, it does not act directly by influencing the consciousness or subjectivity of actors, but operates more in the mode of affectation: "The docility it produces is not the effect of injunctions to productivity and/or enjoyment, but of affectation—in the mode of reflex rather than reflection—of individual and collective behaviors."[26] Thus, instead of relying on a process of subjectivation that leads individuals to think about themselves as businesses, algorithmic governmentality "bypasses and avoids reflective human subjects. It feeds on infra-individual data that is insignificant in itself, in order to shape supra-individual behavior patterns or profiles without ever appealing to the subject."[27] This form of power is thus embodied in the form of behavioral automation. While neoliberal governmentality assumes the existence of a rational subject, algorithmic governmentality focuses on the management of drives and desires. It thus produces an impulse-driven digital subject profoundly different from the neoliberal *homo oeconomicus*.[28]

But the relationship between neoliberal and algorithmic rationality is in some instances more complex and bidirectional. This is well illustrated, for example, by the Quantified Self movement. In this encounter between the two rationalities, data extraction and the computing power of algorithms are put in the service of neoliberal subjectivation, and vice versa. The call for the "quantified self," launched in 2007 by two editors of *Wired* magazine, has since then spread to more than a hundred cities around the world. Its aim is to promote "self-knowledge" through numbers, personal data, and self-tracking (self-monitoring of physical

26 Sadin, *L'intelligence artificielle ou l'enjeu du siècle*.

27 Antoinette Rouvroy and Thomas Berns, "Gouvernementalité algorithmique et perspectives d'émancipation: Le disparate comme condition d'individuation par la relation?," *Réseaux* 177, no. 1, 2013, pp. 163–96. See also the discussion in Antoinette Rouvroy and Bernard Stiegler, "Le régime de vérité numérique, de la gouvernementalité algorithmique au nouvel État de droit," *Socio*, no. 4, 2015, pp. 113–40.

28 See also Laurence Barry, "The Rationality of Digital Governmentality," *Journal for Cultural Research* 23, no. 4, October 2019, pp. 365–80. In Theses 16 and 17, we offer theoretical avenues into understanding algorithmic subjectivation.

activity, sleep, diet, mood, and so on), in order to be healthier and more productive.[29] The quantification of everyday life is made possible by the rapid democratization of the tools able to capture, analyze, measure, and share personal data. Self-quantification is combined with an injunction to constantly self-improve. Dardot and Laval note that neoliberal rationality had already made use of quantification and performance indicators to manage organizations and guide people's lives. In algorithmic capitalism, individuals continue to live and work in a world based on competition, and they find new devices to help them measure their performance and become more efficient. The injunction to become an "entrepreneur of the self," so dear to the neoliberal ideal, does not mainly consist in starting a business, but in continually *working on oneself*, in measuring one's personal efficiency the better to increase that efficiency.

If it is true that the algorithmic governmentality generated by machines acts on individuals "from the outside" in an automated and impersonal way, algorithmic subjectivation also consists of a certain relationship to oneself, one that is mediated by the quantification of performance in different spheres of activity. From this point of view, algorithmic governmentality serves to exacerbate the earlier neoliberal rationality. The quantification of the self fuels the datafication of everyday life and the training of algorithms, which in turn reinforce the automated management of behavior: "In this sense, practices of self-quantification directly and actively contribute to the digitization of the world [and] previously untouched sectors of activity: health, emotions, travel, etc."[30] Here we can clearly see the possibilities for a complex intertwining between the two types of power, even though they are each distinct.

But there is a second way in which algorithmic governmentality differs from neoliberal logic. While the latter stems from an extension of economic rationality (*homo oeconomicus*), from market logic and the form taken by a normativity internalized by individuals, algorithmic governmentality is instead based on an "objectivity" that seems to be perfectly in tune with reality. The automated processing of information, coupled with the use of this "objective knowledge" to guide behavior, confers an almighty power on algorithms that seem to dictate rules of conduct based on an impartial and exhaustive analysis of reality. Here,

29 Emily Singer, "The Measured Life," *MIT Technology Review*, June 21, 2011.

30 Maxime Ouellet et al., "Big Data et quantification de soi: La gouvernementalité algorithmique dans le monde numériquement administré," *Canadian Journal of Communication* 40, no. 4, November 2015.

"norms seem to emerge directly from reality itself."[31] Multiple algorithms are thus endowed with "injunctive power," which can be based on an incentive structure (apps for sports coaching, healthy eating, responsible consumption), but also a prescriptive one: automated medical diagnosis, loan approvals, job interviews. In each case, the dynamic, horizon, and main threat posed by algorithmic governmentality appear to be the gradual relegation of human judgment in favor of fully automated decision-making.

In short, in algorithmic capital understood as a power relation, behavior is no longer ultimately guided by the market, competition, or the entrepreneurial ethos, but rather by automated decision-making processes and algorithmic governmentality enacted within companies and public institutions, and within everyday life as well. Algorithmic governmentality encompasses and supplants neoliberal rationality, which does not suddenly disappear, but rather becomes integrated into the circuit of algorithmic capital accumulation. Neoliberal subjectivity becomes a moment in the datafication of the world, which is accelerating along with the other dimensions of algorithmic capitalism that we will soon discuss. Finally, the term "algorithmic capitalism" draws attention to the form of power, regulation, dominant rationality, and subjectivity within this regime of accumulation. In this new phase of capitalism, it is not primarily an industrial or financial logic that guides economic activities and social practices, but instead an algorithmic logic. We will now see how this logic is spreading throughout the state, the public administration, and the police.

---

31 Rouvroy and Berns, "Gouvernementalité algorithmique et perspectives d'émancipation."

Thesis 10

Algorithmic logic extends to the state, the public administration, and the police

> *"Our authorities as far as I know, and I only know the lowest grades, don't go out looking for guilt among the public; it's the guilt that draws them out, like it says in the law, and they have to send us police officers out. That's the law. Where d'you think there'd be any mistake there?"*
>
> —Franz Kafka

The rapid spread of algorithmic logic now encompasses state power and the public administration of society. We have already discussed how an algorithmic governmentality guides human behavior—and this is not limited to the economic sphere, but rather extends to politics, reorganizing public governance by developing new possibilities for managing society. We are seeing a new algorithmic logic take root as the successor to the New Public Management proper to the neoliberal era. The neoliberal logic of public administration favored privatization, competition between agencies and organizations for funding, the use of (quasi-) market mechanisms, and the implementation of business management methods within the state apparatus. Algorithmic public administration is based on the efficiency goals seen under the previous model but also emphasizes the use of algorithms and AI to automate decision-making, program administration, and the management of social complexity.

The algorithmization of state administration went through an acceleration phase in the US during the rather grotesque DOGE moment in 2025.[1] Wearing a dark MAGA hat, Elon Musk and his team barged into

1 The Department of Government Efficiency, an initiative of the second Trump

the data gold mine that is the US federal government and deployed AI systems in a number of agencies. The process of integration of algorithmic systems in the administrative state has, however, been underway for more than a decade, in numerous countries. In this thesis, we will first examine this logic of public administration and some of its effects. We will then address the transformation of policing practices toward a model of "algorithmic policing" (also known as predictive policing or Big Data policing) and the trends at work in the algorithmization of prison systems.[2]

Algorithmic power and public administration

In Thesis 9, we discussed the new modalities of algorithmic power and the emergence of a new power dynamic which is "pitting those who produce, control and manipulate algorithmic power against those who are subjected to it." As the state adopts algorithmic decision-making systems, a new algorithmic paradigm for the exercise of state power is emerging. Within this paradigm, power is exercised less through the traditional tools of regulation, legislation, and incentives and more through algorithms that centralize information and then coordinate actors in decentralized fashion, using a mode of regulation through information. Algorithmic public governance is linked to the refinement of social control technologies (surveillance, nudges, recommendation systems, feedback loops) and thus aims to guide behavior and administer society not only by establishing policies based on data input, but also by using information as a policy tool, that is, by configuring "choice architectures," information environments that encourage various actors in society to act in certain ways rather than others.[3] Even alongside the process of integrating algorithms into decision-making mechanisms, various new and complex algorithmic regulation systems are thus being put into operation (see Table 9.2 in the previous chapter).

A growing number of areas of government activity are being entrusted to algorithmic systems, whether in the administration of criminal justice,

---

administration, was a cross-departmental organization under the Executive Office of the President (EOP), which took over the US Digital Service in January 2025.

2 The authors would like to thank Jérémi Léveillé for his help with the research for this thesis.

3 Yeung, "Hypernudge: Big Data as a Mode of Regulation by Design."

Table 10.1 Applications of algorithmic technologies in the state sphere

Field	*Examples of algorithms being used*
Child protection	Calculation of current or future risk of harm to a child
Social programs	Access to government benefits Detecting fraudulent applications Calculating benefit amounts
Access to housing	Determining eligibility for subsidized housing Prioritizing applications
Education	Predicting the risk of violence by a pupil Admissions Teaching evaluations
Police	Surveillance Predicting the risk of a crime being committed, where, and by whom Systems management
Criminal justice	Assistance in drafting presentence reports Calculation of sentences, bail amounts, risk of reoffending, required conditions of imprisonment, access to parole and its conditions
Immigration	Recommendations on eligibility or immigration status Prioritization of applications Refugee camp management and monitoring
Healthcare	Use of algorithmic technologies in medical practice Systems management Care access
Military	Military AI/militarization of AI (see Thesis 12)
Prison system	Datafication of prisons and prisoners Facial recognition of prisoners 24/7 prisoner surveillance Optimization of prison administration Phantomization of prison-guard work

Source: Adapted from LCO Forum on AI in Ontario's Criminal Justice System, "Automated Decision-Making in the Criminal Justice System, Background Information," March 2019, p. 8

education, transport, healthcare, or social programs.[4] The Canadian federal government, for instance, uses AI in more than 300 projects and initiatives across its agencies.[5] The automation of decisions affects an impressive number of legal decisions and administrative processes

4 Pascal König, "Dissecting the Algorithmic Leviathan: On the Socio-Political Anatomy of Algorithmic Governance," *Philosophy and Technology*, no. 33, 2020, pp. 467–85; Karen Levy, Kyla E. Chasalow, and Sarah Riley, "Algorithms and Decision-Making in the Public Sector," *Annual Review of Law and Social Science* 17, no. 1, October 2021, pp. 309–34.

5 Anja Karadeglija, "Federal Government Use of AI in Hundreds of Initiatives Revealed by New Research Database," CBC News, April 11, 2024.

within the state. Figure 10.1 presents some examples of the use of algorithms in the state apparatus.

The integration of automated decision-making mechanisms creates new challenges for public sector decision-makers and state employees related to the problem of human–AI interaction. Indeed, the introduction of algorithms creates a new human–machine symbiosis in decision-making processes. Public employees who experience these new technologies will react in different ways to these new interactions. Several such scenarios, which may range from over to underconfidence in algorithms, raise different degrees of "algorithmic bias," and public employees or decision-makers will show a "selective adherence" to the algorithmically generated advice.[6] The normative framework for decision-making in the public sector is being disrupted, often in the absence of standardization mechanisms, a common framework for assessing the impacts of AI use in the public sector, or a consensus model of "best practice."[7]

Algorithmic public governance and the automation of inequalities

The introduction of algorithmic technologies into the administration of social programs is a particularly interesting case, one that has raised several concerns in the specialist literature. Recent years have seen a rapid increase in the use of databases and algorithms in the management of public services, especially in the United States, but also in Australia, China, Canada, and across Europe. Massive investments and public tenders to automate delivery systems and administrative and decision-making processes in these areas of government have been undertaken largely in the name of efficiency and saving resources. This, after all, is taking place in the context of over a decade of austerity policies that have severely undermined public institutions' ability to do their jobs properly. Today, predictive algorithms, risk models, and automated eligibility and benefit calculation systems govern access to social services and benefits for millions of people.

6 Saar Alon-Barkat and Madalina Busuioc, "Human–AI Interactions in Public Sector Decision Making: 'Automation Bias' and 'Selective Adherence' to Algorithmic Advice," *Journal of Public Administration Research and Theory* 33, no. 1, January 2023, pp. 153–69.

7 Maciej Kuziemski and Gianluca Misuraca, "AI Governance in the Public Sector: Three Tales from the Frontiers of Automated Decision-Making in Democratic Settings," *Telecommunications Policy* 44, no. 6, July 2020: 101976.

In her book *Automating Inequality*, Virginia Eubanks meticulously documents, based on in-depth field surveys, how the delegation of decision-making processes to algorithms has already become firmly established in US society. She shows how this algorithmic decision-making is used to control, monitor, regulate, and discriminate against poor and marginalized populations.[8] These algorithmic technologies control the provision of services and access to resources. For example, they determine which children will be taken into care by youth protection services (based on an algorithmically calculated risk score) and decide whether to continue or discontinue coverage for a medication that is essential to the survival of a chronically ill patient. They form an automated and extremely opaque technological apparatus for managing poverty and social programs. In this administrative use, algorithms classify and decide, and it is very difficult to establish.

The use of algorithmic technologies to administer social programs produces three main types of effects: exclusion, inclusion, and prediction. According to Eubanks, these consequences mainly impact marginalized populations. We can speak of exclusion mechanisms when algorithms are used to determine eligibility for programs and benefits, as well as the amounts or services provided. The exclusion effect stems from changes in eligibility determined by an algorithm, or from erroneous exclusion, such as when an algorithm flags a potential case of fraud—which may not really be one—and requires the recipient to prove once more that they are eligible for some program or service. Complex interactions with an automated bureaucracy, in which it becomes almost impossible to speak to a human being, can discourage people from taking the necessary steps to obtain a service or benefit, thereby heightening the mechanisms of exclusion. These exclusionary effects can have the most dangerous consequences. For evidence of this, we need only look to those cases where seriously ill people have had their drug coverage or access to food stamps withdrawn because of a missing signature on a form or a data entry error interpreted by the algorithm as a "lack of cooperation," punishable by a complete denial of service.[9] While exclusion may operate at the individual level, the systemic exclusion of a population group through arbitrary algorithmic decisions is also a significant ethical concern.[10]

8 Virginia Eubanks, *Automating Inequality: How High-Tech Tools Profile, Police, and Punish the Poor*, New York: Picador, 2019.

9 Ibid., pp. 41–3, 55–7.

10 Kathleen Creel and Deborah Hellman, "The Algorithmic Leviathan:

Algorithmic public governance thus raises grave problems of accountability and legitimacy. Once these algorithmic systems are in place, human oversight is limited, because the employees of a ministry or agency are not high-level coders, and the functioning of the algorithms quickly exceeds anyone's ability to understand their calculations and outputs. The algorithms used are so complex that it becomes difficult to explain their outputs to the concerned parties or to the general public—and their operations are also often protected by commercial secrecy, which makes it even more difficult to discover how they work. In this sense, the opacity of algorithms creates a lack of accountability that affects the legitimacy of the government institutions that use them.[11] This is the problem of "explicability," widely discussed in scholarly literature, here applied to the sphere of state decision-making.[12]

The experiment with algorithmic public governance in Indiana was a dismal fiasco, as Eubanks recounts. The state government had hired IBM to integrate algorithmic technologies into the administration of several public programs and services. This resulted in a series of erroneous benefit denials and serious management problems that even IBM was unable to remedy, due in part to the opacity of the algorithms. Within a few years, the same Republican governor who had campaigned to introduce this system—all in the name of making the Indiana government "more efficient"—had to resort to legal action against IBM for breach of contract. The State of Indiana claimed $437 million from IBM, which responded with a counterclaim of $100 million. IBM won the legal battle and Indiana ultimately had to pay it $52 million. The judge ruling on the case could only conclude that "both parties [were] to blame" in this "'perfect storm' of misguided government policy and overzealous corporate ambition."[13]

Inclusion mechanisms serve the purposes of surveillance and have criminalizing effects of their own. The inclusion effect is not least perverse when it results from the extensive collection of data on marginalized populations. Take the case of homeless people targeted by an algorithm meant to connect them with social housing provision appropriate to their situation. These same databases are accessible to police

Arbitrariness, Fairness, and Opportunity in Algorithmic Decision-Making Systems," *Canadian Journal of Philosophy* 52, no. 1, January 2022, pp. 9–13.

11 Kate Vredenburgh, "The Right to Explanation," *Journal of Political Philosophy* 30, no. 2, 2022, p. 210.

12 The terms "explicability" and "explainability" are used interchangeably in the scientific literature.

13 Eubanks, *Automating Inequality*, p. 72.

forces, which use them to monitor, control, and target marginalized individuals and populations. We will return to this problem below.

Risk-prediction mechanisms are used to control behavior. This involves the use of statistical models to provide a predictive risk index to the employees of the various social services. For instance, in child protection, such indexes will influence these employees' decisions on whether to investigate reported cases of abuse. These algorithms, which attribute a predictive score, are also widely used in the justice system. Algorithms are now used to calculate the risk of reoffending, sentences to be served, and bail amounts. Judges no longer do more than determine whether an accused person is guilty; the rest of the justice process is increasingly automated, with AI assistance. This raises issues of access to justice, equality, and respect for procedure. In just a few years after their first use, many of those who had wanted algorithmic decision-making processes to be introduced into the justice system were calling for their use to be suspended.[14]

Ultimately, these mechanisms of exclusion, inclusion, and prediction ensure that disadvantaged people become the targets of algorithmic technologies. The result is that they face new risks, risks that may even be fatal. The automation of eligibility systems can discourage many people from using the public services that they need, while predictive models can also disqualify them from obtaining loans, identify them as parents in need of surveillance, or label them as potential criminals. Their personal information often ends up in low-security databases with no privacy protection, where the government, police, and commercial interests can monitor them. Since poor or vulnerable people generally interact more with public services and social programs, they are also the ones who are subject to this data collection, which adds to their marginalization when this same data is used for surveillance. This creates a constantly renewed cycle of surveillance and marginalization, tantamount to "automating inequality."[15]

Algorithmic policing

In his short story *The Minority Report* (1956), the great sci-fi author Philip K. Dick portrays three rather unusual humans. Gifted with astonishing precognitive abilities, they can predict the future up to two weeks

14 Law Commission of Ontario, "The Rise and Fall of AI and Algorithms in American Criminal Justice: Lessons for Canada," Toronto, 2020, p. 4.

15 Eubanks, *Automating Inequality*, pp. 7, 11.

in advance. Chained by the police to machines that transcribe their predictions, these "precogs" are used by the "Precrime" unit to predict murders and prevent them by preemptively arresting future murderers. This premise was still the stuff of science fiction when Steven Spielberg made it into a successful feature film in 2002. But the reality now far exceeds fiction, with the exception that machines no longer need *human* precogs and instead use algorithmic technologies to predict future crimes. Welcome to the era of algorithmic policing.

Nowhere in the realm of state power have algorithms made such a breakthrough as they have in policing. What is now known in the United States as Big Data policing, and what we refer to here in Canada as "algorithmic policing," is now well-established in most major US cities.[16] This concept refers to the predominant role now played by Big Data and algorithms in police work. Los Angeles, Chicago, and New York have been at the forefront of this revolution in the exercise of police power, where algorithmic capital, its technologies, and forms of power are transforming policing institutions.[17] At the federal level as well, the US Immigration and Customs Enforcement (ICE), whose violent tactics toward targeted racialized and immigrant populations have triggered acts of resistance and demonstrations across the US, uses AI in its operations in various ways, including facial recognition and data analysis.[18]

Today, the model of algorithmic policing is being rapidly, if not systematically adopted, in several regions and countries around the world: The city of Amsterdam has implemented the Crime Anticipation System (CAS), and the Krimpro system is in use in Berlin. Similar systems are used in Norway, Switzerland, the United Kingdom, China, and elsewhere.[19] In Canada, although it is difficult to get information from police

16 Andrew G. Ferguson, *The Rise of Big Data Policing: Surveillance, Race, and the Future of Law Enforcement*, New York: New York University Press, 2020.

17 On the algorithmization of the Los Angeles Police Department, see Sarah Brayne's remarkable book *Predict and Surveil: Data, Discretion, and the Future of Policing*, Oxford: Oxford University Press, 2021. It should be noted that there are still huge difficulties in obtaining information on police practices. Even in the case of the LAPD, some claim that algorithms such as Predpol have already been abandoned. These difficulties in accessing information should be kept in mind when reviewing this literature.

18 Department of Homeland Security, *United States Immigration and Customs Enforcement—AI Use Cases*, dhs.gov (accessed November 11, 2025).

19 Wim Hardyns and Anneleen Rummens, "Predictive Policing as a New Tool for Law Enforcement?: Recent Developments and Challenges," *European Journal on Criminal Policy and Research* 24, no. 3, September 2018, pp. 201–18; Lukas Lorenz, Albert Meijer, and Tino Schuppan, "The *Algocracy* as a New Ideal Type for Government Organizations: Predictive Policing in Berlin as an Empirical Case," *Information Polity* 26, no. 1, 2021, pp. 71–86; Mareile Kaufmann, Simon Egbert, and Matthias Leese, "Predictive Policing and

on the implementation of algorithmic policing, the most comprehensive study thus far does reveal a creeping algorithmization.[20] The Vancouver and Saskatoon police forces have confirmed that they use algorithmic policing technologies, while algorithmic surveillance and prediction tools for the areas and individuals with the highest propensity for crime have been deployed by police forces in Calgary, Toronto, Waterloo, and the Ontario Provincial Police (OPP).[21] It is also worth noting that the Royal Canadian Mounted Police uses spyware in its surveillance operations. We might wonder if the datafication of the Global South—which, as we will see in Thesis 12, is already well underway—will result in an outright globalization of algorithmic policing.

The core of algorithmic policing is so-called "predictive policing," which involves "the collection and analysis of data about previous crimes for identification and statistical prediction of individuals or geospatial areas with an increased probability of criminal activity."[22] These predictions are also used by the police at all levels of intervention planning. This predictive analysis method was initially developed by anthropologist Jeffrey Brantingham and mathematician Andrea Bertozzi. It is an adaptation of the algorithms used to predict aftershocks following an earthquake. While computerized statistical analysis models have been in use by US police forces since the 1990s, the predictive logic that has developed with the rise of algorithmic capital has been of increasing interest to police leaders since the late 2000s. In 2009, Los Angeles police chief Charlie Beck, a pioneer of algorithmic policing, published an article with the revealing title "Predictive Policing: What Can We Learn from Wal-Mart and Amazon About Fighting Crime in a Recession?" Since then, the predictive policing model has continued to make inroads.

Predictive policing consists of four main stages or phases. First, it is part of the datafication of society, as it requires enormous amounts of

---

the Politics of Patterns," *British Journal of Criminology* 59, no. 3, April 2019, pp. 674–92; Ana Brandusescu et al., "Comments on the Toronto Police Services Board Proposed Policy on AI Technologies," Montreal, Montreal Society and Artificial Intelligence Collective (MoSAIC), December 15, 2021; Daniel Sprick, "Predictive Policing in China: An Authoritarian Dream of Public Security," *Navein Reet: Nordic Journal of Law and Social Research*, no. 9, 2019, pp. 299–324.

20 Kate Robertson, Cynthia Khoo, and Yolande Song, *To Surveil and Predict: A Human Rights Analysis of Predictive Policing in Canada*, Citizen Lab Toronto, University of Toronto, 2020, citizenlab.ca.

21 Mélanie Marquis, "La GRC armée de logiciels espions," *La Presse*, July 6, 2022.

22 Albert Meijer and Martijn Wessels, "Predictive Policing: Review of Benefits and Drawbacks," *International Journal of Public Administration* 42, no. 12, September 2019, p. 1037.

socioeconomic, demographic, geospatial, and crime data. The data that feeds into the police algorithms comes from three main sources. Governments and public authorities are the first source, and it is important to remember that it is not only corporations that extract data; public authorities and different levels of government also have large databases on individuals' education, health, property, civil status, criminal history, and other categories. Many police organizations now share their access to these databases. Second, police officers themselves are collecting ever more data in the course of their duties, whether using connected devices, or by manually entering data during patrols and investigations. Police forces also draw on data from private companies. Like any other customer in the data market, the police can access the data extracted by applications, connected devices, and other elements of the algorithmic capital extraction architecture. For example, Amazon has developed its Ring camera technology: Private surveillance cameras connect to the Neighbors app and allow users to become freelancers for the wider apparatus of surveillance, reporting suspicious activity captured by neighborhood cameras. Amazon offers discounts to police forces for its Ring cameras, as well as access to a portal that aggregates feeds from all cameras installed in the area.[23] Police forces are also routinely customers of data brokers. With a warrant, the police can also obtain data from digital companies, large and small.

In this second phase, algorithmic surveillance and prediction technologies developed by private companies specializing in surveillance and high technology are leased or purchased by police forces to process and analyze this vast quantity of data. In Thesis 12, we will discuss the deep connections between certain Big Tech companies, intelligence services, and the US defense sector. The rise of predictive policing is supported by numerous "police–private sector partnerships" in which police forces and other public authorities enter into contracts with high-tech companies to develop surveillance and prediction tools. This is one salient aspect of a major convergence between the private sector and the state in establishing a large-scale surveillance regime. Various levels of government and police forces are outsourcing functions and hiring private security and surveillance companies simply to track citizens. In this vast market, established companies such as Microsoft, Hitachi, Motorola, IBM, and Google are competing with start-ups for lucrative contracts.

23 Crawford, *Atlas of AI*.

As a result, a commercial/state surveillance complex has now taken root in workplaces, in homes, on the streets, in public spaces, and in the social world in general. This is all feeding into the development of algorithms by specialized companies that supply crime prediction models to police. Several technologies once developed for military applications or for espionage abroad are now being deployed against citizens, spinning off from this vast outsourcing process. Palantir offers a striking example: Cofounded by Peter Thiel in 2003, this specialized firm develops surveillance applications, having started out with its Gotham system, which was used by US intelligence services and the Department of Defense in counter-terrorism operations in the 2000s. Palantir, which has since diversified its clientele, sells its services for tracking undocumented immigrants to ICE and its facial recognition services to the LAPD. The technology used by Walmart to monitor its employees also comes from Palantir, and the Palantir Foundry automated decision-making system counts Morgan Stanley, the British government, and Ferrari among its users.

In the third phase, the predictive policing algorithmic system, which feeds on all this data, produces outputs in the form of crime predictions. With reference to either locations (neighborhoods, street corners, vacant lots, buildings), times (of the day, or days of the week), or people (individuals or groups defined as at risk of committing a crime), predictive policing uses algorithms that share the same logic we analyzed in the previous theses. These algorithms, too, are predicting behavior—but in this case, potentially criminal behavior. When it concerns individuals or groups, the prediction often takes the form of lists. Much as an Amazon algorithm, for example, can assess based on a customer's purchase history whether they are pregnant, predictive policing can predict who will be the potential next victim of a violent crime, who is trafficking drugs, and which criminal network is most likely to hit a community.[24] Depending on their gender, age, family history, friends and acquaintances, previous connections to violent events, and other data points, a person may potentially find themselves on the "heat list" produced by a Chicago police algorithm that calculates who is the likely victim or perpetrator of a shooting. More than 1,400 young men—the vast majority of them racialized and disadvantaged—have been placed on this list without their knowledge, some of them visited by police officers to inform them that they are being actively surveilled, without them having any record of

24 Ferguson, *The Rise of Big Data Policing*, pp. 3, 19.

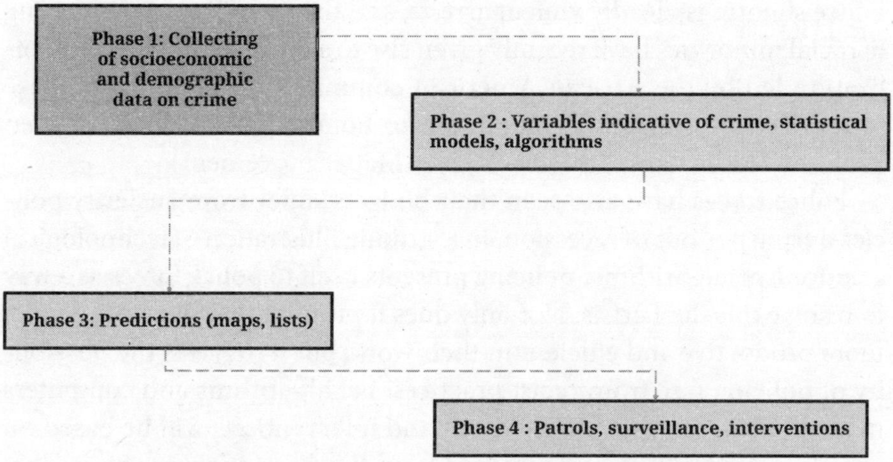

Figure 10.1 The four phases of predictive policing

Source: Inspired by and adapted from Hardyns and Rummens, "Predictive Policing as a New Tool for Law Enforcement?," p. 205

violent crimes.[25] Prediction can also take the form of a geographical map when it uses field risk modeling. The prediction thus focuses on "red square boxes," areas with a high probability of crime.[26]

Finally, in the fourth phase, personnel are deployed and a police operation (a patrol, or an intervention) is launched in order to preempt some future crime. In the practical context of police patrol work, for example, at the start of their shift officers might be issued a digital map of the day's crime predictions. Patrol officers will then obviously focus their attention on these areas, the famous red square boxes, with a view to prevention and deterrence. Spyware is also used to track the phones of people said to be "at risk" or under investigation. This means a profound transformation in the policing model: The police who once operated largely *reactively* now intervene *proactively*.[27]

The advent of algorithmic policing in the US stands at the intersection of two social dynamics. On the one hand, the establishment of the extractive infrastructure of algorithmic capital installs a surveillance apparatus and creates a new abundance of data that offers new possibilities for handling crime. On the other hand, American police forces have faced a prolonged crisis of legitimacy stretching back to the 1990s. Corruption scandals and, above all, evidence of (especially racial) discrimination in the exercise of police power have now been building up for many years.

25 Ibid., p. 37.
26 Jackie Wang, *Carceral Capitalism*, South Pasadena, CA: Semiotext(e), p. 241.
27 Ferguson, *The Rise of Big Data Policing*, p. 58.

Police shootings, deadly violent arrests, and the patrolling and targeting of racial minorities have recently given rise to a far-reaching social mobilization led by the African American community, which resists these forms of oppression and demands top-to-bottom police reform, or even abolition (for instance, the Black Lives Matter movement).

Police forces have also seen their budgets suffer from austerity policies during periods of recession. In a striking illustration of technological solutionism, algorithmic policing presents itself to police forces as a way to resolve this dual crisis. Not only does it promise to make police units more productive and efficient in their work, but it suggests the possibility of policing free from racist practices: Let algorithms and computers manage crime, so that our methods and interventions will be based on objective, neutral facts, far removed from the discriminatory biases that some officers may have. By substituting machine learning and AI for human bias, a now-algorithmic policing can aspire to overcome its crisis of legitimacy.

However, it seems that, despite this change, algorithmic policing actually perpetuates historical discriminatory practices within a new paradigm. Although its proponents seem to believe that a new era has arrived, the temporality inherent in predictive policing strongly encourages the reproduction of past crime patterns and models in the future. It is quite typical for new technological developments to be seen as forces for change. When it comes to the algorithmic technologies used in predictive policing, however, we are much more likely to see the same results recurring without end. Indeed, the model's epistemology comes straight out of David Hume's treatise on inductive reasoning: It is based entirely on the belief that the future will resemble the past. Since the predictions are a reflection of past data, this is necessarily the case. A red square box is red, among other things, because many crimes have already been recorded in this area. The simple act of predicting that this will happen again leads to increased surveillance, which increases the chances of a crime being observed in that area and decreases surveillance outside the red boxes. Choosing to deploy an anti–drug trafficking unit near a street corner in a disadvantaged neighborhood—that is, instead of in the corridors of a reputable law firm, in a Harvard dormitory, or at a party at Elon Musk's house—will inevitably increase the likelihood of apprehending the perpetrators of a transaction taking place in that poor neighborhood where racialized communities live, while lowering the chance that the perpetrators of a comparable transaction will be arrested in a space typically occupied by wealthy whites.

The same logic applies to the management of individuals already on a list. They are there because of sociodemographic data inputs that faithfully reflect the racist biases of the criminal justice system through its history. Being or having been monitored by the police is one of the factors that increases an individual's risk score. Thus, the police make someone more suspect, simply because they are suspicious of them. And, of course, when someone is subjected to greater surveillance, the probability that they will be apprehended increases.[28] As Sandra G. Mayson summarizes in an influential article, algorithms are thus just a new version of an old problem concerning prediction itself:

> The deep problem is the nature of prediction itself. All prediction looks to the past to make guesses about future events. In a racially stratified world, any method of prediction will project the inequalities of the past into the future. This is as true of the subjective prediction that has long pervaded criminal justice as it is of the algorithmic tools now replacing it … Algorithms, in short, shed new light on an old problem.[29]

The problem of self-fulfilling prophecy is thus inherent in the epistemological horizon of predictive policing. Seeing it in this light, we understand that this logic does not have to do with crime reduction, but rather with the prediction and reproduction of crime: "Crime prediction algorithms should more accurately be called crime *production* algorithms," as Ruha Benjamin aptly puts it.[30] And *predicting* does not mean *reducing*. Ultimately, it should come as no surprise that a thorough review of studies on the subject concludes that it is impossible at this stage to say whether predictive policing actually reduces crime.[31] Chicago's heat list may well predict the victims and perpetrators of violent crimes, but the fact remains that violence in that city is on the rise.[32] It is also interesting to note that, despite the rollout of predictive policing in the United States, which is supposed to predict and reduce violent crime, the country is still grappling with a scourge of mass shootings that shows no sign of abating, occurring at an alarming rate of almost 2 (1.92) per

28 Ibid., p. 103.

29 Sandra G. Mayson, "Bias In, Bias Out," *Yale Law Journal* 128, no. 8, 2019, pp. 2218–300.

30 Ruha Benjamin, *Race After Technology: Abolitionist Tools for the New Jim Code*, Cambridge: Polity, 2019, p. 83.

31 Meijer and Wessels, "Predictive Policing," p. 1037; Ferguson, *The Rise of Big Data Policing*, p. 69.

32 Ferguson, *The Rise of Big Data Policing*, p. 39.

day in shopping centers, workplaces, schools, and other locations in the last five years (2021 to 2025).[33] Algorithmic policing predicts, but these predictions remain silent on the real causes of violence and the resources needed to confront it.

Beyond the eternal reproduction of the same results, predictive policing raises a whole series of other problems that are still without solution. The five most important problems are racial discrimination, accountability, individual rights, data errors, and "data gaps."[34] First, and depressingly, algorithmic policing reproduces the racial discrimination it was supposed to eliminate. We have discussed the main reason for this: Although the algorithms do not include data on "race," algorithmic policing relies on data (past arrests, contacts with the police, and so on) that is directly correlated with practices that reflect the racist and discriminatory history of social control, and policing in particular. It thus reproduces discrimination in its predictive models: If police forces have historically targeted a population because of racist bias, this bias is reflected in the data, and the algorithm that uses this data can generate biased predictions. The individuals and groups targeted by the predictions are disproportionately racialized people, and the red squares reflect not so much the history of crime as the history of police interventions, concentrated in disadvantaged neighborhoods where racialized communities live. Algorithmic policing thus reifies many systemic inequalities inherited from traditional policing.[35] The use of algorithmic predictions in policing and the criminal justice system produces a "net burden" for racialized communities.[36] This is unfortunate—and seems to be inherent in the logic of algorithmic policing. After all, the problem of systemic racism lies within the police force itself, not in the populations being policed. Yet algorithmic technologies are deployed by the police to monitor the populations being policed, and not the other way around. *Surveillance of the police* might be a more effective way to understand its biases and confront them.

Second, the same problem concerning accountability and transparency that we mentioned earlier in the discussion on automated decisions by public bodies also applies here. Indeed, neither the police

---

33 See massshootingtracker.site for a database on mass shootings in the United States.

34 Ferguson, *The Rise of Big Data Policing*, pp. 177–85; Meijer and Wessels, "Predictive Policing"; Brayne, *Predict and Surveil.*

35 Ferguson, *The Rise of Big Data Policing*, pp. 47, 132.

36 A. Z. Huq, "Racial Equity in Algorithmic Criminal Justice," *Duke Law Journal* 68, no. 6, 2018; Benjamin, *Race After Technology*, pp. 80–4.

nor anyone else (courts, academics, citizens, journalists) can explain the exact reasons why, for example, a particular young man has ended up on the Chicago Police Department's heat list. Automated decision-making processes remain opaque even to police officers themselves; the commercial secrecy that protects the private algorithms of firms like Palantir and Microsoft only adds to the near impossibility of explaining the reasoning behind a prediction. Given how important the consequences of these predictions are—they are quite literally a matter of life and death—the inability to understand the reasons behind a decision and to hold the decision-maker accountable is extremely problematic from a moral and legal standpoint.

The third major problem concerns individual rights. This is not the first time we have raised the issue of privacy in this book. This problem is particularly acute with the rise of general surveillance enacted by a commercial–state complex and sweeping police access to personal data. There are two further problems related to respect for individual rights. Predictive policing significantly distorts the meaning of reasonable suspicion and the presumption of innocence. For example, a police officer can arrest a "suspect" based on reasonable suspicion, drawing on their experience and an analysis of the circumstances of the situation. This reasonable suspicion must itself be demonstrable. Predictive policing greatly muddles this concept. Can the algorithm's prediction, which is the result of an opaque process, be considered reasonable suspicion when it is impossible to ascertain the "reasons" that guided the algorithm in its prediction? Is being on a list or presence in a red square box sufficient grounds for apprehension or arrest? We might similarly ask what has happened to the presumption of innocence for an individual classified as being at high risk of committing a crime? These concepts, which apply to individual rights, come from a world before Big Data. This world has now gone, and such principles seem to collapse in the world of algorithmic policing.

A fourth problem: errors. According to the principle of "bias in, bias out," an algorithm fed with biased or incorrect data will produce biased or incorrect predictions. This is, of course, a commonplace, since data can be incorrect for several reasons (incorrect inputs, erroneous processing, data produced by mistake, false information, data produced with deceptive intent, and so on). One Chicago police lieutenant can attest to this, as his own son was mistakenly placed on a list of gang members. The California auditor general reported multiple errors in the CalGang list: This count of street gang members included forty-two babies.[37] Of course, erroneous data in the advertising world also has economic consequences, though the effect is negligible: Someone will receive an advertisement for a product they have no interest in, which ultimately means that part of the marketing budget will have been invested in poorly targeted advertisements. In the field of police surveillance and the justice system, however, data errors can lead to enormous, life-changing consequences for individuals, ranging from undue surveillance and unfair arrest to unfounded charges and the wrongful imprisonment of an innocent person. There is no room for error here, yet mistakes are commonplace in the world of Big Data. This discrepancy poses a serious danger to citizens.

The fifth problem is that of data gaps. The algorithmic policing model relies on an abundance of data, and the accuracy of its predictions depends on the representativeness of all individuals and all spaces in the data. However, there are significant disparities: Some people "produce" much less data than others, and some spaces are much less covered by the architecture of data extraction than others. These disparities and the existence of areas and individuals for which there is little or no data create an imbalance in the systems: An algorithm trained on incomplete data produces erroneous outputs.[38] In this sense, social movements aimed at limiting access to personal data, or data extraction in general, or advocating "disconnection," may in fact reduce the effectiveness of predictive policing.

Toward an automated prison system?

The intertwined roles of public authorities and private companies in developing algorithmic technologies can also be seen in prisons and jails. Several prison systems around the world use smart technologies to monitor inmates and administer correctional facilities. Multinational companies such as Johnson Controls, GTL, and Motorola's subsidiary Avigilion offer algorithmic solutions for prison management, thus helping to open up the closed space of prisons to inroads by algorithmic capital.

Some states in Asia and Oceania are exploring the "guardless prison" model. With this "prison automaton,"[39] algorithmic technologies are

37 Ferguson, *The Rise of Big Data Policing*, p. 49.

38 Ibid., p. 182.

39 Carolyn McKay, "The Carceral Automaton: Digital Prisons and Technologies of Detention," *International Journal for Crime, Justice and Social Democracy* 10, no. 4, 2022.

used not only for surveillance processes, but also for the various functions performed by guards. This includes automated video detection of abnormal occurrences, facial recognition, tracking bracelets, and even an automated check-in and check-out service for prisoner admission and release. In Hong Kong, inmates must wear a Fitbit that records their vital signs. Taiwan recently launched a five-year project to integrate smart technologies into the management of its prisons, in partnership with private high-tech companies such as NEC Taiwan and Syscom Group, "that will deliver significant benefits to officers and inmates alike."[40] Researcher Carolyn McKay has documented the embedding of smart technologies into the infrastructure of Australia's prisons. According to McKay, these technologies have increased the tendency to isolate and dehumanize prisoners. Kaun and Stiernstedt agree that while prisons are becoming algorithmized, prisoners remain largely excluded from access to digital technologies, which adds to their isolation from society:

> The prison environment is … characterized by an ambivalent relationship to digital technologies, which could be described as a paradox of a data-intensive environment observing and documenting prisoners constantly and consequently producing large amounts of data combined with often low digital technology access for prisoners themselves.[41]

The guardless prison model is less popular in the United States, where the focus is instead on platformizing prison guards' work and integrating algorithmic technologies to assist with the various tasks related to the management, surveillance, and discipline of prisoners. The company GUARDIAN RFID, for example, offers a series of "warrior technologies," intelligent data processing solutions, smart cameras, automated patrols, and other devices. Its Spartan application is a "weapon of mass data collection" that guards can use to "gain operational dominance," but which also provides managers with a sophisticated disciplinary tool for monitoring guard performance.[42]

Ultimately, the expansive logic of algorithmic capital is already well established in the administrative and police structures of the state. Based on the datafication of society, we can see that the algorithmization of

---

40 Sophie Yang, "Taiwan's First Smart Prison Unveiled," *Taiwan News*, December 4, 2020.

41 Anne Kaun and Fredrik Stiernstedt, "Doing Time the Smart Way?: Temporalities of the Smart Prison," *New Media & Society* 22, no. 9, 2020, pp. 1580–99.

42 Ibid.

state power tends to create symbioses of interest between public organizations and private companies in the surveillance and management of social complexity. Indeed, in the era of algorithmic capital, the line between public and private, and between the state and the market, is being reconfigured. But what about the impact of algorithmic capital on democratic life and public debate? We will address this issue in Thesis 11.

Thesis 11

Algorithmic capital disrupts and erodes democratic life

Over this kind of men stands an immense, protective power which is alone responsible for securing their enjoyment and watching over their fate. That power is absolute, thoughtful of detail, orderly, provident, and gentle.

—Alexis de Tocqueville

Algorithmic power and democratic politics

What specific consequences does algorithmic capitalism have for the world of politics, public debate, and democratic life?[1] As discussed in Thesis 9, it would seem that the organizations best placed to benefit from the new possibilities offered by algorithmic technologies are private companies (particularly digital platforms) and the state with its bureaucracy, police, and army. This does not mean that individuals, families, and community organizations cannot also leverage Big Data and smart machines to their own benefit. Still, their ability to use these resources to exercise or accumulate social power is bound to be much more limited compared to large organizations (both public and private), which already have a head start in terms of the real power they hold. As Jamie Susskind points out, the overall impact of new technologies must therefore be assessed in terms of the forms of power they allow to be deployed on a

1 This thesis contains and adapts elements from chapter 9 of Jonathan Durand Folco, *Réinventer la démocratie: De la participation à l'intelligence collective*, Ottawa: Presses de l'Université d'Ottawa, 2023.

large scale: Force, surveillance, and control of people's perceptions will increasingly be based on algorithmic technologies. The consequence is that power will tend to be concentrated in the hands of those who control these technologies, whether that means Big Tech or the state.[2]

While these forces are each among the frontrunners in accumulating algorithmic power, it is also true that one is gaining significantly more than the other. The relationship between algorithmic capital and the state is based on highly complex relations of dependency and influence, which currently place governments in a relatively weak position compared to the digital giants, especially in countries outside of the US where large US-based companies occupy dominant market positions. All manner of major companies—whether in the fossil fuels, finance, manufacturing, or services—are often able to impose their will on local authorities and governments. They do this in many ways: by exercising moral leadership; through their ability to finance an army of lobbyists; through their economic might (making them an indispensable creator of jobs and sources of growth and tax revenue); or through their ability to forge strategic partnerships with the military, the police, and the public administration.

If we want to consider the role of business strategies bound up with coercion, influence-seeking, and manipulation, the Uber Files scandal revealed July 2022 offers a case study of the enormous power wielded by a company emblematic of algorithmic capital. Its 124,000 pages of leaked documents exposed hundreds of schemes that Uber has used in dozens of countries around the world in recent years.[3] From privileged access to French President Emmanuel Macron, to promoting chaos and violence in order to tarnish the image of anti-Uber protests, to seeking to enfeeble existing bodies of legislation and transform them to its own advantage, we see the firm pursuing an aggressive strategy of conquest. Uber was ready to use all means at its disposal to achieve its goals.

A softer, more conventional strategy is to create a series of incentives that will steer public policy decisions in one direction rather than another. Today's Big Tech companies do use these traditional tools of political, economic, and moral influence, but the immense technological power they enjoy helps to place them even further beyond governments' regulatory reach. The public authorities are not only increasingly

2 Jamie Susskind, *Future Politics: Living Together in a World Transformed by Tech*, Oxford: Oxford University Press, 2018, p. 89.

3 Harry Davies et al., "Uber Broke Laws, Duped Police and Secretly Lobbied Governments, Leak Reveals," *The Guardian*, July 11, 2022.

dependent on the dissemination of these technologies within the state itself in order to exercise power, bolster the military, and carry on their activities. They also rely on these technologies' dissemination across society, whose administration now depends on its mass datafication. The sheer size of the major platforms, and their quasi-monopoly on digital technologies on which the state and public authorities rely, creates a situation where the incentives for public authorities to tightly regulate the high-tech sector to protect the public are systematically surpassed by other considerations. Indeed, this objective is overshadowed by these same public authorities' own interests in taking advantage of the most dynamic possibilities of algorithmic capital. As a result, the state and governments are often active participants in the dissemination of algorithmic technologies within society and limit their regulatory function to establishing a legal and regulatory framework that promotes the growth of algorithmic capital. This prevents them from curbing some of its harmful effects on society and ensuring that large technological firms contribute their fair share to existing tax systems.

Yet this is not only a question of the rise of capitalist platforms' power over states. Tech giants, social media platforms, and other mobile apps also have a major impact on individuals. Shoshana Zuboff has illuminated how the extraction of personal data, invasions of privacy, and algorithmic control of individual behavior are becoming the norm. In Thesis 3, we saw how designers of social media platforms, online video games, and apps use neuropsychological research to create technologies that aim to make us literally addicted to our screens in order to capture as much of our attention and personal data as possible. Yet privacy and individual autonomy are key values in contemporary democracies, and free time and control over one's life are prerequisites for enjoying any real political freedom. Hence, some of the important developments in algorithmic capitalism are also undermining the practical conditions for democratic life at an individual level.

One major political impact of algorithmic technologies relates to the poor democratic quality of the digital public sphere, and of online information and discussion. The vast majority of information and news today passes through the filter of social media, the use of which continues to grow, having almost doubled since 2015. Today, 5.66 billion people are social media users—equivalent to the total world population over age eighteen.[4] Social media is now the main source of news in the US, ahead

4 Christy Tila, "Number of Internet and Social Media Users Worldwide 2025," Statista, November 19, 2025.

of TV and news sites and apps, a development also noticeable in many other countries.[5] This significant trend is becoming more pronounced, especially since the COVID-19 pandemic, during which the influence of social media grew further. For an increasing share of the population, relations with the outside social world are ever-more mediated by these digital interfaces. Instead of reading the newspaper, listening to the radio, participating in a citizens' assembly, or discussing politics in a coffee shop, people are increasingly getting their information from an evanescent stream of news, images, and videos whose algorithmically filtered content—whether sponsored or not—seeks to generate maximum engagement, that is, to draw clicks and attention time.

At first glance, this does not seem much of a problem. After all, the advent of the Internet and social media has greatly democratized access to information and allowed for greater public interactivity, compared to traditional media where the average individual was a passive recipient. Social networks in the 2000s, like the Internet in the 1990s, were initially a great source of hope for the reinvigoration of democracy and increased citizen power in the world of information.[6] However, these democratic hopes based on the abundance and accelerated circulation of information quickly faded, as it became clear that this technology also facilitates the rapid spread of misinformation. The avalanche of fake news, the rapid spread of conspiracy theories, and the challenges of the "post-truth" era—brought to the fore during Donald Trump's first term as US president, among other things—mark a turning point. They raise far-reaching questions about the role of social media in a democracy.

The infodemic and the ambivalent role of social media

How can we explain this reversal, where the vision of an interconnected "global village" and the utopia of a borderless world driven by openness, collaboration, and "the wisdom of crowds" has seemingly given way to concerns about an increasingly disoriented world plagued by aggression, polarization, and all-consuming irrationality? We should say from the outset that a key issue concerns algorithms, or at least a certain configuration of them in capitalist digital platforms. Take Facebook, the world's

5 Ian Youngs, "Social Media Now Main Source of News in US, Research Suggests," BBC News, June 16, 2025.

6 Robert McChesney, *Digital Disconnect: How Capitalism Is Turning the Internet Against Democracy*, New York: New Press, 2013.

most popular social media platform, with more than 3 billion monthly users worldwide. Every time a user logs on, an algorithm filters the millions of pieces of content available based on their preferences, friendships, areas of interest, past interactions—and a good dose of targeted ads. Each user's news feed is thus the result of an extremely complex, dynamic, and evolving algorithmic curation process. This hyperpersonalization of the news feed can create a "filter bubble," where each user is enclosed in a bubble of information filtered by a combination of individual choices and automated, algorithmic selections. Each bubble is, by definition, unlike others, and so each user interacts within an information environment that will differ, to varying degrees, from that of other users.

These filter bubbles can contribute to the emergence of "echo chambers." In these spaces, many users who share similar opinions, beliefs, and values will occupy increasingly overlapping information environments and be prodded into discussing their common concerns, thus hardening their worldview. These are, of course, complex algorithmic processes, but since a core aim of the algorithm is to maximize engagement, there is a tendency to present users with information, viewpoints, and opinions that resemble others that usually elicit some reaction. This is done without any real regard for the veracity of the information being conveyed, resulting in echo chambers where users navigate familiar waters and are systematically shielded from ideas, opinions, or facts that do not reinforce their own views.

The formation of these echo chambers is one of the factors contributing to the accelerated spread of fake news, sometimes produced by individuals but also sometimes fabricated by bots. As three researchers from MIT showed in a study of Twitter, fake news circulates "significantly farther, faster, deeper, and more broadly than the truth in all categories."[7] This is not to say that a content will circulate more *because* it is untrue, but rather that fake news tends to be unusual and trigger more emotional responses, which might lead to more engagement and sharing.

With the pandemic, we experienced a global public health challenge, but also a massive disruption of the information environment. This complex process of growing disinformation today seems so amplified that we can speak of an "infodemic," an "overabundance of information, of highly variable veracity," which, when it concerns politics or public issues, can "hinder their resolution by preventing the general

---

7 Soroush Vosoughi, Deb Roy, and Sinan Aral, "The Spread of True and False News Online," *Science* 359, no. 6380, March 2018, p. 1145.

population from finding reliable information and acting accordingly."[8] This is a perfect example of a perverse effect: An abundance of information turns into overabundance and then misinformation, as the accelerated production of ideas, knowledge, and commentary paradoxically contributes to a far-reaching and involuntary dumbing-down, with potentially harmful consequences for public life. The recent rise of massive quantities of AI-generated content all across the Internet only intensifies the problem.

Of course, collective intelligence and real knowledge can still find spaces for themselves. Yet even the enormous mass of collective wisdom and knowledge behind such an invaluable free, collaborative project such as Wikipedia seems like a foothill compared to the Everest of stupidity, superficiality, violence, hatred, and sordid conspiracy theories spread on social media, rapidly undermining their epistemic claims. Ultimately, as Cass Sunstein's analysis shows, people's confinement in echo chambers and misinformation can heighten polarization and foster political radicalization, sometimes turning into outright hatred and violence. The mechanisms by which social media generates cascades of information and accelerates the spread of rumors amplify the power of online influencers and the rise of extreme views that can undermine public debate and push democracies toward fragmentation and polarization.[9]

We should note that in the context of the US, as Benkler et al. have shown, the phenomenon of political polarization does not affect the left and the right in an equal manner. Rather, the development of a distinct media ecosystem of the right, fed by a "propaganda feedback loop" of far-right ideas and fake news, and increasingly disconnected from traditional journalistic norms, has resulted in a phenomenon of "asymmetrical polarization," where radicalization is much more apparent to the right of the political spectrum than to the left.[10]

In any case, no clear consensus exists among scholars on the extent and impact of phenomena such as echo chambers, filter bubbles, and the spread of fake news. And we should avoid attributing political consequences simply to technological causes, as that would amount to a form of technological determinism. Rather, we can say that informational

8 Office Québécois de la Langue Française, "Infodémie," terminology sheet, 2020.

9 Cass R. Sunstein, *#Republic: Divided Democracy in the Age of Social Media*, Princeton, NJ: Princeton University Press, 2017.

10 Yochai Benkler, Robert Faris, and Hal Roberts, *Network Propaganda, Manipulation, Disinformation and Radicalization in American Politics*, New York: Oxford University Press, 2018.

phenomena that are rooted in complex social causes can be accelerated or exacerbated by algorithms, with deleterious effects for democracy. All in all, there remains little doubt that social media has altered significantly the information landscape and the way people relate to news over the last two decades, creating new challenges for democratic life.

Yet we should not overgeneralize. While criticism of misinformation and the negative effects of social media is surely warranted, we should also recognize how social media may, in certain circumstances, help enliven the public sphere, energizing citizen participation, collective action, and social movements. The Arab Spring, the #MeToo movement, and #BlackLivesMatter were each expressions of popular participation that used social media to mobilize people in support of social causes. While some commentators have doubtless exaggerated things by calling these social movements "Twitter revolutions," as if these protests were caused by the digital tools they used, the fact remains that digital platforms can indeed increase the impact of social struggles in the public arena. But this is a double-edged sword, and the effect of digital platforms on citizen participation and activism remains ambivalent. Many social justice movements (such as feminist, antiracist, LGBTQ+, and environmentalist struggles) can draw on these tools to find a greater echo and organize. But it is also true that certain religious fundamentalist, masculinist, racist, xenophobic, transphobic, conspiracy theorist, and other currents use these platforms to amplify their voices, target and attack people, and spread reactionary ideas.

Moreover, while social media does sometimes help feed social mobilization in certain sociohistorical contexts (Occupy Wall Street, the 2012 Printemps érable in Quebec, or the Gilets Jaunes movement in France), it can also hinder collective action, undermine concerted and disciplined efforts, and weaken solidarity within these movements. We are today seeing civic engagement become increasingly volatile and ephemeral compared to older forms of activism, which were rooted in more lasting organizations. Various factors serve to give platform activism this ambivalent character: the rise of what could be described as "armchair activism" limited to online actions (hashtag activism, slacktivism); the culture of denunciation or canceling (call-out or cancel culture), as expressed in actions such as mass reporting, disinviting, canceling events, smear campaigns, doxxing, intimidation, and sometimes even death threats. As Benjamin Loveluck explains, these various forms of moral indignation "sometimes slide into punitive actions, which places them in the realm of *digital vigilantism*," which means "taking justice into one's own hands" through "targeted surveillance, repression or deterrence practices."[11]

Acts such as denouncing sexual assault, hate speech, and forms of police brutality are legitimate and useful forms of collective action, particularly when the judicial system proves incapable of delivering justice to victims and trivializes the actions of perpetrators, for example, in so many cases of sexual violence. And yet, a climate of mutual surveillance seems to be surreptitiously creeping into the world of social media. Hyperpolarized discussions, hasty accusations, and interpersonal conflicts reinterpreted as acts of aggression create an unhealthy climate for public deliberation. Social networks promote various forms of "lateral" or "participatory" surveillance, which "is also exercised horizontally between individuals" and brings new "forms of social control."[12] A "culture of humiliation" is thus taking hold. In her 2022 book,[13] researcher Cathy O'Neil examines how this "shame machine" exploits the mechanisms of this social emotion:

> Big tech is making use of shame to profit from our interactions. I think that is a new development that relies quite directly on the ability to hijack our pre-rational, triggered reactions and the existential threat that shame represents to our psyche … I think algorithms are optimised to service that which will arouse us the most. That usually means outraging us so we perform shame. In our filter bubble, our in-group, the algorithm serves to us the most outrageous thing that some other filter bubble has managed to arrive at, so we have the opportunity to be righteous and lob shame on to that other group, and to create this shame spiral.[14]

Thus, the crisis of public space fragmented by algorithms and certain social pathologies associated with online activism undermine the deliberation and forms of participation that any healthy democracy needs. We do not have room here for an exhaustive overview of all the problems that arise in the digital spaces mediated by algorithmic capital. We will return in more detail to certain algorithmic "vices" and propose possible solutions to them in Thesis 19. Table 11.1 presents an overview of some of these issues.

---

11 Benjamin Loveluck, "Le vigilantisme numérique, entre dénonciation et sanction: Auto-justice en ligne et agencements de la visibilité," *Politix* 3, no. 115, 2016, pp. 127–53.

12 Ibid.

13 Cathy O'Neil, *The Shame Machine: Who Profits in the New Age of Humiliation*, New York: Crown, 2022.

14 Cited by Andrew Anthony, "Cathy O'Neil: 'Big Tech Makes Use of Shame to Profit from Our Interactions,'" *The Guardian*, August 9, 2022.

From representative democracy to algorithmic regulation?

Algorithmic capitalism also threatens certain central institutions of representative government: the electoral process and political decision-making by elected officials. This is not new, and was evidenced already by the scandal around Cambridge Analytica, a data analysis company that played a role in the Brexit referendum in the United Kingdom, the 2016 US presidential contest, and the 2017 elections in Kenya. Cambridge Analytica successfully used sophisticated tools to extract and analyze the personal data of 87 million individuals through psychological quizzes on Facebook, which were disguised as innocuous games. Combined with other strategies to extract information about undecided or reluctant voters, Cambridge Analytica then sought to influence these users using hyperpersonalized targeted advertising. The aim: to alter users' perception of reality, get them to vote, or sway their choice one way or the other. It is impossible to measure the exact impact of these measures, or whether the actions of Cambridge Analytica really did make a difference in voting results. Yet a number of experts have proposed that the victory of the "Leave" vote in the British referendum, Donald Trump's first election in the United States, and Uhuru Kenyatta's election in Kenya in 2017 were at least partly attributable to Cambridge Analytica's operations. Disinformation is thus not only the work of blind, automated algorithmic filters, but also a deliberate strategy of personalized propaganda, designed to sway the vote by targeting certain parts of the electorate. These are not trivial examples, but large-scale processes of manipulation carried out on major political powers worldwide, including long-standing constitutional democracies.

A small, cunning private company—a kind of mercenary in the age of algorithms—thus might have succeeded in hacking or short-circuiting the electoral system in some of the world's most powerful countries. The question arises as to whether representative democracies are threatened by this type of political-algorithmic intervention, which may happen again in the future. Even though Cambridge Analytica was forced to close its doors in 2018 in the wake of the global outcry over this scandal, other companies or countries could easily use similar processes to further their own agendas. Recent research on the use by various political interests of "troll farms" to distribute propaganda during electoral contests and fan the flames of polarization and disinformation online shows the extent of these challenges.[15] For example, Brazilian far-right candidate

15 See, for example, Maksim Markelov, "I Investigated Millions of Tweets from the

Table 11.1 Glossary of problems created or amplified by algorithmic capitalism

Problem	*Explanation*
Filter bubbles	Hyperpersonalized content causing a lack of exposure to perspectives other than the user's
Clickbait	Framing method using exaggeration or omission of information to incite users to click on content
Surveillance capitalism	Extraction and monetization of personal information
Algorithmic radicalization	Hypothesis according to which recommender systems can orient users toward extreme content on social media
Slacktivism	Forms of public support for a cause or public criticism made solely through minimal efforts on social media or through online petitions, etc.
Concentration of media	So-called independent media bought by larger media companies, giving an illusion of choice to information consumers
Social bots	Social media accounts piloted by AI or humans with an aim to manipulate discussions or amplify specific messages and discourses
Ad clutter	When the use of a website is rendered difficult by a large number of popups, automated videos, and intrusive advertising
Astroturfing	Publication of content creating the illusion of a citizens or grassroot interest for an issue or a cause
Deplatforming	Ban of individuals or groups from a social media platform
Context stripping	Mass sharing of stories on social media causing more convincing or intuitive content being selected, this digital "natural selection" contributes to masking elements of context or facts
Infotainment	"News" which is optimized to hook users in by focusing on entertainment instead of facts
Disinformation	Deliberate distribution of false information
Dogpiling	Large volume of messages or targeted harassment of an individual by a group on social media
Rumor cascade	Rapid propagation of misinformation through social media posts

Jair Bolsonaro spread fake news through the massive use of WhatsApp during his successful 2018 bid for the presidency, and a growing number of political parties are now using Big Data, Facebook Analytics, and other AI tools to personalize their content and increase their impact.

Kremlin's 'Troll Factory' and Discovered Classic Propaganda Techniques Reimagined for the Social Media Age," *The Conversation*, October 1, 2024.

This use of AI tools in electoral politics by political parties is often done in a way that remains opaque to public scrutiny, while electoral regulations remain vague on the topic. In an age of what Luciano Floridi calls "hypersuasion," AI "can and will be used, evermore commonly and successfully, to manipulate people's views, preferences, choices, inclinations, likes and dislikes, hopes, and fears."[16]

Of course, we should remember that propaganda techniques are nothing new, but today they are amplified by the power of algorithms that can analyze personal data and influence behavior in an ultra-individualized way, in contrast to traditional tools that were primarily aimed at the masses in an undifferentiated manner. Algorithms can successfully predict and influence individual consumer behavior, and there is a very fine line between this and influencing individuals' political views and voting choices. In close elections or votes, a slight change in voter turnout in a given county, for example, can change the national-level outcome. The power of algorithms and the organizations that use them for manipulative purposes does indeed pose a threat to the integrity of the electoral process—that is, the central institution of representative government at the heart of this system's legitimacy. How can we be sure that these processes will retain their integrity, when any hostile company, party, organization, or government in the world can now try to hack elections?

Another dynamic that threatens to erode representative democracies is the prospect of certain human decisions being gradually replaced by more or less automated ones. In Thesis 10, we saw how algorithms are replacing human judgment in administrative, police, and judicial decision-making within the state apparatus. Added to this is the concept of the "smart city," which has rapidly gained popularity around the world since the 2010s. The term refers to the desire to build "connected cities" equipped with networks of sensors, algorithms, screens, smart traffic lights, surveillance cameras, the Internet of Things, and other devices and apps. This paradigm above all embodies a technocentric vision of the city—conceived as a complex system that can be optimized by myriad algorithmic mechanisms. But it is also sincerely intended to improve central aspects of urban life as an aid to governance, the environment, quality of life, inclusion, and the local economy.

The idea of using algorithmic technology to optimize various types of urban services is not new, but it is now finding unprecedented resonance

16 Luciano Floridi, "*Hypersuasion*—on AI's Persuasive Power and How to Deal with It," *Philosophy & Technology* 37, no. 64, May 2024.

through the rapid dissemination of technologies produced by algorithmic capital. This materializes in new forms of data management and urban planning models using contextual computing, heralding a more sustainable and eco-efficient city.[17] Smart cities can take different forms depending on the local and national context. In Canada, the Sidewalk Toronto project sought to entrust Google's Sidewalk Labs with the redevelopment of the Quayside neighborhood. Launched in 2017, the project was abandoned in 2020 due to economic difficulties, but also due to strong local opposition surrounding surveillance, data protection, and the privatization of urban development. Elsewhere in the world, the city of Masdar in the United Arab Emirates and the Songdo international business district in South Korea are two examples of urban projects based on this paradigm. The promise of renewable energy, smart mobility, waste management via underground vacuum systems, and Wi-Fi-connected devices and sensors paints a picture of an omniscient city that can be monitored continuously. It heralds, ultimately, a "real-time city," one powered by Big Data and realized through smart urban planning.[18]

This "colonization of the lifeworld" by algorithmic rationality takes form in the utopian vision of a "responsive city"—a city able to respond to individuals' concerns through interactive systems, networked citizenship, and a data-responsive municipal administration. In short, this means a city designed on the model of digital platforms.[19] This idea of expanding community engagement by means of smart, datafied governance is driving the platformization of political life. Connected objects, open data, algorithms, and cybernetic systems come together in a technological ecosystem that connect a plurality of users (citizens, businesses, community organizations, elected officials, civil servants, and so on). While this platformization holds out the dream of a "data democracy" or "AI democracy," it also raises numerous social and political problems.[20] To conceive of the city or the municipal administration itself as a kind of gigantic "digital platform" may well have its logistical advantages. Yet it also reflects a certain political conception of local democracy, a

---

17 Pervaiz Akhtar et al., "Data-Driven and Adaptive Leadership Contributing to Sustainability of Global Agri-Food Supply Chains Connected with Emerging Markets," *International Journal of Production Economics*, no. 181, 2016, pp. 392–401; Simon Bibri, *Smart Sustainable Cities of the Future*, New York: Springer, 2018.

18 Rob Kitchin, "The Real-time City?: Big Data and Smart Urbanism," *Geojournal* 79, no. 1, 2013, pp. 1–14.

19 Stephen Goldsmith and Susan Crawford, *The Responsive City: Engaging Communities Through Data-Smart Governance*, San Francisco: Jossey-Bass, 2014.

20 Susskind, *Future Politics.*

consumerist one that reduces citizenship to being a user "connected" with services, much in the manner of a Spotify subscription.

The use of algorithmic technologies to optimize certain urban services and infrastructure does not pose an ethical problem in itself, especially if these technical transformations remain subordinate to democratic political decisions. However, problems arise when the increased use of data and intelligent machines is accompanied by the emergence of an "algorithmic governance" paradigm. This is a management model primarily focused on data extraction, predictive analysis, feedback loops, and automated decision-making systems, which are increasingly becoming essential intermediaries in political and administrative life. For example, Montreal City Hall's report on the use of automated decision-making systems notes that various data- and algorithm-driven technologies (such as crowd-density sensors, smart parking, automated vehicle registration plate readers, systems for targeting homeless people, and smart bus shelters including emotion recognition) are already being used by dozens of cities in Canada. This raises many ethical and political issues.[21]

Furthermore, elected politicians have now started to use algorithms in a more direct and uninhibited manner. In March 2023, Romania's Prime Minister Nicolae Ciucă proudly announced that he had become the world's first political leader to use a chatbot, which he readily presented as his "new honorary advisor," named Ion. The prime minister confidently stated: "I have the conviction that the use of AI should not be an option but an obligation to make better-informed decisions."[22] The Ion bot was tasked with providing the government with real-time information on the opinions of the Romanian population through automated analysis of the most viral posts on social media. It would also gather ideas and suggestions from citizens who wished to communicate with the bot directly. This could be seen as mere media hype, since elected officials and their advisors have long used various techniques to gather public opinion (as in poll-driven government). Ciucă's announcement may also have had some relation to the fact that Romania, as one of the EU's poorest countries, is trying to push itself as the "Silicon Valley" of Eastern Europe. But it is also true that the broad popularity of chatbots such as ChatGPT and the normalization of their use by political and

---

21 Lyne Nantel, "L'avis sur l'utilisation des systèmes de décisions automatisées par la Ville de Montréal," Conseil jeunesse de Montréal, August 2021, available at ville .montreal.qc.ca.

22 "Romania PM Unveils AI 'Adviser' to Tell Him What People Think in Real Time," *The Guardian*, March 2, 2023.

business leaders as decision-making tools will likely accelerate the use of algorithmic tools among elected officials at all levels. The result: A growing number of political decisions will indeed soon be mediated by automated systems. The hybridization of humans and machines in political decision-making here crosses a new threshold.

While each use of algorithmic technology can be assessed on a case-by-case basis, it is important to ask whether we are undergoing a gradual transition away from the principle of *democratic rationality* (based on citizen participation, public deliberation, and political representation underpinned by principles of responsibility, transparency, and accountability) and toward an *algorithmic governmentality* that is surreptitiously replacing democratic processes and political debate. In this sense, we would like to raise, as a final question on the crisis of confidence in institutions, the issue of "democratic deconsolidation" that has been underway for at least a decade, and liberal democracies' recent authoritarian turn. Are we shifting, in some measure without even realizing it, toward a form of political rationality governed by the power of algorithms (and thus the organizations that control them)? Is the form of the state corresponding to capitalism's algorithmic stage intrinsically more authoritarian?[23]

Finally, what should we think of recent transformations in political life, as it becomes increasingly intertwined with the networks of algorithmic power? Is it possible that one day algorithms, smart assistants, and robots will be called upon to replace—in whole or in part—not only the judgment of civil servants, but also the decisions of elected officials, especially if it can be demonstrated that AI decisions are more systematically "accurate and relevant" than those of flesh-and-blood human beings? We have already mentioned some key studies concerning the biases and injustices that occur within algorithms.[24] Yet the fact remains that the increased use of these technologies in business, insurance, public services, the justice system, policing, public spaces, and the political arena raises an open (and worrying) question about the future of democracy. Is an algorithmic democracy conceivable and, if so, under what conditions?

---

23 Roberto S. Foa and Yascha Mounk, "The Danger of Deconsolidation," *Journal of Democracy*, no. 3, 2016, pp. 5–17; Roberto S. Foa and Yascha Mounk, "The End of the Consolidation Paradigm: A Response to our Critics," *Journal of Democracy Web Exchange*, April 28, 2017.

24 Safiya Umoja Noble, *Algorithms of Oppression: How Search Engines Reinforce Racism*, New York: New York University Press, 2018; Virginia Eubanks, *Automating Inequality: How High-Tech Tools Profile, Police, and Punish the Poor*, New York: Picador, 2019; Benjamin, *Race After Technology*.

Or is algorithmic authoritarianism what lies ahead? Do the rapid development of AI and the spread of platformization risk overshadowing the central role of human decisions, which will increasingly become mere inputs or auxiliaries to algorithmic calculations that are at most partly subject to human supervision?

This brings us to the heart of a reflection on the political dynamics of the present era. For today, we face a multidimensional democratic crisis that affects participation, deliberation, representation, and collective intelligence, which is being challenged by machine intelligence. How can we democratize institutions, states, local government, and the spaces in which we live? How can we overcome the shortfalls in citizen engagement and the polarization of public debate and regain the ability to solve collective problems through democratic processes? In the final chapters of this book we will explore some possible solutions. But before that, we must continue our critical reflection on another level. Next, we must analyze how algorithmic capital is not only reshaping democracies at local and national levels but is also becoming one of the structuring dynamics of the global geopolitical order.

Thesis 12

Algorithmic capital is helping to reshape the international order

Those who play with fire will eventually get burned.

—Xi Jinping

The rise of algorithmic capital is taking place during a truly historic moment in global capitalism's international relations. For there are today significant indicators of a process of hegemonic transition in the international order. This concept refers to a shift of the focus of power and accumulation in the global system from one dominant state (hegemon) to another. This change is already the object of a significant body of critical literature in the field of international relations: If the United States consolidated its global hegemony within a globalized capitalism after World War II, are we now witnessing the early stages of a hegemonic transition toward China in the twenty-first century?[1]

More mainstream analysts are also studying this question, as evidenced by recent discussions of the "Thucydides trap." This term refers to a situation in which an established power starts a war in order to prevent the rise of a potential rival. Such debates are a variation on the same theme of the growing Chinese threat to US hegemony.[2] Nowhere

1 Giovanni Arrighi, *The Long Twentieth Century: Money, Power and the Origins of Our Times*, London: Verso, 2014; Arrighi, *Adam Smith in Beijing: Lineages of the Twenty-First Century*, London: Verso, 2008; Mark Beeson, "Hegemonic Transition in East Asia?: The Dynamics of Chinese and American Power," *Review of International Studies* 35, no. 1, 2009, pp. 95–112; Lin Chun, "China's New Globalism," *Socialist Register*, no. 55, 2019, pp. 150–72.

2 See, among others, Graham T. Allison, *Destined for War: Can America and*

is this unique and complex conjunction between the advent of a new regime of capitalist accumulation and a process of hegemonic transition more clearly revealed than in the rivalry between the two countries' high-tech sectors.

We must here examine the coordinates of this transition process as they directly relate to algorithmic capital. It is also important to look at the algorithmic reconfiguration of the military sphere, which is giving rise to a new race for "intelligent" weapons. Looking beyond the US–China rivalry, we will also consider the position of other Global North actors, of Russia and of other emerging countries, and of the Global South in this new configuration of international relations in the era of algorithmic capital.

Hegemony and transition

For scholars of international relations, the concept of hegemony is not limited to describing the political domination of one state over others. It also refers to a form of power that combines coercion and consent, expressed through international leadership on multiple levels: economic, financial, political, diplomatic, cultural, moral, and military. Hegemony is therefore always multidimensional. Inspired by Antonio Gramsci, who examined hegemony in its national context, the neo-Gramscian school of international relations offers the most important critical contribution to the concept of hegemony and makes the concept particularly useful for our purposes. Defined as the expansion of the internal hegemony of a dominant state, established by its ruling class, beyond its own national borders, international hegemony is consolidated when "the economic and social institutions, the culture, the technology associated with this national hegemony become patterns for emulation abroad."[3] Hegemonic periods thus correspond to an international order that is dominated by a particular national hegemony, whereas periods when this hegemony is contested, eroded, or ceases to exist are referred to as nonhegemonic.

China Escape Thucydides' Trap?, Boston: Mariner Books/Houghton Mifflin Harcourt, 2018; G. John Ikenberry, "From Hegemony to the Balance of Power: The Rise of China and American Grand Strategy in East Asia," *International Journal of Korean Unification Studies* 23, no. 2, 2014, pp. 41–63; Denny Roy, *Return of the Dragon: Rising China and Regional Security*, New York: Columbia University Press, 2013.

3 Robert Cox, "Gramsci, Hegemony and International Relations: An Essay in Method," *Millennium* 12, no. 2, 1983, p. 171.

From 1945 onward, the postwar order was consolidated into a system of international organizations and a regime of US-centric norms that enabled the United States to reproduce its hegemony. US hegemony was backed up by economic, political, and diplomatic leadership, cultural and moral influence, and, of course, an imposing military supremacy. Consolidated in the Bretton Woods institutions and punctuated by major phases of economic globalization led by US and allied capital, post–World War II American hegemony enabled a new type of postcolonial imperialism to establish itself. This imperialism would be based on market mechanisms for extracting and transferring value to the center of accumulation—that is, the Global North and its center, the United States—whether through the international division of labor, unequal exchange, market coercion, or the mechanism of debt.[4]

The US model of postcolonial imperialism thus relies on market mechanisms, and the exercise of its hegemonic power has been aimed primarily at promoting the spread and institutionalization of market relations on a global scale. While initially exercised through multinational corporations and the actions of the international institutional mediators of US capital (the World Bank, International Monetary Fund, World Trade Organization, and so on), it has recently been recalibrated in parallel with the rise of algorithmic capital. This more recent process has given rise to a host of new players, new companies, new goals, and a rethinking of US imperialism's systemic logic. If we want to understand the particular form of US hegemony and the imperialist practices that have characterized it since the 2000s and 2010s, we must look not only to the Pentagon, Wall Street, and traditional US multinationals, or even to the influence of Hollywood, but also to Silicon Valley: Google, Apple, Meta, Instagram, Microsoft, YouTube, Uber, Airbnb, Netflix, and the like. The major players in US algorithmic capitalism are global forces—Facebook is the number-one social network in 153 countries; Apple and Amazon exploit workforces and labor processes stretching around the globe; Netflix is present in 190 countries; and Google and Meta make most of their profits outside the US. We could go on. These companies have now become the vehicles of a new regime of accumulation, which in

4 Harvey, *The New Imperialism*; Ellen Meiksins Wood, *Empire of Capital*, London: Verso, 2005; Andrew J. Bacevich, *American Empire: The Realities and Consequences of U.S. Diplomacy*, Cambridge, MA: Harvard University Press, 2002; André Gunder Frank, *The Development of Underdevelopment*, New York: Monthly Review Press, 1966; Leo Panitch and Sam Gindin, *The Making of Global Capitalism: The Political Economy of American Empire*, London: Verso, 2012.

turn requires a redeployment of US imperialism and an attempt to reset certain pillars of its hegemony. Thus, the mechanisms of value transfer that characterize US imperialism are now complemented by a double imperative of *data extraction* and of a *transformation of the world's political-economic and technological ecosystems* in order to ensure the sustained growth and dominance of the US algorithmic giants.

While the imperatives of algorithmic accumulation are extending their reach, this does not mean an end to the previous forms of imperialism. Rather, it involves a process of reconfiguration. This is seen, for instance, in the reorientation of investment, in the securing of access to critical infrastructural resources, and data neocolonialism (where algorithmic capital deploys algorithmic technologies in emerging countries in exchange for control over data—a point to which we will return). Culturally, while the "Americanization of everywhere" in the second half of the twentieth century spread through cultural industries (music, films, clothing styles) and the export of major brands (McDonald's, Coca-Cola, Nike, and so on),[5] in the twenty-first century it is more characterized by the promotion of "lifestyles" linked to algorithmic capital, based on social media, the smartphone, influencers, and the values, codes, and cultural references of Web 2.0. All this can be expressed in a variety of languages and national, regional, and local characteristics. The "siliconization of the world" has thus now climbed the ladder to become the archetype of cultural hegemony, which we will explore in Thesis 13. In short, the central pillars of US hegemony are in the process of being algorithmized.

This raises a question. Is the algorithmization of American hegemony sufficient to ensure its long-term global domination in the twenty-first century, preventing the onset of a hegemonic transition? Nothing could be less certain. In our view, it is entirely plausible that the rapid algorithmization of Chinese society in recent years could even accelerate the transition to a period in which US hegemony over the international order comes under serious challenge. Since hegemony is multidimensional, any investigation into a possible hegemonic transition must itself be multidimensional. The growth of new cultural norms or the rise of an economic or technological rival is not alone sufficient cause to identify a complete transition. More fundamentally, the thesis of hegemonic transition refers to new international leadership in economic, political,

5 Peter Conrad, *How the World Was Won: The Americanisation of Everywhere*, London: Thames & Hudson, 2014; Benjamin R. Barber, *Jihad vs. McWorld: Terror's Challenge to Democracy*, New York: Times Books/Random House, 1995.

financial, diplomatic, cultural, and military terms. However, what we are seeing at present is a major trend, dating back at least ten to twenty years, in which the center of global economic accumulation is shifting away from the United States and the North Atlantic region and toward China and Southeast Asia.[6] The other dimensions (in terms of diplomatic, political, cultural, and military leadership) are far from having reached the same stage of transition. Yet they are also showing signs of the beginning of a profound shift away from the US-centric order built in the post-1945 period, which concerns the most central areas of international power. Against a linear reading of the thesis of hegemonic transition, we note that this shift need not necessarily lead toward another unipolar world, centered on China, for example. It is as likely to lead to a protracted phase of rivalry between regional blocs in a multipolar world, the outcome of which is as open-ended as any other historical phase of the international system.

Of course, we cannot systematically address all dimensions of the hegemonic transition here, but we will maintain our focus on the spheres of technology and algorithmic political economy. The symptoms of this underlying condition of the system are becoming increasingly apparent elsewhere and cannot be reduced to the anti-Chinese mood swings of particular US presidents. The tensions over TikTok, quantum computing, the Huawei affair, Deepseek, GPUs, geopolitical debates on 5G infrastructure, as well as the rise of trade rivalries, indicate the tensions arising from the redeployment of US imperialism and hegemonic domination on a new level: the level of algorithmic capital, and the emergence of a powerful Chinese rival in this specific field.

The rise of China and Chinese algorithmic capitalism

Tensions between China and the United States take many forms, but all stem from the same root cause: the emergence of a Chinese economic and political model, indeed a "national hegemony," which has the capacity to project its power beyond Chinese borders and present itself as an alternative to US hegemony and, more broadly, to the model of Western liberal capitalism. This model stems, of course, from the Chinese revolution of 1949 and the merger between the Chinese state apparatus and Communist Party. However, it was in 1979 that major trends began to

6 McNally, *Global Slump*, pp. 35, 84.

take shape that would propel China to the forefront of global capitalism and international relations. The reforms launched under Deng Xiaoping's leadership moved away from orthodox Communist positions, and a number of market principles were accepted and integrated into the Chinese economy. The creation of "special economic zones," based on an integrated industrial development strategy aimed at attracting foreign direct investment, bore fruit. China, especially since the 1990s, has rapidly become not only the center of the world's manufacturing, but is now a leading industrial power.[7] China's GDP growth rates over the period from 1990 to 2010 were among the highest in the entire history of capitalism anywhere in the world. In 1981, 88 percent of the Chinese population lived in extreme poverty, but today that figure stands at just 2 percent, although social inequalities have also exploded over the course of these transformations.[8] Over the past four decades, the size of the Chinese economy has grown from $195 billion to $18.75 trillion. China ranks second in the world in terms of GDP, and first in terms of GDP expressed in purchasing power parity. It is home to the second-largest number of Fortune 500 companies, has more billionaires than all of Europe combined, and every indicator suggests that in the near future it will become the world's leading economy at every level.[9]

During these forty-five years of profound transformation, China has become a fully capitalist country. But the Chinese system is often misunderstood in the West, especially by some of the right-wing intelligentsia and the mass media, who stubbornly see China as a "communist enemy." They associate the authoritarianism of the Chinese state with a Cold War–era idea of communism and thus refuse to recognize the development of what is in fact virulent *authoritarian capitalism*. In a "capitalism with Chinese characteristics," a model is developing that combines state ownership and planning with a dynamic private sector and the fundamental role of market imperatives. Profit-seeking is the main driver of the economy, and Chinese people from all social classes are largely dependent on the market. The private sector now accounts for the vast

7 Ezra F. Vogel, *Deng Xiaoping and the Transformation of China*, Cambridge, MA: Belknap Press, 2013; Julian B. Gewirtz, *Unlikely Partners: Chinese Reformers, Western Economists, and the Making of Global China*, Cambridge, MA: Harvard University Press, 2017.

8 We should always be cautious in using measurements of poverty, as they often equate poverty not with lack of access to resources, but with lack of access to resources that are accessible only with money.

9 Laurie Macfarlane, "A Spectre Is Haunting the West: The Spectre of Authoritarian Capitalism," *Open Democracy*, April 2020.

majority of production, employment, and investment, and there is no democratic or socialist control of the economy by workers. The Chinese economy is now resolutely capitalist, in the sense that the private sector is responsible for most production, most workers are salaried employees, and most decisions concerning production and prices are made in a decentralized manner.[10]

However, the Chinese model of capitalism differs from the Western one in many important respects.[11] As economist Laurie Macfarlane explains, the most striking difference concerns macroeconomic management by the Chinese state, through gigantic institutions such as the state-owned Assets Supervision and Administration Commission. Formed in 2003, this body has since then supervised Chinese companies, controlled large industrial enterprises, and reported directly to the State Council. The companies under its umbrella have combined assets of $26 trillion and revenues of more than $3.6 trillion (more than the UK's entire GDP). No similar institution exists in the United States or in other Western capitalist countries. Such a body would have to control a cluster of dominant companies in each of the major industrial sectors—for example, in the United States, GE, GM, Ford, Boeing, US Steel, Dupont, AT&T, Verizon, Honeywell, and so on—whatever their main shareholder, and be able to make important decisions in each of them: hire and fire management, order the sharing of resources, ensure cross-compatibilities, and so on. The Chinese state also exercises significant macroeconomic management over the financial sector (the world's largest) and banking. The highest authority, the National Development and Reform Commission, oversees the entire Chinese economy. Such an organization promotes the coordination of development projects and conducts a degree of macroeconomic management that has no equivalent in the Western model.[12]

However, over the past fifteen to twenty years, Chinese capitalism has adopted a development model focused on the high-tech sector and has developed an algorithmic accumulation regime of staggering economic, political, and social dimensions. A comparative analysis of the development of the AI sector in the United States and China clearly illustrates the differences between the two models of transition to an

10 Branko Milanovic, *Capitalism, Alone: The Future of the System That Rules the World*, Cambridge, MA: Belknap Press, 2019.

11 Minxin Pei, *China's Crony Capitalism: The Dynamics of Regime Decay*, Cambridge, MA: Harvard University Press, 2016.

12 Macfarlane, "A Spectre Is Haunting the West."

algorithmic capitalist accumulation regime. In the United States, it is more the market and its big players that are spearheading innovation; whoever lags behind in the race for profit or is less effective in the game of market competition and at attracting investment will soon be eliminated. In China, conversely, various levels of central, regional, and local government have implemented a far-reaching policy of massive investment and support for start-ups. This development strategy relies on the abundance of available capital rather than on the optimal profitability of its use. Both models have their advantages and disadvantages, but ultimately they compete to generate what Kai-Fu Lee calls the four main ingredients of an algorithmic superpower: an abundance of data, a savvy and combative entrepreneurial class, skilled scientists, and a political environment conducive to the development of algorithmic technologies. China has already caught up with and even overtaken the United States in the race for the first two elements; there is every reason to believe that it can compete at a very high level in the other two as well, and in the near future.[13]

With regard to AI in particular, it is increasingly obvious that there are powerful trends toward the concentration of power in this sector around a US–China rivalry. The rapid rise of China in the development of algorithmic technologies is based on multiple factors, as Lee explains. First, there are the specific traits of the Chinese Internet, which is a relatively closed digital ecosystem allowing Chinese platforms to extract and analyze a wealth of data in a hypercentralized manner. This runs in contrast to US platforms in Silicon Valley, which remain relatively separate from one another. This specific institutional and technological environment has developed through various phases. China began from an initial phase that could be described as a process of imitation, where a myriad of entrepreneurs unscrupulously copied Silicon Valley products and business models and adapted them to the Chinese market. But far from limiting itself to this dynamic of imitation, "counterfeiting," and perpetual catch-up, the Chinese Internet took a new step forward around 2013 by creating its own infrastructure, devices, and apps, as well as establishing huge programs to support innovation and start-ups that are now on a par with their US counterparts.

In just a few years, China has gone from a world where credit cards were largely absent and only a minority of citizens had access to a laptop (about one-third of the population in 2010) to a hyperconnected one

13 Lee, *AI Superpowers.*

where virtually all economic and social transactions in daily life are now carried out on low-cost smartphones. The speed at which these technologies have penetrated the market, effectively skipping the PC and Web 1.0 stages, means that Chinese platforms now have access to a greater volume of personal data and an unprecedented capacity for centralization compared to Silicon Valley companies. The WeChat super-app developed by Tencent, installed on more than 75 percent of Chinese phones, is a prime example. It is much more than just a social media or messaging service: It has become a universal remote control for various everyday tasks, including paying the bill at a restaurant, hailing a cab, unlocking a self-service bicycle, managing investments, making doctor's appointments and having medication delivered to your home, or ordering a meal. These features have multiplied, blurring the line between reality and virtuality in the use of WeChat. These two dimensions are now intertwined in the daily lives of the majority of the Chinese population. WeChat is almost omnipotent: It handles tasks that, for users of non-Chinese algorithms, are performed separately by Facebook, iMessage, Uber, Expedia, Instagram, Skype, PayPal, Grubhub, Amazon, LimeBike, WebMD, and many others. Even if it does not offer the full range of these apps' features, it effortlessly performs the main ones, all with a seamless, integrated payment system that requires neither cash nor cards.[14]

This vast mesh of apps and data is combined with the social credit system that has been implemented on a mass scale since 2020. This, in fact, involves multiple and multitiered evaluation systems, which have been rolled out unevenly at the national, regional, and local levels. They institutionalize forms of surveillance and social control with dystopian overtones, using mechanisms for quantifying "social reputation" in correspondence with the distribution of (and access to) certain social and economic resources. True, mechanisms for surveillance, prediction, and behavior control are also at the heart of the algorithmic technologies developed by Silicon Valley. However, China seems in many ways to be ahead of the United States, where digital platforms do not enjoy this same degree of vertical integration, data concentration, and direct power over everyday life. Just as China seems to have undergone its industrial revolution at an accelerated pace, becoming the world's leading exporter in less than three decades, it is now embarking on its digital and algorithmic revolution at breakneck speed. It represents arguably the most advanced embodiment of the algorithmic society in terms of integrating

14 Ibid.

human life with the Internet of Things, hyperconnectivity, and other AI devices.[15]

Over the last decade, China has thus become the world's leading laboratory for the utopias and dystopias of algorithmic capital, experimenting on a large scale and within a hybrid institutional environment. In that country, a hyperinterventionist state is rushing to create hundreds of miniature Silicon Valleys by massively promoting "entrepreneurship for the masses." It thus combines the more innovative aspects of the capitalist system with the *dirigiste* power of a single-party political regime endowed with an unrivaled repressive apparatus, in a political culture less concerned with individual freedoms. Faced with Silicon Valley's "liberal algorithmic capitalism," a new form of Chinese-style "authoritarian algorithmic capitalism" is emerging, which could undoubtedly become dominant in the coming decades during the hegemonic transition process we have described. As Macfarlane points out, a specter is haunting the West: the specter of "authoritarian capitalism." This haunting might not, however, come only from China, inasmuch as recent developments in US politics surely denote a tendency toward more and more authoritarian forms developing endogenously in the US itself. Could the competition in the future turn out to be one between two *authoritarian* algorithmic capitalist giants?[16]

Hegemony, norms, and algorithmic capital

In terms of norms, the Chinese and US models of algorithmic capitalist society differ in their distinct cultural and moral values. The Chinese model has fewer legal guardrails to protect minimal privacy and individual freedoms, which are more foregrounded in US society. China's more hierarchical social relations and values, which are less focused on individualism, seem to eliminate much of the friction that results from the development of algorithmic modes of social organization, particularly with regard to issues of political freedom. It must be noted, however, that thus far the US constitutional system, which is supposed to protect

15 Jean-Pierre Cabestan, "The State and Digital Society in China: Big Brother Xi Is Watching You!," *Issues & Studies* 56, no. 1, March 2020; Lee, *AI Superpowers*.

16 On the development of authoritarianism and tendencies toward "technofascism," see Jonathan Durand Folco, *Fascisme tranquille: Affronter la nouvelle vague autoritaire*, Montreal: Écosociété, 2025; Quinn Slobodian, *Crack-up Capitalism, Market Radicals and the Dream of a World Without Democracy*, New York: Metropolitan Books, 2024; Quinn Slobodian, *Hayek's Bastards: Race, Gold, IQ, and the Capitalism of the Far Right*, Brooklyn: Zone Books, 2025.

individual freedoms and privacy, has fallen far short of significantly altering or blocking the deployment of a vast surveillance infrastructure by a complex institutional network of commercial actors and state agencies. Still, China's cultural and moral influence internationally seems limited compared to the traditional US model. Although certain Chinese Communist Party policies, such as the (uneven) transition to green energy or health diplomacy during COVID-19, may have drawn praise from the international community, it remains difficult to foresee any rapid and drastic change in the values and norms that govern the international system, or a major realignment of the principles that guide the UN. In this sense, Chinese norms and values could spread more widely with the establishment of a network of institutions alternative to those of the Bretton Woods system, which remain dominated by a strongly North Atlantic liberal-internationalist ethos, even though it's unclear whether this ethos will weather the Trumpist storm. The creation of new Chinese development banks and the "economic diplomacy" of the new Silk Road project can also be seen as efforts by the Chinese state to spread its hegemony.

These multiple transition processes naturally involve their share of contradictions, as researcher Sean Starrs has insightfully pointed out.[17] China's ambitions have provoked major reactions from several other Asian countries (Vietnam, Malaysia, Indonesia, the Philippines), which have been eager to reassert their alliances with the United States. It is also true that China's economic growth has slowed over the last ten years. But the most significant contradiction undoubtedly lies in the stifling presence of US capital in China and its domination of several Chinese markets. In fact, most of the upper echelons of the value chain in China are controlled by foreign capital. If China wants to achieve a hegemonic transition, it will have to exercise greater control over the drivers of value creation on its own national soil. Reconfiguring the power structures of the international political economy, into which China is so intricately integrated, could prove to be a double-edged sword and jeopardize the foundations of the Chinese state's power. Moreover, the short-term prosperity of China's new middle class could suffer from any major economic upheavals. However, the sustained growth of the high-tech sector and algorithmic capital, combined with economic nationalism supported by the Chinese state's regulatory arm, which protects this sector from the US giants, could well enable Chinese capital, driven by its algorithmic sector, to regain greater control over its national economy.

17 Sean Starrs, "Can China Unmake the American Making of Global Capitalism?," *Socialist Register* 55, 2019, pp. 173–200.

At the heart of these processes of hegemonic transition, US imperialism—described above as a postcolonial economic imperialism recently rearticulated in a more algorithmic mode—now finds itself facing a powerful rival. With China having developed its own algorithmic capital base, the two superpowers are increasingly competitors in terms of the socioeconomic model that they can export around the world, and in terms of the transfer of value (and data) to either US or Chinese centers of accumulation. For its part, the US government is seeking to curb China's technological rise: for example, by trying to convince its allies such as Japan, South Korea, and the Netherlands to stop selling strategic technologies to China; through tariffs or legal mechanisms to limit exports of these technologies by other third countries; and by exerting diplomatic and commercial pressure to prevent China from becoming self-sufficient, particularly in advanced chip technology.[18] China also possesses its own strategic advantages, being for example the world's largest producer and refiner of rare earth minerals, key to the material infrastructure of algorithmic capital.

While American technological imperialism is represented by the acronym GAFAM, China has its own BATX—referring to its digital giants, which have joined the ranks of the world's largest companies: Baidu, Alibaba, Tencent, and Xiaomi—to which we can add, of course, the tech giant Huawei.. The BATX are increasingly exporting algorithmic technologies to countries in the Global South in exchange for the data generated in these countries being transferred to the Chinese giants. China is also investing heavily in building data centers as part of its development projects, such as the gigantic New Silk Road. The forms of imperialism unfolding in this phase of hegemonic transition thus remain strongly economic and postcolonial in character. Yet imperialism is also becoming algorithmic, meaning that it is increasingly exercised by means of new technologies, surveillance, and competition for the extraction of (and access to) vast amounts of data.

All this poses a bigger question: Is the world order of the twenty-first century bound to be an algorithmic one, whether under Chinese or US hegemony or marked by bipolar competition? Regardless of which hegemon is spearheading it, can we really envisage an alternative to the spread of a culture and norms governing international relations that is not tightly aligned with the imperatives of algorithmic accumulation?

18 Evgeny Morozov, "Histoire secrète de l'intelligence artificielle: Une guerre froide 2.0," *Le Monde diplomatique*, May 2023.

The arms race over smart weapons

One particular field is likely to see some of the most lasting effects of the rise of algorithmic capitalism in terms of the exercise and distribution of power in the international system: the development of military applications of AI. Vladimir Putin's widely quoted statement that "whoever dominates artificial intelligence will dominate the world" reveals a deep conviction shared by military thinkers not only in Russia, but also in China and the United States.[19] States are seeking to deploy AI militarily, in order to quantify the "global battlefield" and minimize the uncertainties generated by conflict situations. The possibilities for applying algorithms to military affairs are varied. They range from administrative tasks to the use of autonomous vehicles, and also include algorithms for decision support in warfare scenarios, classification and targeting, intelligence, surveillance, automation of weapon systems, and so on.[20] The day when robots will fully replace soldiers has not yet arrived, but this is a potential development that is widely discussed in the literature.

Just as the development of nuclear weapons led to a frantic arms race during the Cold War, a race for "smart" weapons, despite its aim of reducing uncertainty, actually carries a destabilizing potential of its own. This danger appears on three levels, according to James Johnson: an amplification of the uncertainties and risks posed by existing threats; a transformation of the nature and characteristics of these threats; and the introduction of new threats into the field of international security. AI-controlled autonomous weapons systems (including nuclear weapons) and the proliferation of drones and autonomous weapons belonging to state and nonstate actors will make the task of predicting attacks all the more difficult, requiring the development of strategic "counter-AI" tools. The automation of the attack/counterattack process and lethal weapons systems may well destabilize the handling of military affairs. They bring to mind retaliatory dynamics reminiscent of the Cold War's Mutually Assured Destruction (MAD) doctrine, now automated with AI, with a potential for deterrence still unknown.[21]

Academics have debated the importance of maintaining "meaningful human control" over lethal autonomous weapons and the presence of a

---

19 Amandeep Singh Gill, "Artificial Intelligence and International Security: The Long View," *Ethics & International Affairs* 33, no. 2, 2019, pp. 169–79.

20 Istvan Szabadföldi, "Artificial Intelligence in Military Application—Opportunities and Challenges," *Land Forces Academy Review* 26, no. 2, 2021, pp. 157–65.

21 James Johnson, "Artificial Intelligence and Future Warfare: Implications for National Security," *Defence & Security Analysis* 35, no. 2, 2019, pp. 147–69.

"human element" in the context of intelligent warfare.[22] But it remains difficult for researchers to gather information on these developments, given the veil of secrecy that envelops the military sphere. Nevertheless, we do know that the US military-industrial complex is already mounting a vast algorithmic retooling of its means of violence. In 2017, the US Department of Defense set up the Algorithmic Warfare Cross-Functional Team, which oversees, among other things, the now famous Maven project. This latter aims to integrate AI and machine learning into the DoD's operations in order to maintain an advantage over other countries. It is part of a complete overhaul (known as the JEDI project) of the department's entire information infrastructure, now converted into an algorithmic mode. The Pentagon-led Maven project made headlines in 2017: It had called on the services of algorithmic technology companies to, for example, analyze images filmed by US drones and geolocate potential targets. Google was awarded the contract, but after protests by Google workers who opposed the use of its algorithmic capabilities for military purposes, the firm had to abandon the project, which was taken over by Amazon in 2018.[23] At the end of 2022, the Department of Defense awarded a new $9 billion contract to centralize all its cloud services, the Joint Warfighting Cloud Capability, an initiative designed to replace the JEDI project. The beneficiaries of the contract are Amazon Web Services, Google, Microsoft, and Oracle.[24] This symbiosis between the Pentagon and GAFAM is hardly surprising, given the history of the development of these technologies in the United States. Indeed, the armed wing of the US government (defense and intelligence) has since the 1950s played a pioneering role in research into AI and the algorithmization of violence and surveillance. Telling, for example, is the career of someone like former Google CEO Eric Schmidt, who today wears several hats. Schmidt is director of the Pentagon's committee on AI development and a venture capitalist in military AI through his Innovation Endeavors fund, while also being a philanthropist who, through his Schmidt

22 Ingvild Bode and Hendrik Huelss, *Autonomous Weapons Systems and International Norms*, Montreal: McGill-Queen's University Press, 2022; Avi Goldfarb and John R. Lindsay, "Prediction and Judgment: Why Artificial Intelligence Increases the Importance of Humans in War," *International Security* 46, no. 3, 2022, pp. 7–50.

23 Crawford, *Atlas of AI*, pp. 189–92; Billy Mitchell, "Google's Departure from Project Maven Was a 'Little Bit of a Canary in a Coal Mine,'" FedScoop, November 5, 2019; Adam Frisk, "What Is Project Maven?: The Pentagon AI Project Google Employees Want Out Of," *Global News*, April 5, 2018.

24 Mark Pomerleau, "Pentagon Awards AWS, Google, Microsoft and Oracle Spots on $9B Joint Warfighting Cloud Capability Solicitation," DefenseScoop, December 7, 2022.

Futures foundation, legally remunerates government employees responsible for developing strategic AI policies. He is likewise a founder of the Special Competitive Studies Project (SCSP), a think tank dedicated to artificial intelligence, and cosigned AI propaganda books with Henry Kissinger just prior to Kissinger's death.[25]

In the United States, the militarization of AI is already well underway, and this is having a competitive effect on other military powers around the world. The perceptions of the actors involved, especially China, Russia, and the United States, regarding the development of military AI reinforce what Johnson calls an algorithmic security dilemma. China's People's Liberation Army is working with civilian partners to develop military applications of AI, aimed at a transition to a new military-strategic paradigm branded "smart warfare."[26] Chinese leaders see the AI revolution as an opportunity to catch up with the United States in military terms, and the more the United States militarizes AI, the more China and Russia will seek to do the same.

One of the most disturbing materializations of algorithmic militarism to date is the authorization of so-called "signature strikes." In such actions, lethal drone attacks target "signatures," devices identified by an algorithm as belonging to enemies of the state. When these devices rear their head in the flood of data collected by the US surveillance apparatus, drones shoot down whoever is using them—often causing "collateral" casualties. These are nothing less than algorithm-based murders. The "Skynet" algorithm (its name is inspired by the film *Terminator*), which zones in on targets for lethal drone strikes on the basis of metadata, is one such example of technologies that allow military leaders to override important ethical issues by reducing assassination decisions to quantifiable metrics. In the ongoing genocide in Gaza, the Israel Defense Forces (IDF) has made use of algorithmic technology, an AI-powered database called "Lavender," to target Palestinian men linked by AI to Hamas. Another AI system called "The Gospel" has been used by the Israeli Air Force to determine bombing targets. Such a "technologization of ethics," which delegates central issues of accountability to a technological apparatus, is surely highly problematic.[27] Beyond military ethics,

25 Morozov, "Histoire secrète de l'intelligence artificielle"; Henry Kissinger, Eric Schmidt, and Daniel Huttenlocher, *The Age of AI and Our Human Future*, New York: Little, Brown, 2021.

26 Justin Haner and Denise Garcia, "The Artificial Intelligence Arms Race: Trends and World Leaders in Autonomous Weapons Development," *Global Policy* 10, no. 3, 2019, pp. 331–7.

27 See the excellent article by John R. Emery, "Probabilities Towards Death:

this automation of mass murder should be seen as one actual, real, and present "existential risk" posed by AI.

The convergence between algorithmic capital and the state, both in the United States and China, complicates the relationship between the market, politics, economic nationalism, and algorithmic surveillance. On the one hand, this convergence allows the state to apply algorithmic tools to its own population, when government agencies or police services acquire these militarized technologies and integrate them into their operations. The experiments in mass surveillance and the deployment of control technologies during the COVID-19 pandemic are a telling example of this. On the other hand, this convergence allows monopolistic companies such as Google and Meta to justify their practices to politicians on the basis of national security. Such companies can argue that dismantling them to deal with their monopolistic practices would open the way to global domination by Chinese companies such as Baidu and Tencent, and that this would threaten US national security. Instead, the government should finance and subsidize the US development of AI and the high-tech sector to counter the Chinese threat. This is in line with a recent noticeable trend among US tech giants to pursue military contracts with the Pentagon, while several Silicon Valley executives have even joined the US army in various positions.

Algorithmic capital's global connections

Although Chinese–American rivalry dominates the international order in the age of algorithmic capitalism, algorithmic geopolitics is not confined to this sphere alone. Indeed, the "unequal and combined datafication" of the world articulates a global geopolitical dimension at each stage of its development, as Amaël Cattaruzza has explained.[28] At the stage of data production and collection, actors who produce algorithmic technologies gain a strategic advantage and can leverage their relative geopolitical strength. Indeed, access to a mass of data is crucial for the development of algorithms. At the stage of data transport and transmission, an inequality emerges between actors, in terms of the density of their algorithmic infrastructures. In other words, data does not flow

Bugsplat, Algorithmic Assassinations, and Ethical Due Care," *Critical Military Studies* 8, no. 2, 2022, pp. 179–97.

28 Amaël Cattaruzza, *Géopolitique des données numériques: Pouvoir et conflits à l'heure du Big Data*, Paris: Le Cavalier Bleu, 2019.

equally in all directions, but converges toward centers of accumulation, in this case, the United States and China. The general direction of global data flows mostly overlaps with that of value transfers from the Global South to the Global North, which characterized postcolonial market imperialism, with the notable exception of a peripheralization of Europe in the data economy.[29]

Even within the countries of the Global North, there is a sharp asymmetry. Europe is in a striking position of weakness and dependence, with a massive flow of data from the continent to the United States every second. In France, for example, the three most visited websites are Amazon, Google, and Facebook: The data generated by French users on these sites thus migrates in great measure to the United States. Europe is largely dependent in terms of algorithms, and European experts do not foresee any reversal of this situation. Cattarruzza refers to a "lasting asymmetrical relationship" between Europe and the two major poles, the United States and China. Chinese data, on the other hand, remains in that country, mainly because Google and Facebook are absent and because the BATX companies occupy this space. The inequality between global players is also notable in terms of data retention and storage. Data centers are concentrated in certain countries (the United States, China, Russia), and relationships of dependency are developing between client countries and cloud platforms that are almost all concentrated in the dominant countries. Finally, data processing capacities are also deeply unequal, and at this stage there is almost total dependence on GAFAM and BATX.

We can thus see a territorialization of the international space emerging in terms of the development of algorithmic infrastructure. For examples we could look at anything from the submarine cables that data is transmitted along to the geographical distribution of host servers—an infrastructural network heavily centered on the United States. A political and territorialized digital space, and thus also a geopolitical one, overlaps with international spaces that are structured by other flows, of capital, money, goods, services, investments, ideas, and culture. All this is creating relationships of dependency between various countries. The result is that increasing numbers of states are now attempting to create national digital territories and to control the internal data flows within their own borders. Several countries are now adopting data localization

29 Giuliana Facciolli and Jonathan Martineau, "L'intelligence artificielle et les algorithmes: Au cœur d'une reconfiguration des relations internationales capitalistes," *Nouveaux Cahiers du Socialisme*, no. 31, Winter 2024, pp. 68–78.

laws, and we are seeing the end of the era of neutrality in routing algorithms (that is, algorithms directing data flows to certain storage or processing centers). Geopolitical factors now play a role in the global data-flow routing space, with some countries wanting to prevent their data from being accessed by others. The politics of "digital sovereignty" largely occur in the geopolitical context of algorithmic capitalism.

The most striking example in this regard is undoubtedly Russia, which is doing everything in its power to develop national algorithmic autonomy. Kevin Limonier's work on the construction of data centers in Siberia clearly illustrates the framework of Russia's digital sovereignty policy. It aims to store its data on its own territory, avoid dependence on China or the United States, and exert algorithmic influence in Central Asia.[30] In 2014, Russia's parliament passed a law requiring all Russian companies that process data connected to Russian citizens to host this data on national soil. However, digital sovereignty is only feasible for a very small number of states: Russia is indeed the only country, along with China, where Google and Facebook are not the most widely used search engine and social media platforms (Baidu and WeChat are most used in China, and Yandex and Vkontakte in Russia). At this stage, digital sovereignty seems but a vain hope for other countries around the world.

Although largely dominated by a Chinese–American duopoly, algorithmic capital flows still cross international borders, and certain international players are attempting to find an advantageous position for themselves in this new global power play. Middling powers such as Saudi Arabia and Japan have an important role in the creation of investment funds that will attempt to acquire a certain degree of control over algorithmic appropriation through international finance. One example is the establishment of the SoftBank Vision Fund, born of a partnership between the Japanese conglomerate SoftBank and the Saudi public investment fund, which is currently the world's largest equity fund; it also includes investments from Apple, Foxconn, and Oracle CEO Larry Ellison. The fund focuses its investments on AI, algorithmic technology, and robotics companies, and shows that middling powers want to position themselves in the value chain of the new world order of algorithmic capitalism.[31]

---

30 Kevin Limonier, "La Russie dans le cyberespace: Représentations et enjeux," *Hérodote*, nos. 152/53, 2014, pp. 140–60.

31 Evgeny Morozov, "Billion-Dollar Debts Control the Future of Tech Industry," *The Guardian*, March 11, 2018.

Algorithmic neocolonialism

American hegemony in the second half of the twentieth century was accompanied, as we noted earlier, by unequal relations of exploitation and transfers of value from the Global South to the Global North, particularly the United States. This was expressed through the international division of labor, unequal exchange, market coercion, and the mechanism of debt. The era of algorithmic capitalism continues these exploitative relationships, but it adds a new dimension: the transfers of value that characterize North–South relations now also take the form of an "algorithmic neocolonialism," centered mainly in the United States and China.

Algorithmic neocolonialism refers to the exploitation of labor and the appropriation of data produced by individuals and institutions in the Global South by algorithmic capitalists from the Global North and China.[32] This phenomenon is not necessarily new; IBM, for example, has been appropriating data passing through its computer systems installed in certain Global South countries since the 1970s. The World Trade Organization and most of the free-trade agreements that provide the framework for international trade have, in general, eliminated barriers to the movement of goods and services, but also to the movement of data. The rise of algorithmic capitalism is now reflected in an offensive by both Global North and Chinese companies, which are deploying algorithmic infrastructure in Global South countries and thus appropriating the data produced there.

Indeed, a vast process of "datafication of the Global South" has taken place in recent years.[33] This is visible across numerous Global South countries that are today integrating into the algorithmic capitalist order, and whose consumers are generating new masses of data. Major players in international development (the UN, World Bank, World Economic Forum) now favor "data-driven development" and consider data extraction to be an essential resource and tool in this cause. The UN

---

32 The concept of colonialism figures prominently in Mejias and Couldry's extensive examination of the logic of the data economy. See their illuminating contribution, Ulises A. Mejias and Nick Couldry, *Data Grab: The New Colonialism of Big Tech and How to Fight Back*, Chicago: University of Chicago Press, 2024. Our own concept of algorithmic neocolonialism seeks to grasp how the novel imperative of data extraction is articulated with more traditional forms of colonial extraction and exploitation in the international relations of algorithmic capitalism.

33 Linnet Taylor and Dennis Broeders, "In the Name of Development: Power, Profit and the Datafication of the Global South," *Geoforum*, no. 64, 2015, pp. 229–37.

secretary-general's Global Pulse initiative is spearheading this movement, which seeks to mobilize Big Data and artificial intelligence for development and humanitarian action. So, too, is the US Agency for International Development (USAID), which, before being gutted by the Trump administration, had launched a far-reaching partnership with Google, Amazon, and Apple to develop a new "digital strategy" for international development, which drew on AI in all its forms. China, for its part, is pursuing an active algorithmic diplomacy that consists of offering partnerships to countries in the Global South, notably in Africa, through which China will set up technological and algorithmic infrastructure in exchange for control over the data generated. Thus, the race for investment channels for surplus capital in the Global North and China is accompanied by a rush for data from the Global South.

The datafication of the Global South is also shifting power toward the private sector in the sphere of international development. Algorithmic companies have surpassed other actors in the institutional complex of international development (NGOs, international organizations, states, development banks, humanitarian organizations, and so on) in terms of the information they collect on the populations of Global South countries. Donors and developers are now turning to algorithmic companies to access the data collected by their apps, which is more detailed and extensive than that provided by "traditional" mechanisms for generating information on populations (censuses, government surveys, studies, academic research, and so on). These companies are thus gaining power in terms of understanding, approaching, and defining development issues, the establishment of development priorities, and project governance.[34]

These developments create new power dynamics between public and private actors, and also raise important epistemological and ethical issues.[35] It is true that the data extracted by algorithmic companies focuses more on "connected" populations, users, markets, and profit-generating potential (after all, data is always generated because it does have some potential for profit) than on population groups that are less integrated into the market economy, the digital world, and mass consumption. But it is also important that data ownership gives algorithmic

34 Ibid., pp. 231–2.

35 Rachel Adams, "Can Artificial Intelligence Be Decolonized?," *Interdisciplinary Science Reviews* 46, nos. 1–2, 2021, pp. 176–97. See also the "Decolonial AI Manifesto," available at manyfesto.ai; Jason Edward Lewis, Noelani Arista, Archer Pechawis, and Suzanne Kite, "Making Kin with the Machines," *Journal of Design and Science*, 2018.

capitalists the power to see, even as it denies users the right *not* to be seen. Mark Zuckerberg, for example, speaks philanthropically about his "Internet.org" venture, presented as an international development project aiming to connect disadvantaged populations worldwide to the Internet, "free of charge." Of course, Meta's interest lies in collecting the data that would be generated by such large-scale connectivity, since many countries in the Global South have little or no legislative or regulatory provisions for privacy or data protection. IBM, for its part, wants to use artificial intelligence to solve problems in agriculture, health, education, and sanitation in Kenya. If even a few years ago, algorithmic capitalists in the Global North took the trouble to pass off their thirst for data as a question of corporate social responsibility, aimed at contributing to the global fight against poverty, the accumulation of algorithmic power by those who own the data now exempts them from having to temper their discourse in this way. Ultimately, as Taylor and Broeders summarize, "corporations' data-gathering and analytical advantage means they no longer have to conceal for-profit motives under development discourse … but instead are now free to connect the expansion of markets, and the attendant expansion of digital records, with human development."[36] The global deployment of algorithmic capitalism is well and truly underway.

This algorithmic neocolonialism also concerns the global division of labor, and GAFAM and other algorithmic capitalists' delegation of many of the most alienating and lowest-paid tasks to "click farms" in poorer countries and regions. As noted in Thesis 5, here workers are often crammed together in unpleasant conditions to perform tasks that are presented to customers as if they had been automated. The construction of a digital labor market in the Global South relies on various institutional mechanisms within the development sector. One illustrative example comes in the form of "impact sourcing" organizations, which have set out to "redefine development" around slogans like "Bring jobs, not aid."[37] Impact sourcing is not new: It is part of the same process of offshoring jobs to cheap labor centers, broadly characteristic of economic globalization in the neoliberal era. In the service sector, for example, many companies have relocated call centers and customer services further afield, particularly to India. However, these impact-sourcing

36 Taylor and Broeders, "In the Name of Development," p. 234.

37 Phil Jones, *Work Without the Worker: Labour in the Age of Platform Capitalism*, London: Verso, 2021.

organizations have especially focused on the relocation of digital work. Initially mostly nonprofit organizations, they began to distribute digital work jobs in refugee camps and slums. Supported by the World Bank, several of these organizations have become for-profit companies and are helping to build digital labor markets that draw in communities marginalized by the traditional employment sector. For example, one such firm, Sama, is active in Africa, Asia, and Latin America and has set up low-cost pools of digital labor in Haiti and Pakistan, among other locations. The work performed in these contexts is arduous, often dangerous—and always poorly paid.

These phenomena reproduce the overall dynamics of externalization society, that is, the multiple processes by which dominant actors and countries shift the burden of their economic development onto groups and countries that are somehow disadvantaged, whether economically, socially, environmentally, or now algorithmically.[38] A good example is the externalization of "psychopathological waste," via the tens of thousands of click-workers who have to cleanse the Internet of images and videos of child rape, beheadings, and other sordid content, with all the trauma, anxiety, and mental health issues that this work entails. These "online janitors" concentrated in Global South countries are just one of many examples of North–South inequalities that have been exacerbated by the rise of algorithmic capital.

All in all, what appears as the development of AI in the Global North is based on algorithmic neocolonialism and the exploitation of human labor in the South. In the contemporary global system, capitalism and colonialism are thus rebuilding their historic partnership on new algorithmic foundations. This process of remolding international relations is accompanied by a hegemonic transition, as currently expressed in the growing rivalries between China and the United States. Yet it is also visible in a series of reactions from other countries that are attempting to take refuge in technological nationalism to protect their national interests—aiming for a digital sovereignty which is, in fact, largely illusory for most states around the world. The realm of AI, which now extends across the territories controlled by GAFAM and BATX, is thus taking the form of an arms race by the superpowers seeking to establish their hegemony in a "post-American" world today undergoing what some describe as "repolarization."

38 Stephan Lessenich, *À côté de nous le déluge: La société d'externalisation et son prix*, Montreal: Écosociété, 2019.

The emergence of authoritarian Chinese algorithmic capitalism, as well as the development of algorithmic neocolonialism, thus highlights the reality that algorithmic capital is not solely a Silicon Valley affair. Rather, this siliconization of the world is fraught with tensions and contradictions and is unfolding in complex ways across the planet. We will discuss this further in Thesis 13.

Thesis 13

The siliconization of the world is continuing apace

Human creativity is the ultimate economic resource.

—Richard Florida

The logic of algorithmic capital is advancing in a number of distinct spheres: in the corporate world, in our everyday comings and goings, but also in cities and territories, in culture, and in our imagination itself. Essayist Éric Sadin uses the expression "silicolonization of the world" to describe the multifaceted process by which "technoliberalism" and algorithmic governmentality are intertwining to form a new "civilizational model" that is being imposed across the globe.[1] We will more simply refer to the "siliconization" of the world, which for us involves a twofold process: (1) *the siliconization of people's thinking*, through the ideology that originated in Silicon Valley, the cradle of algorithmic capital; (2) the *siliconization of space* through the hegemony of a model of territorial development inspired by this same coveted region, a real El Dorado that world leaders seek to emulate.

It is thus essential that we describe algorithmic capitalism not only as a mode of production or a new stage of capitalism, but as a means of "making the world." By this term, we mean a set of processes that allow the world to be reconfigured in multiple different ways, be they symbolic or material, cultural or geographical. If algorithmic capitalism is, indeed, an institutionalized social order, this also involves it sinking roots in the

1 Éric Sadin, *La silicolonisation du monde: L'irrésistible expansion du libéralisme numérique*, Paris, L'Échappée, 2016.

collective imaginary and in territories that it sculpts according to its own imperatives. Algorithmic capital produces worlds that embody a certain zeitgeist. Here, Silicon Valley serves as the beacon that lights the way for efforts to establish other such "'Valleys' in many parts of the globe," at different local scales.[2]

Thesis 13 begins with a brief genealogy of Silicon Valley and the worldview emanating from this famous region, the "Californian ideology." We will not address here the TESCREAL ideology we analyzed in Thesis 8 but instead focus solely on the exoteric discourse of algorithmic capital. Next, we analyze the spread of this model at the territorial level. The aim is to build on the work done by geographers in order to highlight the spatial deployment of globalized capitalism in our present era.[3] Finally, we conclude with an analysis of how algorithmic capital seeks to territorially embed itself in a specific territory, namely Quebec.

A brief history of Silicon Valley

The term "Silicon Valley" emerged in the 1970s to describe a region in the south of the San Francisco Bay Area. A unique ecosystem took shape in this region, bringing together numerous tech companies, Stanford University, entrepreneurs, and venture capital funds. This ecosystem, which fostered strong synergies between academic research, public investment, the private sector, and the military, created a unique dynamic. A relatively small geographical area with a population of around 4 million is home to several leading tech and algorithmic capital companies, including Apple, Meta, Hewlett-Packard, Netflix, and Tesla, as well as a wide range of IT and tech start-ups. To date, Silicon Valley is also synonymous with great wealth; it is, in fact, the corner of the United States with the highest number of millionaires and billionaires relative to size.[4]

This "start-up paradise" did not come about overnight. As early as the nineteenth century, this region was already experiencing many of the

2 Ibid., pp. 21–2.

3 Harvey, *The New Imperialism*; Richard Florida, *Who's Your City?*, Toronto: Vintage Canada, 2009; Neil Brenner, Peter Marcuse, and Margit Mayer, eds., *Cities for People, Not for Profit: Critical Urban Theory and the Right to the City*, London: Routledge, 2012; Saskia Sassen, *The Global City: New York, London, Tokyo*, Princeton, NJ: Princeton University Press, 2001.

4 Durand, *How Silicon Valley Unleashed Techno-Feudalism*.

dynamics of the Industrial Revolution, including a spirit of conquest and boundless innovation. In his later years, Karl Marx took a keen interest in California. As he put it to one correspondent:

> I should be very much pleased if you could find me something good (meaty) on economic conditions in *California*, of course at my expense. California is very important for me because nowhere else has the upheaval most shamelessly caused by capitalist centralization taken place with such speed.[5]

In his book *Palo Alto: A History of California, Capitalism, and the World*, Malcolm Harris tells the dark and fascinating history of this prosperous Bay Area city, which has its roots in British colonialism, racial capitalism, and the ravages of extractivism underlying the American Dream.[6] The logic of capitalist accumulation is here embodied in the geography, history, and imaginary of this unique region. For Harris,

> It represents the gold rush and the next gold rush and the one after that, from produce to real estate to radios to transistors to microchips to missiles to PCs to routers to browsers to web portals to iPods to gig platforms to … Not just opportunity but the ceaseless renewal thereof. Silicon Valley is defined by its refusal to stop or even to slow down, which, given the dynamics of finance-led growth, would amount to the same thing.[7]

In the 1870s, Charles Marvin and Leland Stanford (Palo Alto's founder) developed an original method for breeding, training, and marketing trotting horses (used in harness racing). While the traditional method was to wait several years before training horses, Marvin and Stanford decided to target foals with the greatest potential and race them at an extremely young age. This strategy of selecting horses based on their early potential, shortening their production cycle, and optimizing their training according to rigorous data-based metrics, enabled them to break records and revolutionize the horse racing industry.

---

5 Karl Marx, letter to Friedrich Adolph Sorge, November 5, 1880, available at marxists.org.

6 Malcolm Harris, *Palo Alto: A History of California, Capitalism, and the World*, New York: Little, Brown, 2023.

7 Ibid., p. 616.

> Capitalist rationality and the exclusive focus on potential and speculative value was called the Palo Alto System and it worked. Horse-driven chariots were a literally ancient technology, studied for millennia, and the Palo Alto System transformed their production in just a decade. All a man needed to improve the world was an uncompromising dedication to profit and the capital to realize the necessary scale.[8]

The "Palo Alto System"—based on a combination of factors such as data, capital, optimization, techniques for selecting entities with high impact potential, and acceleration of the production cycle—can be found at various points in Silicon Valley's history. We cannot cover all the many historical examples, but it is worth recalling a short set of key moments, inspired by Sadin, who points out five key phases in Silicon Valley's development.[9] The first dates back to the late 1930s, in the context of the US government's planning for approaching war, which sought to establish close collaboration between industry and the military. This early iteration of Silicon Valley was a dynamic arrangement bringing together technology, academic research, the electronics industry, the military, and the spirit of enterprise. The image of the "garage" as a space for incubating and prototyping innovative projects emerged in 1938, when Hewlett and Packard created their first audio oscillator in the garage of their house in Palo Alto. The 1960s and 1970s ushered in the "second Silicon Valley," an experimental milieu that was deeply shaped by the counterculture. It sought to abandon technology deemed too functionalist and centralized in favor of a philosophy pitched toward values of autonomy, horizontality, and creativity. Stewart Brand's *Whole Earth Catalog* in 1972 and the founding of the Homebrew Computer Club in 1975 (bringing together Apple's Steve Jobs and Steve Wozniak) are examples of this strange synthesis of libertarian ideals, psychedelics, entrepreneurial initiative, a freewheeling, piratical spirit, and the desire to create decentralized communication systems in the service of individual freedom.

This unique mindset has been dubbed the "Californian ideology" by Richard Barbrook and Andy Cameron, who seek to describe the hybridization between the deeply antiauthoritarian and utopian hippie counterculture, entrepreneurial values, and enthusiasm for the emancipatory potential of digital technologies.[10] This unprecedented

8 Ibid., p. 71.

9 Sadin, *La silicolonisation du monde*.

10 Richard Barbrook and Andy Cameron, "The Californian Ideology," *Science as Culture* 6, no. 1, 1996, pp. 44–72.

convergence between cultures, which at first glance seem to be diametrically opposed, has given rise to the strange figure of the "libertarian entrepreneur," a kind of emerging synthesis of artist, inventor, mechanic, objective economist, and evolutionary strategist.[11] Several authors describe the emergence of a "dot-com neoliberalism" combined with a techno-utopianism that took hold in the 1990s, when Internet use became widespread and the "knowledge economy" took off.[12] This led to the advent of the "third Silicon Valley," symbolized by the "Declaration of Independence of Cyberspace" pronounced at the Davos Economic Forum in 1996 by John Perry Barlow, the libertarian-minded lyricist for the rock band The Grateful Dead and founder of the Electronic Frontier Foundation. As he put it:

> Governments of the Industrial World, you weary giants of flesh and steel, I come from Cyberspace, the new home of Mind. On behalf of the future, I ask you of the past to leave us alone. You are not welcome among us. You have no sovereignty where we gather … Your legal concepts of property, expression, identity, movement, and context do not apply to us. They are all based on matter, and there is no matter here.[13]

In singing the praises of cyberspace, he drew on themes dear to the immaterial economy, such as distrust of centralized authorities and an embrace of the emancipatory virtues of digital technologies. In his words we can also see the utopian imaginary dear to theorists of cognitive capitalism, for whom immaterial labor may one day free us from the yoke of the "weary giants of flesh and steel." The 1990s saw the emergence of Netscape and Yahoo!, the creation of Amazon, eBay, Google, and *Wired* magazine, the hegemony of Microsoft and AOL, and then the bursting of the Internet bubble in 2001. Sadin then identifies the transition to the "fourth Silicon Valley" in the 2000s, which saw the arrival of social media, the iPhone, the app economy, the mass collection of personal data, and machine learning. It was also in this phase that algorithmic capital took off, with the creation of behavioral databases, the algorithmic organization of collective life, and the creation of applications designed

11 Richard Buckminster Fuller, *Operating Manual for Spaceship Earth*, Baden: Lars Müller, 2013 [1969], p. 98.

12 Barbrook and Cameron, "The Californian Ideology"; Fred Turner, *From Counterculture to Cyberculture: Stewart Brand, the Whole Earth Network, and the Rise of Digital Utopianism*, Chicago: University of Chicago Press, 2008.

13 John Perry Barlow, "A Declaration of the Independence of Cyberspace," Electronic Frontier Foundation, 1996.

for individuals.[14] It was only after the financial crisis of 2007–8 that what Sadin terms the "fifth Silicon Valley" emerged, when this region established itself as a global model—the "Silicon Dream," which countless cities in the United States and other countries would try to replicate, with varying degrees of success. Los Angeles, Miami, and Sydney each describe themselves as "Silicon Beaches," while Boston and its hub at MIT, and indeed the Chilean capital Santiago, also seek to bring this same ideal to life. Cape Town, Lagos, and Nairobi want to become "African Silicon Valleys." Dubai, Bangalore, Dalian, Zhongguancun, and Dublin are each touted as promised lands, seeking to replicate an economic model today synonymous with prosperity. Silicon Valley thus represents a miracle or a "winning formula" that thousands of leaders around the world are seeking to recreate locally in order to re-energize their economies in a context of fierce international competition.

The creativity ecosystem

To fully understand the economic logic of algorithmic capital, we must examine the ecosystem where it is created. This means exploring the world of start-ups, with their incubators, accelerators, entrepreneurs, investors, venture capital firms, universities, research groups, hackathons, living laboratories, innovation valleys, and so on. Before these companies took flight as major platforms, they were conceived in "incubators" known as "technology clusters," where a variety of players from engineers to researchers and entrepreneurs come together to develop what they hope will be highly lucrative innovations.

A key thinker for understanding the geography of algorithmic capital is Richard Florida, a consultant and leading theorist of the "creative class."[15] Florida argues that, before click-workers can perform small routine tasks in the service of algorithms and before Uber drivers can start competing with taxi drivers, a creative class must first take form. It will consist in his view of affable and passionate individuals, guided by the three T's (talent, technology, tolerance), who settle in attractive neighborhoods in large cities to design disruptive innovations that will become the flagships of the "creative economy." In his *Who's Your City?*, Florida tracks down innovators, star scientists, engineers, designers, and other members of the creative class, mainly concentrated in certain

14 Sadin, *La silicolonisation du monde*, p. 72.
15 Richard Florida, *The Rise of the Creative Class*, New York: Basic Books, 2006.

metropolitan areas: Tokyo, Seoul, New York, San Francisco, Boston, Seattle, Austin, Toronto, Vancouver, Berlin, Paris, Stockholm, Helsinki, Osaka, Taipei, Sydney, Beijing, Shanghai, Bangalore, and similar creative hotspots. The concentration of innovation hubs on a global scale is increasingly taking the form of "peaks" and "valleys"—the clusters where high-level research is being pursued. Silicon Valley is the archetype of this economic-territorial model, which hundreds of other cities around the world want to replicate in order to get ahead.[16]

But the rise of metropolises, megaregions, and "global cities,"[17] which comes with the process of neoliberal globalization, is not merely a vast process of standardization. For it is also accelerating competition between territories and an international division of labor, with each city now having its own specialty and reputation in order to attract different segments of the creative class. According to Florida's rough typology, this class spans skilled professions in sectors from science and technology to arts and design, media and entertainment, law, finance, management, health, and education. The geography of creative work brings into relief this functional differentiation of territories, which fuels innovation and nurtures the talented individuals who help to drive the twenty-first-century economy.[18] At the dawn of algorithmic capital, before the rapid spread of social media and smartphones, Florida's writing had already enthusiastically noted the rise of the creative class and its strong propensity to move around in order to find the best employment, in combination with the overall rise in service economy job numbers.[19]

Such a detour into the creative-class narrative is helpful because it helps us see the process of innovation through the eyes of entrepreneurs, engineers, innovators, and investors who played a key role in incubating the technologies and digital platforms that would become dominant in the 2010s. While Wall Street remains emblematic of financialized capitalism, Silicon Valley is the archetype of algorithmic capital. It does so with an optimistic philosophy and sense of adventure, which it shares with the planet through a series of TED talks, each more inspiring and "revolutionary" than the last.

In this sunlit world, all ideas are welcome as long as they have strong potential for impact and growth, and creative individuals are invited to develop their projects within complex and personalized entrepreneurial

16 Florida, *Who's Your City?*
17 Sassen, *The Global City*.
18 Florida, *Who's Your City?*, p. 118.
19 Ibid., p. 109.

frameworks known as "incubators." These offer hosting, advisory, and financing services aimed at helping businesses to take off in their early years. The concept of the business incubator was invented in the United States in 1959 by Joseph L. Mancuso, but it began to spread only in the 1980s, taking various forms around North America and Europe: innovation centers, business incubators, technopoles, and science parks. It was in the 2000s that incubators began to spread like dandelions across the world. While there were barely a dozen incubators in North America in 1980, there were more than 1,400 in 2006. Between 1997 and 2005, the number of incubators in the United Kingdom grew from 25 to 270. Today, the International Business Innovation Association estimates that there are more than 7,000 incubators worldwide.

These incubators are keen on the newest companies, whether they are start-ups, pre-start-ups, or in their infancy, as nothing more than ideas in the minds of carefully selected innovators. These start-ups (or young companies) are mostly guided by the classic ideal of the entrepreneur, which seeks to capture the best ideas that create value and can be quickly disseminated and marketed on a large scale in order to become profitable in the medium and long term. Other start-ups, inspired by social and environmental concerns, may also take form in social-innovation incubators, which combine an entrepreneurial approach with a desire to solve social problems and, perhaps at the same time, try to better the world. But as we shall see, in the vast majority of cases, the most successful and impactful innovations are mainly concentrated in the technology sector and belong to the select club of "unicorns."

Welcome to the unicorn club

What is a unicorn? This legendary creature, present in many ancient traditions, became the most important imaginary animal in Christendom from the Middle Ages until the end of the Renaissance. Endowed with rich symbolism, a source of mystery and inspiration, representing the experience of the divine, strength in difficult situations, and the power of alchemy, love, and protection, the unicorn embodies, according to Carl Jung, an archetype of the dual, changing, and paradoxical nature of the human psyche. In the disenchanted world of late modernity, faced with the failure of financialized capitalism, the myth of the unicorn is strongly re-emerging in venture capitalists' symbolic universe. This new use of the word "unicorn" was first coined in 2013 by the venture capitalist Aileen

Lee, referring to those rare, incredibly dynamic start-ups whose valuation has managed to cross the symbolic $1 billion threshold.

Unlike the value of established companies, which is often based on their economic performance, the valuation of unicorns is derived from their growth potential and anticipated long-term development within their potential market. For example, venture capitalists regularly invest heavily in start-ups that demonstrate significant capabilities for data extraction or already have significant data assets. Capital is thus "risked" as a bet on the young company's ability to monetize its data assets, further develop its data extraction mechanisms, or create new algorithms in the near future.[20] This is, then, not a matter of market capitalization, but rather of the valuation carried out by venture capitalists themselves during the financing rounds for promising start-ups. This purely speculative value may sometimes be corroborated when start-ups are bought up by large companies—for instance, Facebook's purchase of Instagram for $1 billion in 2012, a transaction that effectively transformed Instagram into a unicorn.

The capitalist imaginary is now infused with the magic of unicorns, symbols of the dream of any start-up seeking spectacular growth, capable of reaching the heights of success even before reaching the critical stage of market capitalization. Not all unicorns are companies directly linked to algorithmic capital, but most of them do operate in the digital sector, the collaborative economy, e-commerce, fintech, AI, and other areas related to the high-tech sector.

In 2013, only 39 start-ups could claim the title of unicorn, but 349 companies achieved this classification in the year 2021 alone (more than in the previous five years combined). The global total reached 1,000 unicorns in May 2022,[21] 1,215 in June 2023, and between 1,300 and 1,600 in late 2025—with a total value of more than $4 trillion. One study shows that start-ups founded between 2012 and 2015 are growing twice as fast as those created between 2000 and 2003.[22] How can we explain the sudden proliferation of these creatures, which were so few in number little more than a decade ago?

On the one hand, investors and venture capital firms are adopting the Get Big Fast (GBF) strategy to boost the growth of start-ups, so that

20 Jathan Sadowski, "When Data Is Capital: Datafication, Accumulation, and Extraction," *Big Data & Society* 6, no. 1, January 2019.

21 Russ Zalatimo, "Venture Capital: How the World Reached 1,000 Unicorns," *Forbes*, May 19, 2022.

22 Jim Goetz, "How Unicorns Grow," *Harvard Business Review*, January 2016.

they can occupy a dominant position in the market by eliminating their potential competitors as quickly as possible.[23] It should be remembered that platforms' performance relies heavily on network effects, which allow them to occupy a monopoly position by being first to market. Speed is a key factor for success. Add to this the technological advances in mobile phones, enabling the creation of millions of apps; cloud computing, which reduces fixed capital costs so that companies can focus on platform design; and the rise of social media, which allows innovations to spread at lightning speed, and we have ingredients for an environment conducive to the emergence of unicorns.

Furthermore, start-ups now have access to a vast pool of private investment, allowing them to delay their entry into the stock markets. Whereas in 1999 it took a tech firm around four years to get to its initial public offering (IPO), today it takes more than eleven years to take this "leap of faith." Why delay this crucial step? Because a start-up that is highly valued in venture capital and private markets risks being devalued on the stock markets if investors consider the company worth less than it claims. In other words, venture capital–fueled financing allows young companies to delay their coming of age and postpone confrontation with harsh reality. Uber's chaotic entry into the stock market is a case in point: The company was valued at $76 billion just before its IPO in May 2019, but it was worth only $33 billion in March 2020. It has consistently struggled to generate actual profits.

WeWork offers an even more telling example. This coworking start-up, founded in 2010 by Adam Neumann and Miguel McKelvey, already had forty employees, 2,000 members, and three coworking sites in New York by 2012. The founders decided to step on the accelerator to open sixty spaces and attract 46,000 members in 2014, convincing investors such as JPMorgan Chase and Benchmark Capital to put in extra millions.[24] WeWork became a unicorn in 2013, three years after its creation, and accelerated its rise as it opened ten to fifteen new sites per month in 2017. At its peak in January 2019, WeWork had more than 5,000 employees, at 280 locations in eighty-six cities and thirty-two countries around the world, representing an estimated value of $47 billion. But even as its IPO was being scheduled for September 2019, signs of overheating were becoming apparent. Goldman Sachs valued the company

23 John D. Sterman et al., "Getting Big Too Fast: Strategic Dynamics with Increasing Returns and Bounded Rationality," *Management Science* 53, no. 4, 2007, pp. 683–96.

24 Isabelle Massé, "WeWork: La grosse licorne qui s'est plantée," *La Presse*, October 28, 2019.

at just $20 billion, investors' interest cooled, and Neumann was shown the door by one of his main investors, SoftBank, which accused him of mismanagement. WeWork's downfall quickly followed: Neumann left the company in September 2019, its value plummeted to $8 billion that same October, and SoftBank proceeded with numerous restructurings, closures, and several rounds of layoffs. In addition to its fragile business model and its overly optimistic forecasts, which eventually brought it to the brink of collapse, the example of WeWork perfectly embodies the spirit of the start-up nation.[25] Just like fallen angels, resplendent and seemingly invincible unicorns can also experience the humiliating ordeal of decline.

When they do not collapse of their own accord through misfortune or lack of foresight, most unicorns are ultimately swallowed up by the world's major platforms. While venture capital funds help to increase the value of start-ups on the supply side, on the demand side web giants are keen on new products, and they have huge sums of money at their disposal to buy up unicorns that will further inflate their total value and market dominance. While GAFAM companies invest internally in research and development, from time to time they can also afford to buy certain unicorns when they reach their prime. It is a small act of caprice with high strategic value. When firms like Meta and Apple go spending, they prefer to acquire unicorns and start-ups that help them consolidate their market dominance, rather than solely grow their own fixed capital.[26] The key moment for acquisition comes when the start-up is mature enough and has shown its strong growth potential, but is still young enough to be affordable and attractive. Successful start-ups play hard to get in order to secure the best price, secretly dreaming of being bought out by these wealthy giants; the luxury and prestige of the big players are exchanged for the beauty of a ravishing, promising youth. The GAFAM companies are the sugar daddies of algorithmic capital. As of March 2023, these five giants had acquired hundreds of companies.

We can already see here an initial paradox in the Silicon Valley imaginary: While it claims to be based on the free market, horizontality, and the dynamism of talented small entrepreneurs from the "creative class," it actually promotes the consolidation of monopolies and oligopolies, the hypercentralization of power, and the end of free competition.

---

25 Matthew Zeitlin, "Why WeWork Went Wrong," *The Guardian*, December 20, 2019.

26 Neil Howe, "What's Feeding the Growth of the Billion-Dollar 'Unicorn' Start-ups?," *Forbes*, March 18, 2015.

The five contradictions of the Californian ideology

Let us look again to the spirit of Silicon Valley to examine whether this model of development can actually deliver on its promises of prosperity on a global scale. One of the foundational texts of the Californian ideology, published in 1995 by a group of technology essayists and investors, is entitled "Cyberspace and the American Dream: A Magna Carta for the Knowledge Age."[27] According to its authors, cyberspace opens up a new sphere of universal freedom that everyone can explore; but for this "territory of knowledge" to realize its potential, the state must be rolled back, competition intensified, the creative power of markets unleashed, disruptive innovations propelled forward, and private property reconfigured. The influence of the libertarian philosopher Ayn Rand is deeply felt in this manifesto:

> To create the new cyberspace environment is to create *new* property—that is, new means of creating goods (including ideas) that serve people ... Demassification, customization, individuality, freedom—these are the keys to success for Third Wave civilization ... ownership by the people—private ownership—should govern every deliberation. Government does not own cyberspace, the people do.[28]

Will this utopian vision of the future lead to the promised land of freedom? Cédric Durand's sharp analysis is worth rehearsing here in part. He identifies five pillars of the "Silicon Valley consensus" to show how this worldview produces results that are diametrically opposed to the hoped-for scenario. These pillars are: "(1) the continuous reinvigoration of economic structures thanks to start-uppers' thirst for adventure; (2) the pinnacle of workplace autonomy and creativity; (3) a culture of openness and mobility; (4) the promise of shared prosperity; and finally, (5) the ideal of the state becoming obsolete."[29] Borrowing from Durand's analysis, we can clearly highlight the contradictions of the Californian ideology.

First, as we saw earlier, a start-up is not intended to thrive on a small scale in the local market, providing a living for its owner, but rather to grow as quickly as possible, often in a speculative manner, to become a

27 Esther Dyson, George Gilder, George Keyworth, and Alvin Toffler, "Cyberspace and the American Dream: A Magna Carta for the Knowledge Age," *Future Insight Release* 1.2, August 1994.

28 Ibid.

29 Durand, *How Silicon Valley Unleashed Techno-Feudalism*, p. 30.

large platform or be bought out at a high price by existing digital giants. Because of factors such as the need for gigantic quantities of data and access to immense computer power, algorithmic capital amplifies the monopolistic dynamics of capitalism, reducing competition to a handful of giants. After an initial phase of dynamic competition, the concentration of capital at the company level and its centralization in a given sector contribute to a monopolistic trend. "Socialization of production" refers to the development through which the increasing complexity of the means of production and the greater cooperation between multiple different actors mean that economic units must coordinate a host of interdependent processes, thereby increasing the "collective" nature of the work process. "The result," Durand points out, "is that the size of production units increases, at the same time as their interrelationships in 'networks' of companies become denser at both the domestic and international levels."[30] This process is evolving along an S-curve, illustrating how the optimism associated with cyberspace in the 1990s gave way to a more monopolistic dynamic in the 2000s and 2010s.

Second, Silicon Valley's promotion of autonomy and creativity at work, and the simplistic flattery of the "creative class" narrative, must confront the harsh reality of digital labor, the persistence of precarious salaried work, and the control mechanisms targeting employees of platforms, assembly lines, and factories controlled by algorithms. The utopia of Google and Facebook's playful work environments is accessible only to a select few—the "venture workers" we discussed in Thesis 5—which forms a kind of digital labor aristocracy, while the deterioration of working conditions continues to be seen across many sectors. The dynamics of control and surveillance are penetrating the corporate sphere, with the imperative of competitiveness generally taking precedence over employee well-being. Although the use of algorithms could, in theory, allow for greater freedom in the workplace, algorithmic governmentality is mainly used to increase the value of the company. The power relationship between capital and labor therefore favors the former at the latter's expense.[31]

Third, while it may seem that the culture of mobility and openness promoted by Silicon Valley is a model that can be easily exported to a multitude of geographical contexts, in reality it is extremely difficult, if not outright impossible, to generalize this model. The creation of innovation ecosystems requires a high concentration of human and material

30 Ibid.
31 Ibid., pp. 47–8.

resources in a single area, whether in the form of infrastructure, specialized suppliers, skilled workers, or other related services. Silicon Valley's success stems from a unique combination of circumstances, creating a quite specific territorial and human space that took several decades to bear fruit. Some cities, such as Bangalore, Singapore, Boston, and Tel Aviv, have managed to succeed in developing something comparable, while hundreds of other initiatives around the world have failed. Not every region can become a Silicon Valley. These attempts at imitation cannot simply copy and paste from this success story, because it requires a technological, cultural, economic, and human background that is not easily replicated.[32]

Fourth, while Silicon Valley is often associated with a horn of plenty, in reality it is primarily a source of "innovation without growth." On the one hand, algorithmic capitalism surely is generating far-reaching social changes and technological progress through the tools and applications that it offers to connect and carry out a multitude of projects. Yet we are also seeing economic decline, marked by weak GDP growth, a slowdown in productivity advances, and a financial sector that acts like a deadweight on the economy.[33] There is no doubt that both growth and productivity have been stagnating in most Global North countries since the 1970s. Hence, technological progress has become decoupled from general economic (and social) progress. Algorithmic capital is a new stage of capitalism that does not necessarily perform better than the previous ones in terms of economic growth, profitability rates, productivity gains, or overall economic stability.

Fifth, while the Californian ideology hoped that technological innovations would free humanity from the burden of the state, this has not been the case in practice. It should be remembered that behind Silicon Valley's façade of openness and generosity lies a surveillance culture rooted in the post-9/11 "war on terror." Although surveillance practices by states and their intelligence agencies are nothing new, they have had more leeway since the adoption in the early 2000s of a series of legislative measures and security rhetoric allowing for increased intrusions into privacy. Once a state of emergency was declared, the new technological possibilities provided by the digital giants could be mobilized for reasons of state. These companies forged close collaborations with intelligence agencies, resulting not in a decline of the state and its intrusive practices but in their strengthening.

32 Ibid., pp. 49ff.
33 Ibid., p. 55.

The growing symbiosis between state power and algorithmic technologies, which we explored in Thesis 10, thus accentuates a trend toward what Zuboff calls "surveillance exceptionalism." This goes far beyond concerns about Google: The mass surveillance programs revealed by Edward Snowden in 2013 show that the National Security Agency (NSA), together with the British, Canadian, Australian, and New Zealand intelligence services, was enabling systematic surveillance of Internet users. The PRISM program launched in December 2007 included providers such as Microsoft, Google, Yahoo!, Facebook, PalTalk, YouTube, Skype, AOL, and Apple. As of 2008, 700 XKeyscore servers were installed in more than 150 sites in various countries, while the Tempora project enabled Britain's GCHQ to intercept data transiting via cables between Europe and the United States. These widespread surveillance practices, bound up with the global fight against terrorism, are fueling the development of a "security-industrial complex" that deploys facial recognition techniques, cameras, drones, microphones, and other devices to discreetly collect data. Systemic surveillance optimizes the power of algorithms, which are extremely useful for authoritarian states, prison services, and police departments in major cities. This technology is clearly not freeing us from the state. Instead, we are seeing an unprecedented social contract take form between Silicon Valley and the Leviathan with its ever-expanding security measures.[34]

The spirit of Silicon Valley was meant to enhance market competition, increase greater autonomy in the workplace, and lead to the ever-wider adoption of its winning formula as a source of prosperity for all and the gradual withdrawal of the state. In reality, it is bringing about the exact opposite: a world of monopolies, control, territorial disparities, lack of growth, and increased state surveillance. The Californian ideology acts like a camera obscura—a fully inverted representation of reality.

The St. Lawrence project: The dream of a Silicon Valley in Quebec

The siliconization of thinking and even territory has found a particularly visible—and mediocre—expression in Montreal, Quebec, and Canada as a whole. This can be seen in François Legault's St. Lawrence project, the public funding invested in the AI ecosystem in Canada, and the "artificial intelligence city" in Montreal's Mile-Ex neighborhood. All three of

---

34 Morozov, *Le mirage numérique*, pp. 94–5.

these express the contradictions of the Californian ideology. These cases highlight the paradoxes of the Silicon Valley model, with benefits that mainly accrue to a privileged minority and to foreign interests.

Let us start with the simplest example of the siliconization of the mind. In 2013, the ex-businessman and politician François Legault published a book in which he presented in some detail the economic vision of the Coalition Avenir Québec, the political party he leads, which would come to power in Quebec five years later. The book, entitled *Cap sur un Québec gagnant* ("Toward a Winning Formula for Quebec"), is a summary of the Silicon Valley vision adapted to the context of Quebec. Seeking to propose an alternative to the *Plan Nord* ("Plan North") advanced by Jean Charest's Liberal government, which focused on natural resource extraction, Legault believes that the key to success lies in the development of innovation-focused areas:

> My dream is to turn the St. Lawrence Valley into an Innovation Valley, a place where imagination and creativity will become the driving force behind re-invigorating our economy … Our neighbors to the south created Silicon Valley. We could draw inspiration from them to invent, on our own scale of course, the Innovation Valley![35]

Legault seeks to replicate this "winning formula" in order to create nothing less than the "next major global cluster" of innovation. He draws inspiration in particular from the 22@Barcelona project, an urban regeneration strategy launched in the 2000s in El Poblenou, a run-down industrial district of Barcelona. The city administration there, in collaboration with the Spanish government, promoted the conversion of industrial real estate into lofts, public transport, and dynamic public spaces to make these neighborhoods more attractive to creative workers with discerning tastes. Legault points out that "it took *planned human intervention*, through public institutions, for the revitalization to really take shape."[36] He thus recognizes that economic development does not magically happen thanks to market forces alone but relies on vigorous intervention by public authorities to develop a city that serves innovation (and private interests). The siliconization of cities, aimed at transforming former working-class neighborhoods into new, hip districts, thus represents a broad process of gentrification planned by the state.

---

35 François Legault, *Cap sur un Québec gagnant: Le projet Saint-Laurent*, Montreal: Boréal, 2013, p. 12.

36 Ibid., pp. 87–8. Our emphasis.

An urban regeneration strategy modelled on 22@Barcelona guided the Quartier de l'Innovation (QI) project in Montreal's Griffintown district. Launched in 2013 through a collaboration between the École de technologie supérieure and McGill University, the QI aimed to create a "living laboratory where organizations, researchers, citizens, businesses and students could collaborate, in order to increase the city's potential for creativity and innovation." Encompassing the Cité du Multimédia and the Cité du commerce électronique, the goal was to make this into Quebec's own (smaller) version of Silicon Valley.[37] The Griffintown neighborhood had no schools and little local service provision for families, with various problems of community cohesion and of equality in living standards. Still, this ecosystem, which includes at least twelve incubators and accelerators, has attracted companies such as IBM, Coveo, Nüvü Cameras, and Espace Fabrique, and succeeded in creating nearly 600 jobs between 2016 and 2018. François Legault, showing great enthusiasm for this initiative, is seeking to replicate this experience throughout Quebec:

> The Innovation Valley that I am proposing is a much larger project than the Quartier de l'Innovation. In fact, it would encompass the metropolis and cover the entire St. Lawrence Valley, the great plain that stretches between the Appalachians and the Laurentians. It would include cities located along the St. Lawrence River and others adjacent to it, such as Sherbrooke and Gatineau. In short, it would be the economic heart of Quebec. This Innovation Valley would have enough heft to compete on equal terms with the rest of the world in attracting strategic investments, company HQs and research centers.[38]

Legault's economic vision thus consists of "siliconizing" the St. Lawrence Valley by creating multiple innovation zones, hubs tailor-made to stimulate private investment in the technological innovation sector, "high-quality locations where well-educated and well-connected people live in creative proximity."[39] Collaborations between researchers and entrepreneurs, the commercialization of research, venture capital funds, start-up incubators—all these ingredients must be brought into the mix. The cornerstone of this economic strategy is to strengthen the ties

---

37 "Quartier de l'innovation: Un Silicon Valley québécois," Radio-Canada, May 13, 2013.

38 Legault, *Cap sur un Québec gagnant*, p. 90.

39 Ibid., p. 91.

between Quebec's major universities and the dozen innovation zones accredited under the Zones d'innovation Québec banner, such as the Technoparc Montréal, the Parc technologique du Québec métropolitain, Sherbrooke Innopole, and the Technopole du Québec maritime in Rimouski.

After Legault's Coalition Avenir Québec took office in autumn 2018, the St. Lawrence project began to take shape. In February 2020, Legault's government launched a call for tenders, seeking proposals for the creation of innovation zones. This was accompanied by a detailed guide to the objectives, implementation schedule, and success criteria for initiatives, which had to be industrial, entrepreneurial, based in the knowledge economy, connected, collaborative, and sustainable. Around thirty projects were submitted from thirteen regions of Quebec, and in February 2022, the Legault government announced investments of more than $690 million and the creation of the first two zones: one in Sherbrooke (focused on quantum technologies) and the other in Bromont (Technum Québec, specializing in "smart" electronic systems). The aim was to accelerate "the transition from idea to market" by attracting as much private investment as possible, particularly from abroad. The companies that announced investments included IBM, 1QBit, Pasqal, Eidos, Black Pearl Technology, Teledyne DALSA, and Aeponyx.[40]

Nevertheless, some innovation zone projects are struggling to get off the ground. One example is the Littoral Est project, which aims to revitalize the Maizerets/D'Estimauville port area of Quebec City. This project, developed jointly by the local city hall, Laval University, and Québec International, sought to create an innovation zone focused on smart transport logistics, sustainable health, and urban clean technologies, with a "strategic and connected location thanks to the multimodal nature of the passenger and freight transport network (maritime, rail, motorway, public, and active transport)."[41] The goal is to revitalize 375,000 square meters of municipal land, thanks to $109 million in investments for site development, $1.5 billion in private and public investments, and $4.5 billion in further investments by 2035, in order to create more than 15,000 new jobs. Another objective is to create a "connected living environment" through a network of sensors linked to algorithmic systems in street furniture and municipal facilities, providing a testing ground

40 Jean-François Venne, "Les zones d'innovation sont en plein essor au Québec," *Le Devoir*, May 2021.

41 Ville de Québec, "Projet de Zone d'innovation Littoral Est," Québec, August 2020, p. 10.

for the Internet of Things and 5G networking. The whole project would be steered by a governance model designed both by and for the private sector: The Maison de l'innovation sociale, a nonprofit responsible for the future innovation zone, its operations building, and its incubator, planned to have a board of directors mainly controlled by private companies.

This innovation zone project has sparked a strong public response, organized around the Table citoyenne Littoral Est. Bringing together around twenty resident groups and other organizations from local communities and student, environmentalist, and anticapitalist spaces, their demands oppose gentrification and the privatization of urban development. This opposition is combined with their determination to protect air quality and the environment, finance social housing, ensure genuine citizen participation in planning decisions, and convert a municipal garage into a self-managed social center.

Weak community acceptance of the Littoral Est project, combined with procrastination over the route of a future tramway to this area, has meant that the Legault government has not yet recognized Littoral Est as an innovation zone (the project has since been renamed InnoVitam, giving it a fresh coat of paint).[42] Like the aborted Sidewalk Toronto project, which aimed to entrust the entire management of Toronto's Quayside district to a subsidiary of Google, social mobilizations against gentrification, the privatization of urban development, the Internet of Things, and GAFAM-controlled cities seem to be throwing a spanner in the works of the siliconization of the world. We will return to this contestation in Thesis 18 when we look at the resistance to algorithmic capital, which also takes the form of struggles for the right to the city and the defense of local territories.

Public-funded AI, privatized profits

The siliconization of the world is not the preserve of a few isolated individuals, but the result of a strategy pursued at various levels of government. In addition to the St. Lawrence project led by the Legault government, it is interesting to take a closer look at the flow of finance linked to AI in Canada, and particularly in Montreal, since this city has become one of the industry's global hubs in recent years. In her report

42 Sébastien Tanguay, "Zone d'innovation ou zone d'embourgeoisement à Québec?" *Le Devoir*, January 17, 2023.

published in 2021, researcher Ana Brandusescu offers a clear, critical, and detailed portrait of this booming industry, an "AI ecosystem" that in 2019 included "more than 800 start-ups, 60 public research laboratories, 75 incubators and accelerators, and 60 investor groups based in Toronto, Montreal, Waterloo, Edmonton, and Vancouver."[43]

A huge portion of public funding for AI comes from the federal government, which launched its first pan-Canadian AI strategy in 2017, led by the Canadian Institute for Advanced Research (CIFAR). The development of this strategy pulls in government funds, but it is also funded by Meta and the RBC Foundation. Nonprofit research centers Amii, MILA, and the Vector Institute lead the three main AI clusters in Edmonton, Montreal, and Toronto, respectively.[44] It should be noted that, of the total CA$1.1 billion in Canadian government funding allocated between 2017 and 2020, 72 percent went to the private sector.

This innovation ecosystem in Montreal is currently Canada's most dynamic and prosperous hub for AI development. This is the result of decades of funding for the video game industry, the presence of deep learning pioneer Yoshua Bengio, the city's multiple universities, high-caliber researchers, an attractive image, investment funds, and other factors mentioned above. In 2018, Brandusescu reported on the CA$950 million budget of the Initiatives des Supergrappes d'innovation ("Innovation Superclusters Initiative") for AI research, as well as CA$2 billion for industry and CA$1 billion for university research.[45]

But, Brandusescu insists, we also need to look at something else: Who benefits from all this public funding for AI? Her conclusion is clear: Massive public investment in algorithmic technology mainly benefits the private sector. Even when we consider nonprofit organizations and other research centers, the picture is hardly any brighter, as Brandusescu points out: "The concentration of power provides advantages to a handful of entities with financial resources, data and technologies distributed among a few universities and nonprofit organizations, start-ups and (large) international technology companies associated with them."[46] What is more, the public funding for start-ups is *not* primarily used to encourage job retention and to stimulate the local economy but rather

---

43 Ana Brandusescu, "Politique et financement de l'intelligence artificielle au Canada: Investissements publics, intérêts privés," Montreal: Centre de recherches interdisciplinaires en études montréalaises, McGill University, March 2021, p. 12.

44 Ibid., p. 18.

45 Ibid., p. 3.

46 Ibid., p. 6.

to propel companies that will quickly be bought out by foreign interests. The acquisition of Element AI by the California firm ServiceNow (worth more than $100 billion) is a prime example of this. Cofounded by researcher Yoshua Bengio, Element AI was a true Quebec flagship, financed with tens of millions of dollars from the Caisse de dépôt et placement and the Quebec government. The company, valued at between $600 million and $700 million in its last funding round in September 2019, was apparently sold for a lesser amount—estimated at $500 million, though the exact number remains secret. Although this acquisition stirred discontent in Quebec's tech sector, it was perfectly "normal" and even predictable according to the logic of algorithmic capital: Future unicorns are destined to be bought out by large platforms. As various studies show, there appear to be very few positive side effects locally, even in strictly economic terms.[47]

Public funding for technology ecosystems also has another perverse effect, as seen in the accelerated gentrification of areas surrounding the "AI city" in Montreal's Mile-Ex district—to the detriment of long-term residents, who will soon no longer be able to afford to live in their own neighborhood, now transformed into an innovation hub. AI laboratories, such as those run by MILA, RBC, and the multinational Thales, occupy the new Ô Mile-Ex building complex, owned by a San Francisco real estate investment fund and located a stone's throw from Microsoft's offices. As Myriam Lavoie-Moore notes, gentrification is calculated by the growing number of luxury vehicles and condos in the neighborhood. The gentrification process begins when a tenant population leaves—because rental buildings are being converted into real estate projects, and because of the rise in the cost of living—and is replaced by wealthy homeowners.[48]

In San Francisco, the AI boom has contributed to amplifying the housing crisis and increasing evictions. Importing this development model is fueling the same harmful dynamics in Montreal, which has been experiencing a major housing crisis for several years now. In addition to the Mile-Ex area, which is already almost entirely gentrified, there is Parc-Extension, one of the most multiethnic and disadvantaged

47 Ulysse Bergeron, "La vente d'Element AI à une firme américaine suscite la grogne," *Le Devoir*, December 1, 2020; Myriam Lavoie-Moore et al., "Financer l'intelligence artificielle, quelles retombées économiques et sociales pour le Québec?," Montreal, IRIS, 2019.

48 Myriam Lavoie-Moore, "L'intelligence artificielle contribue à la gentrification des quartiers de Montréal," Montreal, IRIS, 2021.

neighborhoods affected by this "AI-amplified gentrification."[49] The recent construction near the MIL Campus of the University of Montreal, an institution that prides itself on being a new player in this innovation ecosystem, has only added to this real estate boom. At the Global Conference on Machine Learning and Artificial Intelligence, held at the MIL Campus on May 5, 2022, the Comité d'action Parc-Extension disrupted the opening session by setting off a fire alarm and intervening with a public statement of opposition.

The siliconization of the world is, indeed, having destabilizing effects. This is especially clear if we consider the interests of disadvantaged groups and local communities rather than simply adopting the perspective of AI researchers and start-up entrepreneurs and the radiant image projected by Silicon Valley. Algorithmic capital is not just an abstract logic or a dynamic of accumulation. Rather, it is a worldview materially embodied in territories, which it helps mold according to the interests of the "creative class"—a force for the gentrification of neighborhoods and even of thinking itself. The economic strategy of innovation zones, pursued at the local or national level, promises a shared future of prosperity. Yet this is merely a way for algorithmic capital to increase its power through the socialization of costs and the privatization of profits. Unfortunately, the siliconization of the world that began a decade ago is only accelerating, having reached full tilt with the arrival of a new player: COVID-19.

49 Nick Gertler et al., "Digital Divides: The Impact of Montreal's AI Ecosystems on Parc Extension: Housing, Environment and Access to Services," Parc Extension Community-Based Research Network, April 2022, p. 44.

Thesis 14

The global pandemic brought a great leap forward for algorithmic capitalism

The term "shock doctrine" describes the quite brutal tactic of systematically using the public's disorientation following a collective shock.

—Naomi Klein

On March 11, 2020, the World Health Organization (WHO) sounded the alarm, declaring that COVID-19 could be classified as a global pandemic. The WHO then asked countries to implement a series of measures aimed at curbing its spread: elimination of physical proximity; mandatory handwashing; cancellation of social, sporting, and cultural events; shutting borders; preventive quarantines; and the closure of schools, nurseries, and nonessential businesses. A state of emergency was declared in almost every country around the world, giving public authorities exceptional powers to deal with this unprecedented health, economic, and social crisis. This was the Great Lockdown that marked the first wave of COVID-19.

This "health exceptionalism" represented the perfect storm for accelerating the deployment of algorithmic capital, digital platforms, and other technological solutions. After all, such mechanisms had become indispensable for ensuring a certain continuity of social and economic life despite drastic government-imposed constraints. Overnight, the whole world was plunged into a universe of social distancing, compulsory remote working, strict lockdowns, home schooling, queues outside shops, more or less temporary unemployment, and increased police surveillance of cities. Digital devices quickly became essential for maintaining social, economic, and emotional ties during this time of hardship, whether for work, school, entertainment, meal delivery, or communicating with friends and family.[1]

The total time spent in front of screens, which was already high before the pandemic, exploded. The statistics we mentioned in Thesis 3 were already striking, with the time devoted to screens as of 2019 having increased dramatically compared to just twenty years earlier. These pre-pandemic figures for screen time still provide a useful statistical basis, insofar as this crisis can be considered an exceptional event, an anomaly. Still, it is important to note that the COVID-19 pandemic led to historic peaks in screen time that would have been difficult to imagine even extrapolating from the 2019 figures. According to various studies, screen use suddenly increased by 5 hours per day, reaching as high as 17.5 hours per day for heavy users and 30 hours per week for median users.[2] The average daily time spent online now stands at nearly 7 hours (6 hours 54 minutes). All age groups were affected by these trends, including children, teenagers, adults, and the elderly, marking an average increase of 60 to 80 percent compared to the pre-pandemic period.[3] Whether or not a matter of health-crisis exceptionalism, some of these practices—for instance, forms of remote working and reliance on Zoom—are likely here to stay, to some degree. In this sense, the pandemic truly set off a great leap forward in screen time.

This sudden increase in attention time spent online had ambivalent psychological effects. Human beings are social animals, and being able to maintain contact using social media and digital devices during prolonged lockdowns and periods of isolation at home enabled many people to cope, and even to get "closer" to each other. Various social interactions moved online: virtual parties, family conference calls, romantic dates, concerts and shows, virtual tourism, online *apéro*, and other such activities. Yet overuse of digital technologies can also have negative impacts on well-being and mental health: It can drive sleep problems; contribute to emotional instability, increased risk of depression, anxiety, and isolation; and drive impulsive, compulsive, and addictive behaviors

1 Amar Kanekar and Manoj Sharma, "COVID-19 and Mental Well-Being: Guidance on the Application of Behavioral and Positive Well-Being Strategies," *Healthcare* 8, no. 3, 2020.

2 Tobias Dienlin and Niklas Johannes, "The Impact of Digital Technology Use on Adolescent Well-Being," *Dialogues in Clinical Neuroscience* 22, no. 2, 2020, pp. 135–42.

3 Simon Kemp, "Digital 2021 Global Overview Report," Kepios/We Are Social/Hootsuite, 2021; Apurvakumar Pandya and Pragya Lodha, "Social Connectedness, Excessive Screen Time During COVID-19 and Mental Health: A Review of Current Evidence," *Frontiers in Human Dynamics*, no. 3, 2021, p. 4.

related to the Internet.[4] Further studies will be needed to shed more light on the long-term consequences of these phenomena. Still, for many, the significant slowdown imposed by the lockdowns did not translate into a feeling of having more time to themselves. It would seem that the increase in screen time, with its tendency to accelerate the pace of life, often canceled out the decelerating effects of being confined in the home. For many parents, for instance, remote working combined with having to supervise their children's online schooling felt nothing like a relaxed slowing of their pace of life. There even emerged a phenomenon of "Zoom fatigue," which psychologists and neuroscientists are now studying in order to better understand the mechanisms and the negative effects of the mental exhaustion caused by excessive video calls related to work, school, and social life.[5] Overall, the problems of addiction and the transformation of time described in Thesis 3 were multiplied during this great disruption, during which daily and social life was largely confined to digital platforms and devices.

For the platforms, more screen time meant more use of their services, resulting in more accumulation of data … and of profits. Home data consumption was already 16 percent higher in February 2020 compared to the previous year, and households then downloaded 28 percent more data in March, 36 percent more in April, and 30 percent more in May 2020 (during the first wave), returning to an average of around 14 percent more for the rest of the year. Zoom Video Communications, a platform created in 2011 and relatively unknown before the pandemic, reached a market capitalization of $47.8 billion in May 2020, exceeding the value of the world's seven biggest airlines combined.[6] The global ranking of companies by market capitalization clearly shows a hardening of Big Tech's

4 Mark S. Allen, Emma E. Walter, and Christian Swann, "Sedentary Behaviour and Risk of Anxiety: A Systematic Review and Meta-Analysis," *Journal of Affective Disorder*, no. 242, 2019, pp. 5–13; Muhammad Aziz Rahman et al., "Factors Associated with Psychological Distress, Fear and Coping Strategies During the COVID-19 Pandemic in Australia," *Global Health*, no. 16, 2020, pp. 1–15; Daria J. Kuss and Olatz Lopez-Fernandez, "Internet Addiction and Problematic Internet Use: A Systematic Review of Clinical Research," *World Journal of Psychiatry* 6, no. 1, 2016, p. 143; Pandya and Lodha, "Social Connectedness, Excessive Screen Time During COVID-19," p. 5.

5 Brenda K. Wiederhold, "Connecting Through Technology During the Coronavirus Disease Pandemic: Avoiding 'Zoom Fatigue,'" *Cyberpsychology, Behaviour and Social Networking* 23, no. 7, 2020, pp. 437–8; Hadar Nesher Shoshan and Wilken Wehrt, "Understanding 'Zoom Fatigue': A Mixed-Method Approach," *Applied Psychology* 71, no. 3, 2022, pp. 827–52.

6 Carmen Ang, "One Year In: Data Usage Surged During Pandemic," Visual Capitalist, 2021; Iman Gosh, "Zoom Is Now Worth More Than the World's 7 Biggest Airlines," Visual Capitalist, 2020.

Table 14.1. Ranking of companies by market capitalization, 2019–21

Rank	*2019 Q2*	*2019 Q4*	*2020 Q2*	*2020 Q4*	*2021 Q2*	*2021 Q4*
1	Apple $1028B	Apple $1305B	Apple $1576B	Apple $2254B	Apple $2286B	Apple $2913B
2	Microsoft $928.5B	Microsoft $1203B	Microsoft $1551B	Microsoft $1634B	Microsoft $2040B	Microsoft $2525B
3	Amazon $911.2B	Alphabet $922.1B	Amazon $1432.6B	Amazon $1577B	Amazon $1735B	Alphabet $1922B
4	Alphabet $751.2B	Amazon $916.2B	Alphabet $979.7B	Alphabet $999.6B	Alphabet $1680B	Amazon $1691B
5	Facebook $551.5B	Facebook $585.32B	Facebook $675.7B	Alibaba $795.4B	Facebook $985.9B	Tesla $1061B
6	Berkshire H. $521.1B	Alibaba $569B	Tencent $620.9B	Facebook $746.1B	Tencent $721.5B	Meta (Facebook) $935.6B
7	Alibaba $439.2B	Berkshire H. $553.5B	Alibaba $579.7B	Tencent $646.8B	Tesla $654.8B	Nvidia $732.9B
8	Tencent $432.1B	Tencent $461.4B	Berkshire H. $432.6B	Berkshire H. $509.5B	Berkshire H. $637.3B	Berkshire H. $668.6B
9	Visa $379.3B	JPMorgan C. $437.2B	Visa $412.7B	Visa $425.5B	TSMC $623.2B	TSMC $623.9B
10	Johnson & Johnson $370.3B	Visa $416.8B	Johnson & Johnson $370.6B	TSMC $420.4B	Alibaba $615.1B	Tencent $559.9B

Source: Wikipedia, "List of Public Corporations by Market Capitalization"

hegemony between 2019 and 2021. The GAFAM companies continue to battle it out at the top of the rankings, with the arrival of new players such as Tesla (electric cars), Nvidia (processors and AI), and TSMC (Taiwan Semiconductor Manufacturing Company). Chinese algorithmic giants such as Tencent and Alibaba are also climbing into the top-ten list, while Berkshire Hathaway (Warren Buffett's investment company), Visa, and Johnson & Johnson continue to carve out a place for themselves among this same leading group (Table 14.1).

The increase in the overall value of the GAFAM companies during this period is especially striking given that some other sectors were so hard hit, with the drastic fall in financial markets and the overall economic slowdown forcing governments to invest trillions of dollars to prevent collapse. The slowdown in the first half of 2020 did not extend to Apple, Amazon, Alphabet, Microsoft, and Meta—far from it. The value of their shares, which rose by 37 percent between January and July 2020, then represented around 20 percent of the total value of the US stock market. There had not been such a market value concentration in a single sector since 1950.[7] Apple's market capitalization alone almost doubled during the first year of the pandemic.[8] Despite some market corrections in subsequent years, the trends are telling: We witnessed a historical conjuncture where the pandemic of COVID-19 largely benefited the interests of the GAFAM companies and algorithmic capital.

While algorithmic capitalists were able to take advantage of the crisis, the pandemic also amplified social inequalities to a staggering degree, whether in terms of income, gender, ethnicity, race, or status. During the pandemic, "essential workers," the "guardian angels" of healthcare (a workforce that is often female and/or racialized), and other migrant workers indispensable to food production, among other things, went to work in grocery stores, essential businesses, fields, hospitals, and nursing homes, often with inadequate protective equipment and sometimes with meager bonuses to chivvy them along. In this same period, those in quite different professions—the world's billionaires—saw their fortunes grow to an insane degree. Jeff Bezos's net worth grew from $131 billion in 2019 to $159 billion in 2022, but he still lost his title as the world's richest man, as he was overtaken by Elon Musk, whose fortune grew from $26.4 billion in 2020 to $260 billion in 2022. In total, according to Oxfam, "573 people became new billionaires during the pandemic, at the rate of one every 30 hours." Meanwhile, it was expected "that 263 million more people will crash into extreme poverty [in 2022], at a rate of a million people every 33 hours."[9] The pandemic did not invent wealth inequality, but it did amplify it. The digital giants benefited more than anyone else.

Public health authoritarianism and technological solutionism

The COVID-19 crisis also marked a qualitative change in political governance, raising the specter of an "authoritarian turn" in liberal democracies. Once again, the public health crisis revealed processes already at work over the previous decade, notably the "democratic deconsolidation" marked by the crisis of confidence in institutions, the rise of

7 This concentration has continued. In early 2026, the ten largest US companies, eight of which are tech companies, count for more than 40 percent of the total value of the S&P 500.

8 Félix Tréguer, "The Virus of Surveillance: How the COVID-19 Pandemic Is Fuelling Technologies of Control," *Political Anthropological Research on International Social Sciences* 2, no. 1, 2021, p. 16.

9 Oxfam, "Pandemic Creates New Billionaire Every 30 Hours—Now a Million People Could Fall into Extreme Poverty at Same Rate in 2022," Oxfam, May 23, 2022.

authoritarian populism, and growing indifference toward democratic values, particularly among young people.[10] The erosion of democratic institutions was already underway before the arrival of COVID-19, with the election of figures such as Trump, Orbán, Erdoğan, Modi, and Bolsonaro in the 2010s offering various symptomatic cases.

Nevertheless, the arrival of the pandemic created a real shockwave, for it suspended overnight a whole series of counterpowers in parliaments, in the justice system, and at the level of ordinary citizens. It thus fueled an unprecedented concentration of power in the upper echelons of the state. Conspiracy theorists were quick to condemn "health totalitarianism"; this sometimes led to wild claims about microchips being inserted into vaccines linked to 5G and the Gates Foundation, or comparing compulsory mask wearing to the conditions at Auschwitz. Regardless of such wild rumors and absurd claims, several serious studies suggested that representative democracies were taking on new "semi-authoritarian" forms. This order was described in various ways: as a "pandemic surveillance capitalism," "resurgent paternalism," "pandemic mitigation authoritarianism," or "gentle authoritarianism."[11] The increase in state surveillance, the prolonged state of emergency extended across months or even years, the encroachment on individual freedoms and human rights, and the suspension of democratic control mechanisms, were all signs of an authoritarian management of the public health crisis. As Stephen Thomson and Eric Ip put it:

> Authoritarian governance in the name of public health intervention is understood in the present context as being characterized by diverse combinations of governmental and administrative overreach, the adoption of excessive and disproportionate emergency measures, override of civil liberties and fundamental freedoms, failure to engage in properly deliberative and transparent decision-making, highly centralized

---

10 Roberto S. Foa and Yascha Mounk, "The Danger of Deconsolidation," *Journal of Democracy*, no. 3, 2016, pp. 5–17, and Foa and Mounk, "The End of the Consolidation Paradigm: A Response to Our Critics," *Journal of Democracy Web Exchange*, April 28, 2017.

11 Alexei Anisin, "Pandemic Surveillance Capitalism: Authoritarian Liberalism or Democratic Backsliding?," *Journal of Political Power* 15, no. 2, 2022, pp. 262–78; J. Fins, "Resuscitating Patient Rights During the Pandemic: COVID-19 and the Risk of Resurgent Paternalism," *Cambridge Quarterly of Healthcare Ethics* 30, no. 2, 2021, pp. 215–21; Joseph H. Manson, "Right-Wing Authoritarianism, Left-Wing Authoritarianism, and Pandemic-Mitigation Authoritarianism," *Personality and Individual Differences*, no. 167, December 2020; Rémy Bourdillon, "Au temps de l'autoritarisme doux," *Le Mouton Noir*, April 2022.

decision-making, and even the suspension of effective democratic control. In a nutshell, the pandemic has served as a powerful justification for authoritarianization—the process by which state authorities "slowly undermine institutional constraints on their rule," through various combinations of the above—and populations have largely responded with obedience.[12]

This authoritarian approach to public health—generally under a technocratic model of governance in which epidemiologists, public health experts, consulting firms and certain private companies exerted considerable influence over government decisions—was compounded by a technological solutionism based on dubious public–private partnerships.[13] Hundreds of governments and public health authorities around the world made extensive use of private tech firms' resources, data, and apps in order to monitor, quarantine, and track entire populations. In so doing, they prioritized private business interests at the expense of democratic freedoms and the public interest.[14] In Canada, a report submitted to the House of Commons Ethics Committee "shows that Canadians' movements, including trips to liquor stores and pharmacies, were tracked via their mobile phones without their knowledge during the COVID-19 pandemic." The Public Health Agency of Canada used Toronto-based BlueDot, an "intelligent pandemic management" platform, which compiled reports on citizens' movements based on anonymized data.[15]

As public authorities around the world stepped up their use of control technologies, open surveillance of citizens was introduced on a massive scale. In Western Australia, the governor decided to force individuals suspected of having COVID-19 to remain in isolation by wearing an electronic bracelet. The Polish government forced individuals in quarantine to install a surveillance app linked to geolocation and facial recognition, and to send regular selfies from their homes.[16] In New Zealand, the police launched a digital platform encouraging citizens to report their

12 Stephen Thomson and Eric C. Ip, "COVID-19 Emergency Measures and the Impending Authoritarian Pandemic," *Journal of Law and the Biosciences* 7, no. 1, 2020, p. 4.

13 Stefania Milan, "Techno-Solutionism and the Standard Human in the Making of the COVID-19 Pandemic," *Big Data & Society* 7, no. 2, July 2020; Luca Marelli, Katharina Kieslich, and Susi Geiger, "COVID-19 and Techno-Solutionism: Responsibilisation Without Contextualisation?," *Critical Public Health* 32, no. 1, 2022, pp. 1–14.

14 Anisin, "Pandemic Surveillance Capitalism."

15 La Presse Canadienne, "La santé publique a pu suivre les Canadiens à la trace pendant la pandémie," Radio-Canada, May 5, 2022.

16 Tréguer, "The Virus of Surveillance."

neighbors for violating pandemic-related restrictions. In France, police forces in Paris, Nice, and Marseille used drones equipped with loudspeakers to broadcast messages instructing people to stay at home, with the devices also assisting ground teams in identifying breaches of lockdown rules.[17] In Singapore, Boston Dynamics robot dogs roamed parks, counting the number of people and broadcasting prevention messages to passersby. China deployed its entire sprawling arsenal of surveillance technology (drones, thermal-imaging cameras, facial recognition, social credit, the monitoring of social media) to contain the pandemic, setting an example to the rest of the world that Chinese-style "authoritarian capitalism," free from "bothersome" democratic processes, was a superior model for curbing a public health crisis. The Chinese government openly boasted of its ability to monitor people's movements, trace contacts, and enforce curfews, presenting these as an example of its model:[18] good inspiration, perhaps, for *Black Mirror* scriptwriters.

The gleeful use of digital and algorithmic tools in the name of protecting public health was also justified by technological solutionism—that is, the belief that technological innovations can solve and manage complex social problems.[19] The rapid development of contact-tracing apps by several companies and start-ups in the pandemic's early months is one such example. In Quebec, the MILA Institute launched COVI, a mobile health app designed to combat COVID-19. This app, its promoters claimed, had countless virtues. It could do everything from informing individuals to self-isolate if they had been exposed to infected people within two meters of them to predicting the daily risk level based on behavioral data, personalizing public health recommendations, reducing the total number of cases, providing early-warning signs of outbreaks, fostering the creation of new epidemiological models, and predicting the contagion curve. All this was supported by decentralized Bluetooth technology, to ensure maximum privacy protection.[20] However, despite a strong PR campaign and the firm's lobbying of the federal and provincial governments, MILA came away empty-handed: The Canadian government instead turned to Shopify, a private e-commerce company, which had developed the COVID Shield app

17 Jean-Michel Normand, "Confinement: La surveillance policière par drones dénoncée par deux associations," *Le Monde*, May 2020.

18 Laurie Macfarlane, "A Spectre Is Haunting the West: The Spectre of Authoritarian Capitalism," *Open Democracy*, April 2020.

19 Morozov, *To Save Everything, Click Here*.

20 See MILA's website, mila.quebec.

(later renamed COVID Alert), incorporating notification technology from Apple and Google.[21]

What happened to COVID Alert and dozens of other contact-tracing apps around the world? In some Asian countries such as China, Taiwan, South Korea, and Singapore, contact-tracing apps were mandatory and thus used extensively. In contrast, these apps were used much less widely in almost all countries in Europe and North America, where take-up was voluntary. This also reflected a cultural and political context more sensitive to issues of surveillance and privacy. The more sporadic and scattered use of the apps produced a drastic decline in their effectiveness, as well as a series of other issues that slowed their adoption by the masses: fears related to surveillance, lack of trust, social discrimination based on age and group, increased risks of false positives or false negatives, and many other concerns. In general, most studies question the effectiveness and relevance of contact-tracing apps during the pandemic.[22] In Canada, COVID Alert cost the federal government $20 million—including $15.9 million in advertising—for a meager adoption rate. There were 6.89 million downloads, but only 57,704 single-use codes were used.[23] The abandonment, in 2022, of government-provided PCR testing for the general public (ditched in favor of rapid home testing) sealed COVID Alert's fate, and the app was eventually discontinued.[24]

In addition to the dismal failure of contact-tracing applications, the massive vaccination campaign took on a more coercive tone with the arrival of apps such as health or vaccination passports. Through these, access to various sites, social spaces and public services, "nonessential businesses" (restaurants, bars), cultural spaces (museums, theaters),

21 Delphine Jung, "COVID-19: L'application de traçage du MILA mise au placard par Ottawa," Radio-Canada, June 10, 2020.

22 Leonardo Maccari and Valeria Cagno, "Do We Need a Contact Tracing App?," *Computer Communications*, no. 166, 2021, pp. 9–18; A. Akinbi, M. Forshaw, and V. Blinkhorn, "Contact Tracing Apps for the COVID-19 Pandemic: A Systematic Literature Review of Challenges and Future Directions for Neo-Liberal Societies," *Health Information Science and Systems* 9, no. 18, 2021; Steve Jacob and Justin Lawarée, "The Adoption of Contact Tracing Applications of COVID-19 by European Governments," *Policy Design and Practice* 4, no. 1, 2021, pp. 44–58.

23 "For an infection to be recorded, the infected person must be a user of the application and must register that they have contracted the coronavirus, using a single-use code provided by public health authorities or obtained from a website, depending on the province." Michel Bolduc, "L'application Alerte COVID est-elle un échec?," Radio-Canada, February 12, 2021

24 La Presse Canadienne, "L'application Alerte COVID sera bientôt abandonnée," *Le Devoir*, June 4, 2022.

trains, planes, and other areas could be controlled according to individuals' medical status. We cannot enumerate here all the adverse effects, forms of discrimination, and ethical issues associated with the use of vaccine passports. Yet clearly, these policies bring to mind the society of control theorized by Gilles Deleuze, where individuals moving through different spaces find that access is granted or denied by means of extended digital controls:

> The conception of a control mechanism, giving the position of any element within an open environment at any given instant (whether animal in a reserve or human in a corporation, as with an electronic collar), is not necessarily one of science fiction. Félix Guattari has imagined a city where one would be able to leave one's apartment, one's street, one's neighborhood, thanks to one's (dividual) electronic card that raises a given barrier; but the card could just as easily be rejected on a given day or between certain hours; what counts is not the barrier but the computer that tracks each person's position—licit or illicit—and effects a universal modulation.[25]

Furthermore, the distinctive feature of the health vaccine passport is that it *required* people to carry their smartphones with them whenever they left their homes, except for very few people who had their passports printed. It accelerated and, in a way, forced along the transition to a society where the tracking and control of each individual's digital identity has become the norm. It extended the logic of the customs controls enacted at borders and airports to the entire social world. But there was also something else. The passport was initially used as an incentive (or rather a constraint) to maximize vaccination coverage among the population. Ultimately, however, it was as if the vast vaccination campaign were itself a means of forcing people to convert to this system of digital control, to carry their phones with them at all times, and to allow themselves to be monitored without too much complaint. As the anonymous authors of the *Manifeste conspirationniste* noted:

> The "health pass" is also anything but health-related. It is a *police* pass that allows the population to be sorted into docile and rebellious groups and ultimately ensures their voluntary tracking. It is a *behavioral* pass

25 Gilles Deleuze, "Postscript on the Societies of Control," *October*, no. 59, Winter 1992, pp. 3–7.

> that can be used to force anyone to do anything under threat of having it taken away from them. It is a *financial* pass aimed at taking a major step towards individual digital identity. Without it all the data produced by electronic interactions, by all the sensors and connected objects that 5G promises to saturate our daily lives with, would be almost worthless because they would lack the necessary medium … In this sense, the goal of vaccination is indeed the pass, and not the other way around.[26]

We need not subscribe to far-fetched theories that the pandemic was a vast conspiracy planned by the world's powers that be, or assert that the virus is benign and has only minimal consequences. At the same time, it is worth considering the more balanced and plausible claim that large technology companies and governments exploited this unprecedented public health crisis to undermine democratic counterpowers and to expand their control by pushing populations to adopt a fully digitalized way of life. This represents an authoritarian shift, taken by governments in the context of exceptional circumstances, but in hindsight, was it also a milestone in the normalization of authoritarian politics?

The Screen New Deal or the "all-remote" society

Many believe that the excesses and abuses associated with the management of the health crisis are now behind us, having quickly been followed by a return to normality. This is a dangerous illusion, especially when we consider that democratic erosion and the rise of algorithmic capital were already proceeding apace before COVID-19. If we date the onset of the transition from neoliberalism to a new stage of capitalism to the financial crisis of 2007–8, we can say that the post-2020 world is characterized by a much more advanced form of algorithmic capitalism. It has extended its influence to all corners of the globe and to myriad spheres of social life. The pandemic has acted as a real accelerator for the new development model, driven by Big Data, algorithms, and the hegemony of platforms over the social, economic, and political world.

This is why we refer to a "great leap forward" in algorithmic capitalism. History buffs will note that this expression usually refers to Mao Zedong's economic policy implemented from 1958 to 1962. The forced collectivization of agricultural land, the rapid modernization of

26 Anonymous, *Le Manifeste conspirationniste*, Paris: Seuil, 2022, p. 68.

industrial infrastructure, and the implementation of large-scale public works projects led to the "Great Chinese Famine," causing between 15 and 55 million deaths, according to historians. Clearly, we do not wish to compare the advent of algorithmic capitalism to a humanitarian disaster of this magnitude. But in both cases, we can see a similar original intention: a deliberate acceleration of economic and technological development through more efficient use of resources, based on the tools available. Mao might have indeed embraced Mark Zuckerberg's motto: "Move fast and break things."

During the pandemic, algorithmic capitalism's "great leap forward" was largely coercive. During its first decade, it had been embraced "freely and willingly," that is, through the adoption of social media, mobile phones, and the spread of algorithmic platforms and machines. In contrast, the unprecedented pandemic context forced individuals and organizations to adapt overnight to a hyperdigitalized model largely controlled by tech firms. What is more, governments developed multiple partnerships with Big Tech and imposed a series of measures requiring individuals to use their computers to work from home or attend classes, and to leave their homes with smartphone in hand to ensure their access to different spaces. Some described all this as a rapid fusion between "surveillance capitalism" and "state surveillance."[27] We saw in Theses 10, 11, and 12 that this "algorithmic proximity" between the state and the market has been developing since the first stirrings of algorithmic capital—and there is no doubt that the pandemic hardened this partnership. While the algorithmic capitalism of the 2010s was fairly comfortable with the parameters of liberal democracy, that of the 2020s has far fewer qualms about authoritarianism.

This blurring of the boundaries between the state and large technology companies is not a product of chance, but rather a full-fledged political project, openly advocated by certain political and economic elites. It is, in a sense, aimed at strengthening the ties between platforms and governments and institutionalizing partnerships between digital giants and various public services. This project had already begun to take shape before the pandemic, particularly in Canada, as evidenced by certain investigations linked to the consulting firm McKinsey, which signed an open contract for IT services in 2019—one that is valid until 2100.[28] Justin Trudeau's government awarded more than $100 million to

---

27 Anisin, "Pandemic Surveillance Capitalism," p. 265.

28 Mylène Crête, "Ottawa a conclu un contrat jusqu'en 2100," *La Presse*, January 30, 2023.

this firm, which played a strategic role in managing the health crisis with lockdown measures and vaccination campaigns, as well as in the digital modernization of services in various ministries, public services, and the army.[29] While algorithmic capitalism had been subject to growing scrutiny and criticism, including through various revelations and publications about large platforms' intrusive practices, the COVID-19 pandemic allowed the GAFAM companies to secure renewed legitimacy by positioning themselves as essential partners in tackling the crisis. According to Eric Schmidt, founder and former CEO of Google, it was important to take advantage of this global crisis to rethink the "world after." In an open letter published in the *Wall Street Journal* in late March 2020, Schmidt asked:

> How could the emerging technologies being deployed in the current crisis propel us into a better future? … If we are to build a future economy and education system based on tele-everything, we need a fully connected population and ultrafast infrastructure. The government must make a massive investment—perhaps as part of a stimulus package—to convert the nation's digital infrastructure to cloud-based platforms and link them with a 5G network.[30]

Similarly, in late March 2020, New York Governor Andrew Cuomo announced a partnership with the Bill and Melinda Gates Foundation to develop a "smart education system." Such partnerships between states and large technology companies multiplied during the pandemic, and this trend continued afterward. We are gradually seeing the emergence of a kind of new "social contract," which Naomi Klein has dubbed the "Screen New Deal." This is a reiteration of her "shock doctrine" theory, with elites taking advantage of the global pandemic to advance their agenda. For Klein, "the future that is being rushed into being as the bodies still pile up treats our past weeks of physical isolation not as a painful necessity to save lives, but as a living laboratory for a permanent—and highly profitable—no-touch future."[31] Symptomatic of the breakneck speed of the "great leap forward," what Klein here describes as

---

29 Romain Schué and Thomas Gerbet, "Les contrats octroyés à McKinsey dépassent 100 millions de dollars sous Justin Trudeau," Radio-Canada, January 14, 2023.

30 Eric Schmidt, "A Real Digital Infrastructure at Last," *Wall Street Journal*, March 27, 2020.

31 Naomi Klein, "Screen New Deal: Under Cover of Mass Death, Andrew Cuomo Calls in the Billionaires to Build a High-Tech Dystopia," *The Intercept*, May 8, 2020.

"the future" in this text is in fact already the present, a present dominated by algorithmic capitalism:

> It's a future in which our homes are never again exclusively personal spaces but are also, via high-speed digital connectivity, our schools, our doctor's offices, our gyms, and, if determined by the state, our jails … a future in which, for the privileged, almost everything is home delivered, either virtually via streaming and cloud technology, or physically via driverless vehicle or drone … It's a future that claims to be run on "artificial intelligence" but is actually held together by tens of millions of anonymous workers tucked away in warehouses, data centers, content moderation mills, electronic sweatshops, lithium mines … It's a future in which our every move, our every word, our every relationship is trackable, traceable, and data-mineable by unprecedented collaborations between government and tech giants.[32]

This "tele-everything society" or "contactless society,"[33] fostered by social distancing and the widespread fear of contracting a potentially fatal disease, intensifies trends that were already at work in the algorithmic capitalism of the 2010s. In this sense, it is instructive to look at phenomena such as the growing intermeshing of public authorities and Big Tech companies, the widespread adoption of remote working in both the private and public sector, the normalization of online education, the increased use of telehealth, and public administrations' rapid moves to shift their technological infrastructure to the clouds controlled by Amazon, Microsoft, Google, and IBM. These developments all represent a condensed version of processes that were pursued on a large scale beginning in the early months of the pandemic. They have themselves produced a series of disruptions and crises that point toward a possible transformation of algorithmic capital.

The emergence of cryptocapitalism, the metaverse, and other mirages of Web 3.0

Not all remained rosy in the land of algorithmic capital. The golden age of its first decade of growth and its consolidation during the pandemic

32 Ibid.

33 François Saltiel, *La société du sans-contact: Selfie d'un monde en chute*, Paris: Flammarion, 2020.

gave way to a crisis of legitimacy that hit some of its flagship firms, and strange new "innovations" around the time of the pandemic. Indeed, since 2017 a series of scandals, investigations, and media leaks have tarnished its shining image, including the Facebook/Cambridge Analytica controversy, sustained criticism of surveillance practices and addictive technologies, and the Uber Files scandal. Facebook is doubtless the platform most damaged by this series of tremors. Mark Zuckerberg even had to testify before the US Congress in April 2018 following the theft of personal data from nearly 87 million users, then a second time in July 2020 alongside the heads of Apple, Amazon, and Google to answer elected officials' questions about GAFAM monopolistic practices. During a hearing lasting almost six hours, the House Antitrust Subcommittee portrayed them as a threat to the free market and democracy. It was not a flattering image for the grand masters of algorithmic capital.[34]

Facebook's legitimacy crisis escalated in October 2021 when whistleblower Frances Haugen, a former head of the company's civic integrity department, released tens of thousands of pages of internal documents. These showed that the platform had deliberately lied to the public and authorities about its practices for combating hate speech, violence, and misinformation, as well as the psychological impact of Instagram on young women and teenage girls. The company had concealed strategic information from the US Congress and was even planning to launch a version of Instagram for users under age thirteen. This affair was further ammunition for critics who argue that the company prioritizes its growth and its own interests at the expense of the safety and well-being of its users. Also in 2021, an iPhone update allowing users to activate a new privacy protection mechanism for certain apps dealt a severe blow to Facebook's advertising revenue, causing it to lose approximately $10 billion in one year, according to its calculations.[35] To make matters worse, on October 4, 2021, a global outage hit the platform's whole network (Facebook, Instagram, Messenger, WhatsApp) for several hours, highlighting the risks of low resilience when digital tools are overly integrated within a single company.

Faced with these multiple setbacks, Zuckerberg changed tack and announced a new name for his company on 28 October 2021: Meta. The platform sought to revamp its brand image and reposition itself as a

34 Tony Romm, "Amazon, Apple, Facebook and Google Grilled on Capitol Hill over Their Market Power," *Washington Post*, July 29, 2020.

35 Ben Gilbert, "Facebook Blames Apple After a Historically Bad Quarter, Saying iPhone Privacy Changes Will Cost It $10 billion," *Business Insider*, February 3, 2022.

spearhead of innovation through the vast "metaverse" project. A jubilant Zuckerberg presented the metaverse as "the future of the Internet," a coming together of technologies related to virtual and augmented reality, cryptocurrencies, and the introduction of cutting-edge algorithmic tools. Resembling an enhanced version of the online game *Second Life*, the goal was to create an immersive virtual environment combining social life, entertainment, e-commerce, work, school, and other spheres of activity within a new open, decentralized, and innovative protocol. While Zuckerberg acknowledged that there would be many social and technical hurdles to clear before a fully functional metaverse could be achieved, investments of $13 billion over fifteen months left little doubt—at the time—about his determination to make it happen.[36]

The sudden introduction of the metaverse—as if pulling a rabbit out of a hat—served several purposes. First, it provided a distraction from Facebook's woes and, more generally, from the negative aspects of algorithmic capital. But far from being a mere buzzword, gimmick, or marketing concept, it signaled the attempt to launch a new phase in the integration of algorithmic technologies: the Internet of Things, VR headsets, augmented reality glasses, sensors, 5G networks, facial recognition algorithms, and other devices. What is more, it offered a "solution" to various problems experienced during the pandemic: If people were exhausted with Zoom fatigue and all those hours spent in front of screens, it was now time to move on to a new interface allowing richer and more immersive social interactions via an "embodied Internet" capable of "unlock[ing] a lot of amazing new experiences"—as Zuckerberg put it, in his lengthy video presentation of the metaverse. This portended the ideal infrastructure for a contactless or "fully remote" society, which promised to transcend the limitations of current technologies.

Finally, the metaverse seemed to come at just the right time, following the spectacular popularity of economic innovations made possible by blockchain technology, including cryptocurrencies such as Bitcoin, representing new means of speculation. Following the publication of the white paper *A Peer-to-Peer Electronic Cash System* under the pseudonym Satoshi Nakamoto in 2008, Bitcoin saw a remarkable rise: From a value of $0.001 in October 2009, it rose to over $64,800 on April 14, 2021. A marginal phenomenon in Bitcoin's early days, cryptocurrencies rapidly developed into a vast industry—there would be more than 20,000 of them by July 2022, the best-known of which, in addition to Bitcoin, are

36 Maxence Fabrion, "Le métavers a coûté 13 milliards de dollars en 15 mois à Meta (Facebook)," *Les numériques*, May 4, 2022.

Ethereum, Dogecoin, Theter, and Binance.[37] They also have an increasingly problematic carbon footprint, as we will see in Thesis 15. In 2021, El Salvador became the world's first country to recognize Bitcoin as legal tender, in a bid to overcome its public debt crisis. At the same time, the rise of nonfungible tokens (NFTs), created by digital artist Kevin McCoy in 2014, made it possible to monetize a myriad of digital content. These digital signatures allowed a range of cultural goods such as songs, videos, and images to be traded in the same way as hockey cards or Pokémon. Popularized rapidly in 2020 in the wake of the pandemic, NFTs provided a new source of wealth for thousands of unemployed young people, geeks, artists, and venture capitalists.[38] The marketing of Thicc Pokémon—NFTs of Pokémon cards with exaggeratedly large buttocks, sold for several thousand dollars—and the user who spent nearly half a million dollars to build a house near Snoop Dogg's virtual mansion in the metaverse, are two of many examples of this speculative craze.[39]

The result: An NFT bubble rapidly inflated during the pandemic, reaching $17.6 billion traded in 2021. Combined with cryptocurrencies, these assets reached a total market value of over $3 trillion at the time. We have said that algorithmic capitalism faced a crisis of legitimacy, since it was accused of concentrating power, practicing surveillance incompatible with individual freedoms, monopolistic practices, and so on. Yet the world of cryptocurrencies seemed to reaffirm the founding values of the Internet: decentralization; openness and transparency; freedom from banks, governments, and major authorities; increased power for users; and widespread prosperity. Here we see the Californian ideology and the promises of Web 2.0 coming back to life thanks to the advent of "cryptocapitalism," seemingly able to overcome algorithmic capital's various wrong turns.[40]

Blockchain was thus perceived as a disruptive technology with significant potential impact, enabling the decentralization of finance (DeFi),

37 Josh Howarth, "How Many Cryptocurrencies Are There in 2022?," *Exploding Topics*, July 2022.

38 Victoria Zavlayola, "COVID-19 Lockdown and the NFT Boom Fueled the Art + Tech Market. What's Next?," *Yahoo! Finance*, October 8, 2021.

39 Yanis Varoufakis and Evgeny Morozov, "Conversation: Yanis Varoufakis on Crypto, the Left and Techno-Feudalism," *Yanis Varoufakis: Thoughts for the Post-2008 World*, April 23, 2022, yanisvaroufakis.eu; Nicole Carpenter, "You Can Spend Thousands on 'Thicc Pokémon' NFTs," *Polygon*, January 4, 2021; Sean Dickens, "Snoop Dogg Enters the NFT Metaverse in Style," *Yahoo! Finance*, September 23, 2021.

40 Christian Martinez, "What Is Crypto-Capitalism?," *Medium*, 2018; Ethan Fridmanski, "Crypto-Capital: The Political Economy of Cryptocurrencies," PhD thesis, University of Notre Dame, 2021.

the emergence of smart contracts, multiply decentralized autonomous organizations (DAOs), and the extension of the P2P (peer-to-peer) logic to the entire socioeconomic world. All of this converged with the discourse on "Web 3.0," which hopes to overcome the hegemony of GAFAM through a combination of the latest algorithmic technologies, such as blockchain, semantic web, machine learning, NFTs, DAOs, and DeFi. The metaverse thus became a symbol of this vast space, which offers the promise of a "creator-driven economy" and a shared, open, and decentralized prosperity. In 2021, the media hype surrounding the cryptosphere was at full tilt, galvanized by low interest rates, the frenzy of the "techbros," and the craze surrounding this new gold rush.

What should we take away from all this? First, crypto-assets and NFTs accelerated the trend toward the "tokenization of everything" by commodifying a wide range of creative and/or insignificant objects from the digital world while also fueling financial speculation. Second, the crypto universe provided a rapid means of jumping past the contradictions of algorithmic capitalism by introducing new strategies for adaptation, crisis resolution, and technological innovation. "Cryptocapital" appears as a strange outgrowth of algorithmic capital, exacerbating some of its worst tendencies and speculative dynamics. Cryptocurrencies, NFTs, and the metaverse all rely on a material and technological infrastructure that makes increased use of algorithms. Cryptocapital thus merely pursues the same logic of accumulation by other means, even if perhaps driving new configurations of algorithmic capitalism in the post-pandemic world. This includes a resurgence of technosolutionism and even takes such forms as blockchain humanitarianism and crypto-colonialism, with the increased use of these technologies to (supposedly) help Global South countries lift themselves out of poverty.[41]

However, this dream soon came up against harsh reality. At the beginning of 2022, cryptocurrencies suffered a significant drop in value, before crashing that May. Bitcoin lost 57 percent of its value and Ethereum 58 percent, and the overall cryptocurrency market fell from $3.3 trillion to $1.3 trillion between May 2021 and May 2022. While most cryptocurrencies are highly volatile in value, stablecoins (including Terra, USD Coin, Binance USD, and Tether) act as functional "bureaux de change" to ensure the overall liquidity of these assets by maintaining a value similar to the US dollar. But stablecoins were also hit by the crash, with the

41 Olivier Jutel, "Blockchain Humanitarianism and Crypto-Colonialism," *Patterns* 3, no. 1, 2022, pp. 1–8.

collapse of some pillars of the crypto ecosystem such as Luna and Terra.[42] The crypto bubble suddenly deflated, leaving it vulnerable to criticisms long leveled by those who termed it essentially a "giant Ponzi scheme" or even a fraud—"nonfungible bullshit," in the same vein as WeWork, the Fyre Festival, and other avatars of contemporary speculative capitalism.[43]

The bursting of the crypto bubble in May 2022 also coincided with a massive loss in the market value of the digital giants, including Microsoft (down $189 billion), Tesla (-$800 billion), Amazon (-$173 billion), Alphabet (-$123 billion), Nvidia (-$85 billion), and Facebook (-$70 billion)—totaling more than a trillion dollars of market value lost in just a few days.[44] Tech companies had taken on tens of thousands of new hires during the pandemic to service their rapid growth, so they subsequently carried out several waves of mass redundancies, slashing more than 365,000 jobs within just seventeen months.[45] Although the digital giants may seem like invincible powers, they remain subject to market fluctuations and imperatives. In November 2022, the bankruptcy of the world's third-largest cryptocurrency company, FTX—valued at over $32 billion just a few months earlier—led to the downfall of its founder, Sam Bankman-Fried. This fallen king of cryptocurrency was arrested for fraud a few days later, at his home in the Bahamas, and later sentenced to twenty-five years in prison in 2024.[46] The ruin of this widely admired figure in Silicon Valley and the philanthropic community, one associated with "effective altruism" and "longtermism," also stained this wider community, which had been aware of his fraudulent practices for some years.[47]

Finally, the bursting of the tech bubble in 2022 and Meta's falling revenues were accompanied by the first, disappointing glimpses of the metaverse. There were the poor-quality avatars, a "metarave party" attended by only six people, cringe-worthy promo videos, and other results that fell far short of the technological revolution promised just a

42 Romaric Godin, "Les cryptos en plein krach," *Médiapart*, May 16, 2022.

43 Sohale Andrus Mortazavi, "Cryptocurrency Is a Giant Ponzi Scheme," *Jacobin*, January 21, 2022; Luke Savage, "NFTs Are, Quite Simply, Bullshit," *Jacobin*, January 26, 2022.

44 Jordan Novet, "Tech Giants Lost More Than $1 Trillion in Value in the Last Three Trading Days," CNBC, May 9, 2022.

45 Arnaud Leparmentier, "High tech, les leçons historiques d'une vague historique de suppressions d"emplois," *Le Monde*, June 9, 2023.

46 "Arrestation aux Bahamas de Sam Bankman-Fried, le 'roi' déchu des cryptomonnaies," France24, December 13, 2022.

47 Charlotte Alter, "Exclusive: Effective Altruist Leaders Were Repeatedly Warned About Sam Bankman-Fried Years Before FTX Collapsed," *Time*, March 15, 2023.

year earlier.[48] YouTuber Dan Olson compared this ugly virtual universe to a dead mall.[49] There was also the impact of factors such as investor impatience, the demands for rationalization within Meta, and the arrival in November 2022 of ChatGPT, which stole the spotlight in the tech world. Faced with all this, Zuckerberg decided to "quietly bury" his metaverse project and focus instead on generative AI.[50] The many setbacks of 2022 thus represented a "great disillusionment" with the omnipotence of the digital industry, bringing to a close the pandemic-era interlude during which it had carried all before it.

Does this mean the end of algorithmic capitalism is nigh? Not at all. It was a phase, one that shows that this new stage of capitalism is also fraught with its own contradictions and can generate its own forms of crisis, fixes, and dead ends. The "Magnificent Seven" (GAFAM plus Tesla and Nvidia) have recuperated and still dominate rankings of the world's largest companies, and their economic, technological, and cultural hegemony continues to grow. We should remember that capital has more than one trick up its sleeve, reflecting a tremendous capacity for adaptation. The recent demise of the tech bubble that developed in 2020–1 does not represent the end of a whole socioeconomic system, any more than the bursting of the dot-com bubble of the early 2000s did. The opposite may indeed be true. Each crisis can be a golden opportunity for the system to renew itself, eliminate the weakest players, and see new ones come out on top. Post-pandemic algorithmic capitalism is not necessarily identical to that of the 2010s: It draws on new actors, technologies, and horizons. Yet ultimately, the same regime of accumulation is consolidating itself with new players and new global articulations. The legitimacy issues are still plaguing it, however, as tech CEOs including Zuckerberg had to testify again before a Senatorial committee in January 2024 on the issue of online child sexual exploitation, while OpenAI and its CEO, Sam Altman, are facing lawsuits in Canada over the use of ChatGPT to help plan a mass shooting in early 2026, among other instances of continued social pushback against Big Tech.

Major crises in the past such as World War II and the two oil shocks of the 1970s accelerated certain other historical developments, and likewise

48 Kurt Robson, "5 Cringe Metaverse Moments You Might Have Forgotten About," *Verdict*, December 16, 2022.

49 Dan Olson, "The Future Is a Dead Mall: Decentraland and the Metaverse," YouTube, March 26, 2023.

50 Luc Olinga, "Mark Zuckerberg Quietly Buries the Metaverse," *The Street*, March 18, 2023.

the pandemic was a major catalyst for the expansion of algorithmic capital and new peripheral reconfigurations, of which cryptocurrencies and NFTs were only add-ons, now relegated to second-order importance with the rise of Gen AI and the current massive capitalist race for its development and deployment. Algorithmic capitalism has globalized at breakneck speed, becoming a world unto itself, a well-entrenched institutionalized social order. We should now assess how this new form of capitalism relates to the major crisis of our time: the climate crisis.

Thesis 15

Algorithms won't save us from the climate crisis

To save everything, click here.

—Evgeny Morozov

One of the most persistent myths surrounding digital technology concerns its supposedly "immaterial" character. Its information flows are said to reside in a "cloud" infrastructure—as if they were floating in some ethereal space above *terra firma*, disconnected from the real-world processes of exploiting and extracting of natural resources. This myth is fused with a short-sighted outlook that fails to grasp or even obscures outright the reality of the digital world's environmental footprint. What is more, not only does the ecological devastation caused by algorithms too often go ignored, but their champions present them as the magical solution to the environmental and climate crisis. By allowing us to streamline, calculate, and optimize economic processes in order to conserve nature and energy resources, algorithmic capitalism and AI will—it is said—succeed where humans have so far miserably failed. However, this vision of "to save everything, click here"—as Morozov sarcastically sums up this approach—is both naïve and dangerous. Algorithms are not going to save us from climate disaster.

Let us work through this thesis in five steps. First, a word on the relationship between capitalism and nature. Second, we should revisit the role of capitalism in the climate crisis. Third, we will paint a picture of algorithmic capital's own footprint and the physical infrastructure that underpins the material existence of the "cloud." Next, we will critically examine a false solution to the ecological transition—one centered on algorithms—and finally we will look into the different scenarios for energy descent.

Capitalism, society, and nature

What relationship is there between capitalist society and nature, from the perspective of the critical theory formulated in the previous theses? We saw with Fraser's concept of institutionalized social order that capitalism is also a mode of relation between society and nature. In what follows, we adopt a position close to critical realism and define nature in the same vein as Kate Soper, as the combination of

> material structures and processes that are independent of human activity (in the sense that they are not a humanly created product), and whose forces and causal powers are the necessary conditions of every human practice, and determine the possible forms it can take.[1]

Andreas Malm casts this notion of nature in his own terms, by saying that it is that part of the world that humans inhabit but have not built—one that in a sense precedes us, exists with us, and will undoubtedly outlast us. Although we can influence nature, this is not to deny that it has its own distinct reality.[2] As for society, it is defined clearly, succinctly, and soberly in Marx's words: It is "the sum of interrelations" between humans.[3]

While the currently fashionable forms of hybridism and constructivism fuse nature and society into a single whole, we maintain that nature and society are material realities which each have their own distinct characteristics. Society emerges from nature, but is not reducible to it, and the two cannot be identified as any one thing. Nor can nature be reduced to something produced by society: while humans have a real impact on nature, it is not a mere "social construct." Maintaining this dualism allows us to think dialectically and to consider the climate crisis, for example,

---

1 Kate Soper, quoted in Andreas Malm, *The Progress of This Storm: Nature and Society in a Warming World*, London: Verso, 2018, p. 28. See also Roy Bhaskar et al., *Critical Realism: Essential Readings*, London: Routledge, 2015.

2 Malm, *The Progress of This Storm*, p. 42.

3 "Society does not consist of individuals, but expresses the sum of interrelations, the relations within which these individuals stand," Marx, *Grundrisse*, "The Chapter on Capital," available at marxists.org.

precisely as a unity of opposites (of natural processes and anthropogenic causality). We have no need, here, to formulate a comprehensive critique of the influential theories of Bruno Latour, Jane Bennett, or Steven Vogel, who—beyond their various differences, particularities, and subtleties—all tend to blur the boundaries between nature and society. The following thesis simply seeks to escape the theoretical impasse into which far too many works inspired by such authors have fallen—and so, too, their opposites, for instance the many orthodox economists who see nature as a separate entity external to society, a resource to be exploited and a force to be dominated. In the same vein, Kohei Saito proposes a middle way between Cartesian dualism (humans and nature are two separate substances) and socio-constructivist monism (everything is a "social construct"). He does this through an "analytical dualism" where nature and society have distinct characteristics and yet are closely bound together:

> Nature is a precondition through which mediating social activities become possible, and nature and society are not separate and independent entities, but they are the two interrelated poles that encompass an array of impact potentials and contexts that are socially malleable … Contact between nature and society does not represent the "end" of nature but an essential character in the eternal process of social organization of human metabolic interaction with the environment.[4]

We shall thus postulate a dialectic of co-constitution, in which society emerges from nature and in turn alters it, and in which nature delimits the field of societies' material possibilities. The relationship between society and nature follows a processual course, where factors coming from both spheres feed into their mutual development in a concrete, nonlinear temporality. Social changes affect the constitution of nature (for instance, the advent of capitalist agriculture transforms the composition of the soil), and changes to the constitution of nature in turn alter the constitution of society (for instance, changes in the carbon cycle raise sea levels, driving migratory movements on a massive scale).

Approaching the human/nature relationship in this way also offers another advantage: It allows us to conceptualize certain disruptions that emerge in the relations between social activities and the rest of the living world, especially by drawing on the concept of the "metabolic rift." This notion, formulated by Marx in *Capital* and updated by John Bellamy

4 Kohei Saito, *Marx in the Anthropocene: Towards the Idea of Degrowth Communism*, Cambridge: Cambridge University Press, 2023, p. 109.

Foster in his work, aims to describe the ways in which the capitalist economy tends to irreversibly disrupt the metabolism between humans and nature.[5] According to Kohei Saito, this phenomenon appears on three different levels. First, there is a *material* rift that arises from the disruption of the cyclical processes of human/nature metabolism, particularly through the depletion of soils by modern capitalist agriculture.[6] Second, there is a *spatial* rift that takes the form of a "contradiction between town and country" due to the fact that the flows of matter and energy produced in rural society and consumed in cities does not return to the land, since the capitalist economy drives the concentration of waste in urban areas, river pollution, and the destitution of the peasantry combined with soil depletion and the deterioration of living conditions for the working class. Finally, this process also leads to a *temporal* rift, expressed in the loss of synchronicity between the pace of capitalist production and the temporality of ecosystems, which struggle to absorb the products of human activity.

The three dimensions of the metabolic rift (the material, spatial, and temporal) are closely linked and take different forms across geographical regions and historical periods. The capitalist mode of production constantly produces this society/nature contradiction, which it attempts to overcome by various means. This leads to a constant dynamic of displacement in order to find new resources, manage negative externalities, and develop new ways to continue the processes of accumulation despite the tendency toward the depletion of the ecosystem.[7] This constant dynamic of economic, social, and spatial displacement, which Stephan Lessenich terms "externalization,"[8] heightens inequalities between Global North and Global South countries, while fueling the technosolutionist logic that seeks new ways to pursue capitalist accumulation without questioning the system's own contradictions. We should thus ask how the social dynamic of algorithmic capital modifies, alters, and affects the balances of nature, and how the climate crisis interacts with the development of an algorithmic society.

---

5 John Bellamy Foster, *Marx's Ecology: Materialism and Nature*, New York: Monthly Review Press, 2000; John Bellamy Foster, Brett Clark, and Richard York, *The Ecological Rift: Capitalism's War on the Earth*, New York: Monthly Review Press, 2010.

6 John Bellamy Foster, "Marx Theory of Metabolic Rift: Classical Foundations for Environmental Sociology," *American Journal of Sociology* 105, no. 2, September 1999, pp. 366–405.

7 Saito, *Marx in the Anthropocene.*

8 Stephan Lessenich, *Neben uns die Sintflut: Die Externalisierungsgesellschaft und ihr Preis*, Berlin: Hanser, 2016.

Climate crisis and the Capitalocene

Before examining more directly the transformation of society/nature relations in the present era, it is important to recall one key fact. The root cause of the climate crisis is not "humanity" in general, but a particular social order: capitalism. This interpretation calls into question the dominant discourse based on the idea of the Anthropocene, first popularized by chemist Paul Jozef Crutzen. This term refers to the advent of a new geological epoch succeeding the Holocene, an epoch in which human activity literally becomes a geological force and the main ecological constraint on Earth.[9] However, as Malm points out, "Blaming all of humanity for climate change lets capitalism off the hook."[10] The *anthropos* (*homo sapiens*) has inhabited the Earth for 300,000 years and is not *in itself* a cause of global warming. Yet capitalism, especially since the advent of its industrial stage, has developed dynamics that do directly impact the climate.

The Anthropocene framework is in our view too unspecific and abstract. It indiscriminately blames the climate crisis on the entire global population, on all countries, and on all previous generations, even though the main culprits behind the current turmoil are capitalists, large corporations, and highly industrialized countries. The idea of the Anthropocene pushes several myths that contribute to a misunderstanding of the causes of the ecological crisis: It suggests that "the problem is overconsumption!," "the problem is overpopulation!," and "we are all in the same boat"![11] Yet overconsumption does not come from nowhere, or from a moral flaw that we all carry within us, but rather from a tendency toward overproduction linked to the imperatives of capitalist accumulation, which involves the relentless pursuit of profit, fierce competition, continuous innovation (contributing to planned obsolescence), territorial expansion, the overexploitation of natural resources, and infinite growth.[12] Moreover, when we look at the forces in the global capitalist economy that are feeding the current crisis, we find that just eighty-three fossil fuel companies (firms that produce oil, gas, and coal) are ultimately responsible for the vast majority (62 percent) of all global greenhouse gas emissions.[13]

---

9 Alexander Federau, *Pour une philosophie de l'Anthropocène*, Paris: PUF, 2017.

10 Andreas Malm, "The Anthropocene Myth," *Jacobin*, March 30, 2015.

11 Frédéric Legault, Arnaud Theurillat-Cloutier, and Alain Savard, *Pour une écologie du 99%*, Montreal: Écosociété, 2021.

12 Erik Olin Wright, *Envisioning Real Utopias*, London: Verso, 2010.

13 Legault, Theurillat-Cloutier, and Savard, *Pour une écologie du 99%*, p. 32.

These few arguments, which are increasingly permeating the media and scientific literature, lead us to prefer the term "Capitalocene" to describe the current geological epoch. This expression, formulated by Malm among others, identifies the main culprit of the current crisis: capitalism, and especially fossil capital.[14] Although capitalism predates the use of coal to power steam engines, the nineteenth century did indeed see a new departure. The turn to coal was not a simple product of technological development. Rather, it was a strategy used by capitalists to counter the labor movement's opposition to factory work, through the use of "abstract energy" freed from the constraints of time and space—enabling them to produce anytime, anywhere. As Saito points out:

> Coal enabled capital to build new factories in large cities where a larger number of workers were in dire need of jobs. This was basically how the power balance between capital and labour radically changed with the invention of the steam engine. Yet this change also intensified the antagonistic relationship between centre and the periphery, degrading both the living conditions of workers in the former and the natural environment in the latter.[15]

We will not here revisit the complex developments in the history of "fossil capital" over the last two centuries (Malm and Saito offer excellent analyses on this subject). But we believe that it is a useful key to understanding the predominant role of this intermeshing of capital and fossil fuels in the "metabolic rift" between society and nature caused by capitalism. Several scientists and historians note that a great acceleration of the impacts of human activity on nature occurred in the post-1945 decades known as the *Trente Glorieuses*, coinciding with the rise of consumer society, the growth of the car industry, and urban sprawl.[16] This acceleration, which continued with the period of globalized neoliberal capitalism, now seems to be entering a new phase sometimes referred to as the "Digitalocene."[17]

---

14 Andreas Malm, *Fossil Capital: The Rise of Steam Power and the Roots of Global Warming*, London: Verso, 2016.

15 Saito, *Marx in the Anthropocene*, p. 27.

16 John R. McNeill and Peter Engelke, *The Great Acceleration: An Environmental History of the Anthropocene Since 1945*, Cambridge, MA: Belknap Press, 2014.

17 Stéphane Grumbach, "L'Entropie du Numérocène: Quelques réflexions sur la Révolution numérique et l'Anthropocène," in Stéphane Grumbach et al., eds., *Anthropocène, à l'école de l'indiscipline,* Grenoble: Éditions du temps circulaire, 2019, pp. 1–19.

Of course, other factors such as colonialism, productivism, and industrial ideology have also fed into the ecological crisis. If capitalism were alone at fault, then how would we explain that the Eastern Bloc and "actually existing socialism" adopted the same trajectory of rampant industrialism, as noted by several degrowther authors?[18] Industrial logic, the "technical system," or the "megamachine"—to use the expressions of Jacques Ellul and Lewis Mumford—seem to have a certain "relative autonomy." However, these generic expressions seem to miss the mark. The problem is not technology itself, as if it were some completely autonomous factor, but rather a certain historical articulation of technoscientific development with various social relations including capitalism, colonialism, patriarchy, and geopolitical rivalry. These various dimensions cannot be considered as isolated causes. Capitalism, understood as an institutionalized social order, implies not only an economic system, but also a multidimensional social formation encompassing a mode of production, conditions of technological possibility, nature, social reproduction, political institutions, a dynamic international order, and a plurality of systems of domination. Ultimately, this is why the neologism "Capitalocene" is still useful in pointing to capitalism's specific effect in humanity's long journey over the last few centuries.[19]

If we accept the interpretation of capitalism as the "ultimate cause" of the climate crisis, the idea that "fossil capital" is mainly to blame still risks leading us into various myths. For instance, we might imagine that "everything will be fine—new technologies will save us!," "nothing is stopping capitalism from going green," "all we need to do is make the energy transition," and so on.[20] Here lies the problem: For while the contradictions of capitalism in terms of energy and the environment are causing a vast crisis of legitimacy for this system, it is seeking to reinvent itself and find a way out. The growing popularity of discourses on degrowth, political ecology, and environmental and climate justice is leading to a broader questioning of the dogmas of growth and of human beings' undifferentiated responsibility for the climate crisis. It further points to the problematic character of warmed-over rhetoric about sustainable development and the "energy transition"—often nothing more than a fig leaf for "green" capitalism.

---

18 Andre Gorz, *Farewell to the Working Class: An Essay on Post-Industrial Socialism*, London: Pluto Press, 1987.

19 Malcom Ferdinand, *Decolonial Ecology: Thinking from the Caribbean World*, Cambridge: Polity, 2021.

20 Legault, Theurillat-Cloutier, and Savard, *Pour une écologie du 99%*.

Just as algorithmic capital offered a response to the crisis of globalized and financialized neoliberal capitalism, it also presents itself as a practical solution to the deep problems surrounding industrial capitalism. A new brand of prophesy now promotes AI, Big Data, and the technological innovations of the Fourth Industrial Revolution as the best means of bringing the ecological transition to completion. In addition to the spatial shift aimed at overcoming the problems associated with the metabolic rift fueled by capitalism, we are thus faced with a technological shift aimed at circumventing the difficulties generated by fossil capital. Techno-utopianism is the faddish result, from Big Data to green applications promoting responsible consumption, via the Internet of Things as a smart infrastructure facilitating the transition to renewable energies and productivity growth. Cédric Villani, formerly an MP in France for Emmanuel Macron's coalition, provided a good example of this pervasive discourse:

> AI will enable us to grasp the dynamics and the evolution of ecosystems based on the reality of their biological complexity, to optimize the management of our resources, and particularly energy, to preserve our environment and to encourage biodiversity … The development of AI may give rise to new ways of maintaining natural areas that need protection, both on land and at sea. From autonomous robots that can hunt invasive starfish to smart barriers that can divert wildlife in order to preserve it, there are many possibilities for developing new ways of interacting with nature that are more adaptive and more respectful.[21]

If such talk were espoused only by geeks or science fiction enthusiasts, it would be fairly harmless. The problem is that it is promoted by many governments around the world—as well as countless start-ups and tech companies offering applications, innovations, smart objects, autonomous cars, drones, connected trackers, sensors on livestock and wild animals, and ever more sophisticated machines that claim to provide "solutions" to the most serious crisis facing civilization. Surprisingly, the solution to a crisis caused by capitalism and industry seems to be … capitalism producing new machines. We seem trapped in a damaging technological and epistemological circular horizon.[22]

21 Cédric Villani et al., "Donner un sens à l'intelligence artificielle: Pour une stratégie nationale et européenne," Conseil national du numérique, March 28, 2018, p. 124.

22 Vijay Kolinjivadi, Diana Almeida, and Jonathan Martineau, "Can the Planet

We cannot possibly provide an exhaustive list here of the many "green technological innovations" based on machine learning and data. Rather, we will get to the heart of the argument by showing that while we are indeed seeing a historic partnership between fossil capital and algorithmic capital, this does not imply that we can achieve an algorithmically driven energy transition. If anything, the opposite is true: Algorithmic capital and the immense industrial and energy infrastructure on which it relies have a huge environmental imprint and bear a series of other tensions and contradictions that complicate the climate crisis rather than help solve it.

Of course, this does not mean that no applications of artificial intelligence could ever be useful for environmental purposes, or contribute to the transition to a more carbon-neutral economy. Yet even if specific uses of AI prove beneficial, the problem lies mainly in the overall course of development of algorithmic capital, which is subject to the imperatives of extraction, prediction, control, competition, accumulation, and the expansion of the underlying industrial and energy infrastructure. This course of development ultimately amplifies the problems that it seeks to solve. A holistic approach leads to the conclusion that algorithmic capital does not contribute overall to the ecological transition, even though certain specific algorithmic technologies, considered in abstract terms, can sometimes make positive contributions. This sums up, in its simplest form, the *ecological paradox of algorithmic capital*: Whatever the good intentions with which it seeks to help resolve the climate crisis, in the last analysis it makes the crisis worse.

Algorithmic capitalism's environmental footprint

We noted earlier the persistent myth of a digital world that floats above natural processes and has only a minimal environmental footprint. This is imagined to be a "clean" or even "green" world, light years away from the industry that extracts resources and energy from the Earth at an unsustainable rate. However, this myth ought to be deconstructed right away. Algorithmic capital is anything but "green"; rather, it is an energy-guzzling ogre that drives an *ecocidal* extractive economy. Its effect on nature is catastrophic. The only "cloud" metaphor that befits algorithmic capital is the kind of cloud that results from the explosion of an atomic bomb.

Really Be Saved in Time?: On the Temporalities of Socionature, the Clock, and the Limits Debate," *Environment and Planning E: Nature and Space* 3, no. 3, 2020, pp. 904–26.

In the first chapter of her *Atlas of AI*, Kate Crawford provides us with a solid perspective for analyzing the environmental footprint of algorithmic capital. It has effects on "the atmosphere, the oceans, the earth's crust, the deep time of the planet," mainly through its material infrastructure and energy demand. Given the enormous impact that algorithms have on the global ecosystem, Crawford goes so far as to define AI itself as an extractive industry.[23] Two forms of resource extraction are typical of algorithmic capital: mining and energy. It is important to note that the material infrastructure of algorithmic capital is itself immense and far-reaching; it includes not only smart devices, but the batteries that power them, data centers, and the computers that underpin the "cloud." The global computer system also includes, among other things, telecoms towers, thousands of miles of underground and submarine copper cables, satellites, modems, antennas, and systems for electricity generation and distribution. All "cloud" platforms, data centers, and algorithms are ultimately "relentlessly material" mechanical processes that cannot function without minerals, resources, and electricity.[24] The "cloud" relies on a prodigious, fully earthly material infrastructure, an "extraordinary physical substrate"[25] made of lithium, coltan, oil, and labor.

The complete cycle of algorithmic capital's material infrastructure transforms natural resources into electronic waste. This material cycle includes four stages: extraction, manufacturing, use, and finally the "return to nature" in the form of e-waste or greenhouse gases. We should keep this distinction between different stages in mind if we want to appreciate algorithms' environmental footprint. For example, while electronic devices' energy efficiency is tested in the moment of use (an iPhone 14 is said to be more energy efficient than an iPhone 4, for example), their manufacturing generates ever-increasing footprints as they feature increasingly sophisticated circuits, screens, and cameras.[26] This feature development completely erases any gains made during the use phase and instead expands their total environmental footprint.

23 Crawford, *Atlas of AI*, pp. 28, 15.

24 Steven Gonzalez Monserrate, "The Cloud Is Material: On the Environmental Impacts of Computation and Data Storage," *MIT Case Studies in Social and Ethical Responsibilities of Computing*, Winter 2022, p. 3.

25 Peter Buell Hirsch, "Footprints in the Cloud: The Hidden Cost of IT Infrastructure," *Journal of Business Strategy* 34, no. 1, 2022, p. 66.

26 Charlotte Freitag et al., "The Real Climate and Transformative Impact of ICT: A Critique of Estimates, Trends, and Regulations," *Patterns* 2, no. 9, September 2021, sciencedirect.com.

In Thesis 5, we discussed the political economy of labor that underpins the production of coltan, a metallic ore essential to the manufacture of electronic devices. Rare earth metals are at the heart of algorithmic capital, and most of the world's reserves are found in China, which processes up to 90 percent of rare earth metals globally. Among the rare minerals and metals essential to the physical infrastructure of algorithmic capital is lithium, whose main known deposits are found in Argentina, Chile, Australia, and China. Lithium is mainly used in the manufacture of batteries, both for smartphones and electric cars—they contain up to 62.5 kg per vehicle. More broadly, global computing and e-commerce, which depend on batteries, are crucial areas of activity for algorithmic capital.

A major problem with the extraction of rare metals, apart from the slight detail that they are non-renewable resources, concerns the fourth stage in the cycle of algorithmic capital's material infrastructure: handling electronic waste. In fact, almost all the rare earth metals used in smart devices go unrecycled. Estimates indicate a recycling rate of around 1 percent in recent years,[27] and the world's e-waste is piling up in poorly regulated sites in countries such as Ghana and Pakistan. The average North American consumer generates about 20 kg of electronic waste per year, at least a third of which will end up in landfills in the Global South. This also contributes to the growth of the Global North's ecological shadow over the South, that is, the share of the ecological footprint of Global North consumption that is borne and suffered by spaces in the Global South. Finally, the socio-material processes that govern the development of artificial intelligence systems and algorithmic capital affect nature by extracting the result of billions of years of the Earth's geological history to serve a fraction of a second of contemporary technological time.[28]

Not only are these components key to the accumulation of algorithmic capital, but this drive for accumulation also generates political and geopolitical tensions in the places where these resources are extracted. The major manufacturers of the physical infrastructure of algorithmic capital—Philips, Foxconn, Intel, Dell, Nvidia, and the like—have set up vast networks to feed them rare earth metals and other resources, which may include tens of thousands of suppliers. Given the complexity of these networks, no company can guarantee that its supply is free of

27 Peter Dauvergne, *AI in the Wild: Sustainability in the Age of Artificial Intelligence*, Cambridge, MA: MIT Press, 2020, p. 95.

28 Crawford, *Atlas of AI*, p. 31.

resources extracted under conditions of exploitation, violent conflict, or environmental destruction.[29]

In addition to the enormous drain that algorithmic capitalism exerts on the global reserves of certain metals, the automation of the mining industry itself increases large global mining companies' extractive capacities. Multinationals such as Rio Tinto are investing heavily in algorithmic applications to automate production and reduce costs. For example, it is possible to improve the predictive capabilities of blasting models, reduce energy consumption, make greater use of autonomous vehicles, robotize drilling equipment, and manage equipment and labor more efficiently by analyzing data generated by sensors and other connected devices. The result is a transformation of the extraction system itself, based on the "smart mine" model. These capitalist spaces are thus sites where the world's mineral resources are being extracted at an ever faster and more unsustainable rate. This acceleration in turn consolidates the large multinational mining firms' economic and political power.[30]

Algorithmic capital: An energy-guzzling ogre

Calculating the carbon footprint of algorithmic capital is no simple task. However, the literature does provide us with some useful benchmarks to begin to grasp the scale of energy inputs required to support these processes and supply chains, and the emissions they generate. Take, for example, the information and communication technology (ICT) sector, which is at the heart of algorithmic capital accumulation. While estimates vary, these technologies are likely responsible for between 1.8 and 4 percent of global carbon emissions.[31] If this sector were a country, it would count among the world's top-five carbon emitters. The analyses sometimes vary when it comes to defining the boundaries of this sector (hence the variations in measurements). All agree, however, on including telecoms networks, digital devices, and data centers. These three key elements of algorithmic capital might be making significant gains in terms of energy efficiency, but their rapid growth means that these gains are offset in aggregate by the "rebound effect" (also known as the "Jevons

29 Ibid., p. 34–5.

30 Dauvergne, *AI in the Wild*, p. 139; Celia Izoard, *La ruée minière au XXI*[e] *siècle: Enquête sur les métaux à l'ère de la transition*, Montreal: Édition de la Rue Dorion, 2024.

31 Freitag et al., "The Real Climate and Transformative Impact of ICT"; Dauvergne, *AI in the Wild*, pp. 73–4.

paradox")—their total emissions are growing faster than their gains in energy efficiency.

There is no consensus on the limits to a device's or a network's life cycle from design to use. Nor is there a consensus on the analytical approach that looks further upstream in the supply chain. However, existing analyses do converge on certain key points regarding digital technologies. First, global data traffic is increasing. Second, the energy footprint is growing, and demand for networks and data centers will continue to rise. Third, the overall footprint of digital technologies will continue to increase unless there is a significant shift toward renewable energy sources. In this sense, the carbon neutrality strategies of companies such as Google and Meta, which rely heavily on a switch to renewables, are surely crucial.[32] These pledges, however, are being revised or even abandoned altogether in the context of the rapidly rising demand for electricity to power generative AI since 2022. The fact that the data centers and clouds belonging to big players such as Google and Amazon use renewable energy could help stabilize global CO_2 emissions from such infrastructure, even though electricity demand in this sector is set to increase significantly in the coming years (for instance, the European Commission estimates that data center electricity consumption in Europe will increase by 28 percent by 2030). Finally, information technologies could reduce emissions in other sectors; however, this remains far from certain, and as things stand, they could also contribute to increasing emissions.[33]

Meanwhile, two issues have attracted growing attention in the debate on digital technologies' energy footprint: cryptocurrencies and, more recently, data centers. The best-known cryptocurrency, Bitcoin, is based on a vast computer network scattered across the globe; its energy requirements account for about two-thirds of such demand from the entire cryptocurrency sector. Bitcoin mining alone consumes a staggering amount of electricity, more annually than the amount consumed in Argentina or the Netherlands.[34] Bitcoin mining is the process by which new blocks are added to the Bitcoin blockchain in order to validate transactions. This process is like a digital guessing game where the one who guesses correctly completes a block and receives a reward in the form of Bitcoin or transaction fees. These highly complex calculations require several miners to be grouped together within a "mining pool" that networks computers in order to increase the power in service of the

32 Ibid., pp. 73–80.
33 Freitag et al., "The Real Climate and Transformative Impact of ICT," p. 3.
34 Hirsch, "Footprints in the Cloud," p. 65.

"game." The system automatically adjusts so that a new block is created every ten minutes. According to Alex de Vries, writing in 2021, nearly 3 million specially devoted computers worldwide were competing for Bitcoins, generating 160 quintillion (160 followed by 18 zeroes) attempts per second and emitting approximately 65.4 megatons of CO_2 per year, the equivalent of that produced by a developed country such as Greece or Austria.[35] If we also consider the emissions bound up with other cryptocurrencies, the entire sector emits nearly 100 megatons of CO_2 per year, the equivalent of countries such as Chile, Norway, or Czechia.

Clearly, the location of these mining pools is all-important, because that determines whether they are powered by renewable or nonrenewable energy sources. For example, crackdowns on Bitcoin mining in China have forced miners to move to Mongolia and Kazakhstan, where they use electricity sources that include coal, in turn significantly increasing CO_2 emissions. The huge cryptocurrency bubble of 2021–2, especially for Bitcoin, caused significant demand for electricity; many fossil fuel companies took advantage of this to revitalize coal-fired power plants, establishing mining pools on these sites. For example, in July 2021, Black Rock Petroleum Company announced the installation of nearly 1 million Bitcoin mining machines at gas production sites in Alberta. In the United States, several fossil fuel assets in New York State that were no longer profitable have been reactivated in order to power Bitcoin mining operations, while the states of Kentucky and Pennsylvania offered tax credits and subsidies to attract Bitcoin miners, thus reactivating coal mines and creating jobs in this sector.[36] Coal is, of course, the most polluting fossil fuel, with a far greater impact than oil or natural gas. From this perspective, Bitcoin mining is directly linked to the fossil fuel industry, and the two feed off each other, with Bitcoin mining offering an opportunity to revitalize fossil fuel assets. Added to this is an intensifying competition for cryptocurrencies, forcing miners to upscale their equipment with more powerful computers at an accelerated pace. The result: Bitcoin mining generates a quantity of e-waste comparable to that of a medium-sized country.[37]

The energy consumption of data centers has recently come under critical scrutiny, especially in the wake of their rapid proliferation in the

35 Alex de Vries et al., "Revisiting Bitcoin's Carbon Footprint," *Joule*, no. 6, 2022, p. 499.

36 Ibid., pp. 499–501.

37 J. Huang, C. O'Neill, and H. Tabuchi, "Bitcoin Uses More Electricity Than Many Countries. How Is That Possible?," *New York Times*, September 3, 2021.

context of the generative AI boom beginning in late 2022.[38] These centers are immense warehouses housing thousands of server-computers. These obviously require electricity to function, but they also generate heat and need to be cooled down. Data centers use air conditioning for up to 40 percent of their energy consumption. Today in North America, most data centers depend on fossil fuel–powered electricity grids, for instance in the "Data Center Alley" in Virginia, where up to 70 percent of global Internet traffic transits. The carbon footprint of the cloud where global data is stored now surpasses that of the airline industry.[39] The International Energy Agency had already stated in 2022, *before* the generative AI boom, that data centers accounted for 1.5 percent of global energy consumption.[40] Every operation on the Internet, banking transaction, interaction with ChatGPT, Google search, image generation on Dall-E, or video on TikTok generates carbon emissions, up to ten times more using the new generative AI algorithms, which are more energy intensive.

A major problem preventing the precise measurement of the energy consumed by data centers is the very low reliability of the information made public by algorithmic capitalists. Investigation by the journalist Isabel O'Brien exposed in September 2024 that the *real* emissions from the data centers of Google, Amazon, Meta, and Apple between 2020 and 2022 exceeded those reported in the official publications of these companies by a staggering 662 percent.[41] The meteoric rise of Gen AI must only have aggravated the situation since then. In official documents, Google reports that its emissions have risen by 48 percent since 2019. However, this increase must in fact have been much higher. Microsoft, even before it teamed up with Open AI to become its main cloud partner, was in 2022 emitting in its data centers 6.1 million metric tons of CO_2, 22 times more than the 280,782 metric tons listed in its official report. Meta has emitted 19,000 times (!) more CO_2 in its data centers than the official number it reported. Computing generates heat in the present, but also in the future by releasing immense quantities of CO_2 into the atmosphere.

38 Paris Marx documents the growing energy and water consumption of data centers in a series of four episodes of the podcast *Tech Won't Save Us*, aired in December 2024.

39 Steven Gonzalez Monserrate, "The Cloud Is Material: On the Environmental Impacts of Computation and Data Storage," *MIT Case Studies in Social and Ethical Responsibilities of Computing*, Winter 2022, p. 7.

40 Isabel O'Brien, "Data Center Emissions Probably 662% Higher Than Big Tech Claims," *The Guardian*, September 15, 2024.

41 Ibid.

Training algorithms and AI also has a carbon footprint. At present, it is very difficult, if not outright impossible, to measure the total carbon footprint of all AI training and use worldwide. The European Commission has issued a call for action on this problem, since currently the available data is at best anecdotal.[42] However, some studies have focused on the emissions effects of training certain algorithms. For example, a study on the emissions generated by training a language-recognition algorithm reports CO_2 emissions equivalent to a return flight between New York and San Francisco.[43] Yet, it is important to note one important variable involved here: Depending on where the algorithm is trained, the source of electricity will differ and the resulting carbon emissions can vary greatly. For example, training the same algorithm may emit ten times more CO_2 if done in South Africa rather than in France. Different computer equipment may also consume more or less energy for comparable operations.

Another indirect indication of AI's environmental impact comes from the financial costs of each algorithmic query. For example, the arrival of ChatGPT has prompted companies such as Microsoft and Google to integrate LLMs (GPT-4 in Bing and Gemini in Google Search) into their search engines. Alphabet's chairman John Hennessy estimates that a search using the Gemini chatbot costs at least ten times more than a regular one.[44] Traditional Google searches were based on an indexing system, with robots that scan the web, store information, and provide relevant results based on users' keywords. In contrast, generative AI such as Gemini or ChatGPT must generate new responses to each query, using a much more energy-intensive inference system. Back in December 2022, OpenAI's CEO commented on Twitter that ChatGPT would have to be monetized at some point, because "the compute costs are eye-watering."[45] Analyses estimate that Google would have to spend an additional $6 billion if half of its searches (3.3 trillion per year) were combined with fifty-word responses from Gemini, and "the same calculation reaches $9 billion for seventy-five-word responses and up to $12 billion for hundred-word responses."[46] This poses a direct economic conundrum for Google,

42 Guglielmo Tamburrini, "The AI Carbon Footprint and Responsibilities of AI Scientists," *Philosophies* 7, no. 4, 2022, p. 2.

43 Hirsch, "Footprints in the Cloud," p. 67.

44 Aaron Mok, "AI Is Expensive. A Search on Google's Chatbot Bard Costs the Company 10 Times More Than a Regular One, Which Could Amount to Several Billion Dollars," *Business Insider*, February 22, 2023.

45 Sam Altman, Twitter, December 5, 2022.

46 Théo Sire, "L'IA coûterait très cher à chaque recherche: Google lâche des chiffres astronomiques," Frandroid, February 27, 2023.

because its business model relies heavily on ad revenue, and that may no longer be sufficient to cover its operating costs. But from an environmental perspective, if each keyword query produces between 1 and 10 grams of CO_2—as various studies report[47]—and the use of algorithms such as Gemini, Claude, and ChatGPT increases this carbon footprint tenfold, even greener data centers will not be enough to limit the constant increase in demand for energy. Some studies estimate that in the near future, the carbon footprint for training and operating generative AI could exceed that associated with cryptocurrency mining.[48]

While the data available on AI's carbon footprint is insufficient to establish any fully accurate overall picture, we can see that the accelerated spread of algorithms means that AI's carbon footprint is growing. That footprint is sure to grow, unless practical applications of these algorithms serve to reduce emissions in some other sector by improving energy efficiency. This is, indeed, the priority of several research and training programs. In other words, although AI research promises practical applications that can eventually mitigate global warming, it is itself currently one of the factors driving the climate crisis.[49]

Applications of algorithms to environmental management: Promises and limitations

Algorithms and AI have already found many applications in environmental management. Some of these seem, at first glance, to contribute to improving scientific research, particularly through the collection and analysis of data on ecological systems. Others aim to offer practical solutions to certain environmental problems. However, looking beyond these particular types of interactions between AI and the environment, we can see a further set of relations to consider, in which algorithms are used to accelerate the exploitation of natural resources. Here lie the main problems we need to identify.

At the first level of interaction, between algorithmic applications and environmental management, the outlook is quite promising. Peter

47 Anne Quito, "Every Google Search Results in CO_2 Emissions," *Quartz*, May 7, 2018.

48 Sharon Zhang, "Report on ChatGPT's Model Emissions Offers Rare Glimpse of AI's Climate Impact," *Truthout*, April 6, 2023.

49 Josh Cowls et al., "The AI Gambit: Leveraging Artificial Intelligence to Combat Climate Change," *AI & Society*, no. 38, 2021, pp. 283–307; Tamburrini, "The AI Carbon Footprint and Responsibilities of AI Scientists," p. 7.

Dauvergne provides an overview of some of these applications in the fields of environmental science and the protection of certain species, forests, and oceans. He mentions several benefits of AI for global conservation, including machine learning applied to the scientific observation of species or climate change. In this vein, he also points out that "deep neural networks are enhancing high-resolution mapping of forests and the modeling and forecasting of biodiversity loss," and how "AI analytics are supporting efforts to protect wildlife sanctuaries and preserve coral reefs." Intelligent robotics is also being used in conservation efforts: "Smart drones are helping state rangers and community activists combat poaching, illegal logging, and illegal mining. Intelligent robotic devices are helping to eradicate invasive species, clean up pollution, and rewild countrysides."[50] In addition, algorithms are able to measure temperature changes, pollution, and air quality with unmatched accuracy, and predict ocean phenomena, weather systems, extreme weather events, migratory animal movements, and carbon emissions.[51] Initiatives such as Machine Learning for Earth and Climate Sciences at the European Laboratory for Learning & Intelligent Systems (ELLIS) aim to model and understand the Earth as a whole using machine learning. They promise to use these tools to drive breakthroughs in climate research.

Some of these applications do help to improve ecosystem management, and there is no doubt that algorithmic technologies can improve the quantity and quality of environmental data. Dauvergne cites the examples of Ebird, an open-access database that compiles bird observation entries from thousands of professional and amateur ornithologists around the world; EarthCube, which aims to combine Earth science research and observation databases; and the Global Fishing Watch platform, which uses AI to record commercial fishing activities in real time across the world's oceans. Rainforest Connection uses bioacoustic technologies to monitor poaching and illegal logging activities, while the ChimpFace application uses facial recognition to identify chimpanzees in order to prevent their illegal trafficking.[52] In this sense, algorithmic technologies enhance the ability to track, in real time, environmental changes and the activities that cause them, and to produce tools that can be used in conservation efforts.

As well as making pledges toward carbon neutrality goals, high-tech firms have also taken initiatives to marry their algorithmic capabilities

50 Dauvergne, *AI in the Wild*, p. 53.
51 Cowls et al., "The AI Gambit."
52 Dauvergne, *AI in the Wild*, pp. 53–61.

with the efforts of environmental NGOs and nonprofits. Although modest, investments in this area allow small grassroots organizations to access certain cutting-edge technologies in their work. Among the GAFAM companies, Microsoft is probably the most active in this regard, notably with its AI for Earth initiative, which supports projects such as the PAWS program, which tracks illegal hunting and uses algorithmic models to predict areas at risk of poaching.[53]

Despite these commendable and sometimes positive initiatives, we should not expect algorithms to solve the climate crisis or to provide any lasting answers to environmental problems. Environmental science is gaining in accuracy, but this does not undermine the structures of power and exploitation associated with algorithmic capitalism. What AI does help us to see is how quickly we are racing toward ecological catastrophe. As Dauvergne insightfully points out: "AI technology can take us only so far toward lasting, far-reaching solutions, as it has little power to confront the underlying political and socioeconomic forces exploiting natural resources and destroying nature for profit."[54] In any case, the environmental problems we face far exceed the mitigation capabilities of any one algorithmic tool, and the state of the world's ecosystems is continuing to worsen at an accelerated rate.

Applying algorithmic technologies to consumer goods also has potential in terms of increasing energy efficiency and sustainable consumption. One highly desirable side effect often cited by the champions of autonomous vehicles is their potential to reduce the greenhouse gas emissions coming from the global vehicle fleet by an order of 15 to 20 percent. The Internet of Things is accompanied by—often empirically demonstrated—promises of energy savings in the use of all kinds of devices, such as household appliances, televisions, and air conditioning or heating systems. However, several obstacles complicate the situation, as do the undesirable effects of these same innovations. First, energy efficiency often comes with a rebound effect (mentioned above), where more efficient use of an energy source reduces its costs and increases use, effectively creating greater total use of the resource even though it is being used more efficiently.[55] Second, the process of replacing conventional objects can take several years, and have the effect of developing new

53 Cowls et al., "The AI Gambit."

54 Dauvergne, *AI in the Wild*, p. 68.

55 David Owen, *The Conundrum: How Scientific Innovation, Increased Efficiency, and Good Intentions Can Make Our Energy and Climate Problems Worse*, New York: Penguin, 2012.

markets and increasing the supply of goods and overall consumption, which in turn leads to an increase in energy demand for the materials used to manufacture these objects (including the rare earths mentioned above) and other resources needed for their large-scale manufacture and distribution. In the case of algorithmic technologies, the problem of the fourth moment of the material cycle of algorithmic capital—the processing of electronic waste—is particularly alarming. This is especially true when we consider the planned obsolescence strategies adopted by manufacturers to stimulate demand for new product ranges. As Rick Paulas notes, the smart home is likely to quickly become an obsolete home.[56] Third, insofar as targeted advertising, on which algorithmic capitalists rely for much of their revenue, is effective, we can expect an increase in global consumption. Wrapped up in a constant and relentless stream of advertising nudges and prompts, there is no doubt that algorithms fuel a culture of overconsumption that is already highly problematic from an environmental standpoint. Even if each individual product could indeed have a smaller footprint per unit, the explosion in the total quantity of products consumed will quickly make the original problem worse rather than help solve it. We can conclude, with Dauvergne, that smart objects are not a remedy for the unsustainability of capitalism: quite the opposite.[57]

In terms of transport, reliance on the solutions promoted by Silicon Valley, such as Uber or Lyft's brand of "carsharing," autonomous vehicles, or electric cars, leads us to a collective social and environmental dead end.[58] Instead of focusing on viable solutions such as public transport, massive investment in cycling and rail infrastructure, urban redevelopment, and various means of reducing car dependency, capitalist industry is diverting media attention and public resources toward inefficient and environmentally unsustainable privatized solutions. The best such example is surely the gargantuan Hyperloop project proposed by Elon Musk and his Boring Company: a "vactrain" consisting of a double tube carrying capsules that could—theoretically—transport people or cargo at speeds of over 1,102 km/h. This concept, inspired by a short story by Jules Verne, was first announced by Musk in 2013; over a decade later, after half a dozen initiatives, this target remained a distant prospect.[59]

56 Rick Paulas, "All Good Gadgets Go to Waste," *New York Magazine*, July 31, 2018.

57 Dauvergne, *AI in the Wild*, p. 86.

58 Paris Marx, *Road to Nowhere: What Silicon Valley Gets Wrong About the Future of Transportation*, London: Verso, 2022.

59 Sarah McBride, "Startups' Hyperloop Dreams Still Distant, Almost 10 Years

The company Hyperloop One achieved a speed record of 309 km/h in 2017; the first commercially available version of this project is the Las Vegas Convention Center Tunnels, which allow Tesla cars to travel at 50 km/h over a distance of 2.4 km.[60] Even apart from the fact that these results are laughable compared to the revolutionary promise, the main problem with Hyperloop lies in its deeply ideological inspiration. While purporting to create an ultra-high-speed transport solution safer than airplanes, from the outset this initiative was aimed at stifling collective projects that would have brought much better results. As Paris Marx notes:

> The Boring Company was supposed to solve traffic, not be the Las Vegas amusement ride it is now … Musk admitted to his biographer Ashlee Vance that Hyperloop was all about trying to get legislators to cancel plans for high-speed rail in California—even though he had no plans to build it.[61]

Thus, technological solutionism is not only ineffective, but it can in fact delay our taking action in terms of our collective interests.

The agri-food industry is also undergoing significant changes under the influence of algorithms. We briefly mentioned in Thesis 8 the case of John Deere becoming a data company over the last fifteen years. Several productivity trends already set in motion by the industrialization of agriculture are being accelerated by soil and crop management using drones, sensors, and algorithms. Data collection on crop conditions, plant and animal health (applications such as Connecterra, similar to Fitbit, are now used on dairy farms), and production planning using algorithms are combined with the robotization of agricultural equipment and certain work tasks in order to propel agriculture into the age of algorithms. These gains in productivity and management efficiency are surely desirable for businesses in the sector, yet they also serve to perpetuate and in some cases even sharpen the unsustainability of the global agri-food industry.[62] As in previous episodes of technological innovation in this sector, such as the GMO revolution, these trends toward the algorithmization of agriculture are also accompanied by an increase in

---

After Elon Musk Paper," *Los Angeles Times*, February 25, 2023.

60 Jay Ramey, "Las Vegas Loop Launches with Tesla Cars, Minus Some Promised Features," *Autoweek*, June 10, 2021.

61 Paris Marx, "Elon Musk Is Convinced He's the Future. We Need to Go Beyond Him," *Time*, August 2, 2022.

62 Michael S. Carolan, *The Real Cost of Cheap Food*, London: Routledge, 2018.

the power of large agri-food companies, which are currently the only ones with the capital necessary to carry out such innovations. The costs of this equipment may also worsen debt problems for small farmers.[63] The catastrophic ecological consequences of industrial agricultural practices, as well as the serious social injustices and historical and contemporary oppression that underlie this system, are well documented in academic literature.[64] While the transition to algorithms may improve certain methods or streamline certain practices, it risks strengthening the global power relations that underpin these injustices and accelerating the impact of practices that are destabilizing ecosystems.

This transition to Agriculture 4.0 may also weaken farmers' resilience faced with various economic, social, and ecological shocks, particularly through its effect on their ability to maintain agricultural equipment and machinery. For example, the semiconductor shortage during the pandemic directly affected this sector, particularly because capitalist farm machinery manufacturers such as New Holland, John Deere, and Kubota were no longer able to supply certain essential parts due to the lack of electronic chips.[65] Added to this is the capitalist ringfencing of intellectual property, which cuts off the possibility of independently repairing some agricultural equipment without going through the manufacturer's customer service department. A hacker nicknamed Sick Codes managed to jailbreak John Deere's locks on its technology in the interest of making agricultural equipment less dependent on the original manufacturer. Serving the interests of global food security, such actions uphold the farmers' "right to repair," under the slogan "Free the Tractors!"[66]

In short, algorithmic capital does not free us from the problematic model of industrial agriculture, but rather accelerates its worst tendencies, bound up with a productivist vision that seeks to optimize the exploitation of soil, plants, and animals as much as possible. Agriculture 4.0 is essentially another form of automation in industrial agriculture. It amplifies the flaws of a model that fuels the overexploitation of nature by once again replacing humans with robots and other smart machines.

63 Dauvergne, *AI in the Wild*, p. 112.

64 Sébastien Rioux, *The Social Cost of Cheap Food: Labour and the Political Economy of Food Distribution in Britain, 1830–1914*, Montreal: McGill-Queen's University Press, 2019; Jennifer Clapp, *Food*, Cambridge: Polity, 2020.

65 Alayna Harrell, "How the Microchip Shortage Impacts the Agricultural Industry," *Wengers*, July 16, 2021.

66 Lily Hew Newman, "A New Jailbreak for John Deere Tractors Rides the Right-to-Repair Wave," *Wired*, August 13, 2022.

The partnership between algorithmic capital and fossil capital

The connection between algorithmic capital and the fossil fuel industry is becoming more and more apparent, especially since the dawn of Gen AI in late 2022. Far from proposing solutions to limit carbon emissions or developing green technologies allowing for a decrease in the extraction and use of fossil fuels, algorithmic capital accelerates the extraction and consumption of fossil fuels and the accumulation of fossil capital. On the one hand, AI technologies, the cloud, and connected devices generate a gigantic demand for energy. On the other hand, fossil fuel companies are using more and more algorithmic technologies in their operations, establishing partnerships, for example with Microsoft, to use AI to find new oil reserves, optimize extraction operations, automate steps in production, and ultimately generate more profits.

The recent AI boom has triggered a spectacular renaissance of construction and commissioning of new gas power plants in the US. On the one hand, this gas boom has precipitated a decline in higher-emitting coal plants, but on the other hand it meets up to 60 percent of new energy demand, leaving renewable energy sources such as solar or wind far behind. The typical lifespan of gas power plants is forty years, so these new investments are orienting development of energy resources for the coming decades in the opposite direction of a reduction in CO_2 emissions.[67] Furthermore, gas industry infrastructure, like the fossil fuel industry in general, produces methane emissions, a greenhouse gas up to 80 times more potent than CO_2 for global warming.[68] The Stargate Project, launched with great hype in January 2025 by Donald Trump and Silicon Valley "broligarchs" from Oracle, OpenAI, and others, promises investments up to $500 billion in the energy infrastructure and data centers needed for an exponential development of AI.

At this juncture, algorithms are used to boost the profitability of fossil fuel production. This industry now uses a myriad of algorithmic applications to accelerate the processing of the enormous amounts of data extracted by satellites, sensors, drones, and other devices to discover and evaluate oil, natural gas, and coal reserves. Algorithms calculate the most profitable methods to extract these fuels, whether by drilling or fracking, as well as to locate new reserves and speed up management operations

---

67 Josh Saul et al., "AI Boom Is Driving a Surprise Resurgence of US Gas-Fired Power," *Bloomberg*, September 16, 2024.

68 Kate Aronoff, "The U.S. Has a Methane Problem. A.I. Is Making It Worse," *New Republic*, September 23, 2024.

including applying for drilling permits. Deep learning is applied to underground data to improve the profitability and productivity of giants such as ExxonMobil, Royal Dutch Shell, and Sinopec. Big Tech firms such as Amazon, Google, and Microsoft are creating ever more algorithmic products specifically designed for the global fossil fuel industry. This ultimately results in accelerated extraction and combustion of global fossil fuel reserves through increased automation in this sector, thus releasing CO_2 into the atmosphere at an ever-increasing rate.[69] The automation of the fossil fuel industry is thus also "the automation of the climate crisis," as Brian Merchant puts it.

The case of Microsoft is particularly alarming. As Karen Hao has shown, behind its public relations narrative of green technosolutionism and its pledges to reach carbon neutrality, Microsoft is in fact a major contributor to the acceleration of the climate crisis, not only through its own footprint, but now increasingly also through its partnership with fossil fuel companies.[70] This sector represents a potential market between $35 and $75 billion annually for Microsoft. As such, it promotes its own AI products and provides its technologies to the likes of Chevron, Halliburton, ExxonMobil, and Shell in order to automate drilling, find and exploit new reserves, and maximize production. The company preaches that AI can help mitigate global warming, for instance by making production more efficient, or by inventing new technologies that can reduce the amount of CO_2 in the atmosphere. While these benefits remain largely speculative, the real effect of deploying algorithmic technologies in the fossil fuel industry is to accelerate extraction and production. Meanwhile, Microsoft asks its own employees not to use plastic straws in the company cafeterias.

The AI boom, whether it is simply hype or not, generates in turn an increase in expected future energy demand, and thus pushes energy companies to invest in new plants and drilling projects to be able to meet this demand. As Brian Merchant mentioned in an episode of his podcast *System Crash*, the speculation on AI's future feeds speculation on the development of new energy infrastructure. A vicious circle is taking shape, where more AI calls for more fossil fuels and more fossil fuels call for more AI.

---

69 Dauvergne, *AI in the Wild*, p. 137.

70 Karen Hao, "Microsoft"s Hypocrisy on AI," *The Atlantic*, September 13, 2024.

The threefold acceleration of the climate crisis and scenarios for energy descent

Even if algorithmic technologies may have certain positive uses for environmental science and some practical applications, their impact in a capitalist context points to accelerated extraction and consumption that also accelerates global warming. These three forms of acceleration each feed into each other, as we just evoked in the case of the fossil fuel industry.

What role can algorithmic capital play in a world that is sinking into climate crisis? To respond to this central challenge of our century, a growing number of scientists, activists, and politicians are questioning the paradigm of "green growth" and the technocentric vision of "transition." They are instead making the case for a form of "energy descent" that pivots on a socio-ecological transition, citizen action, and local initiatives.[71] Rob Hopkins defines energy descent as

> the continual decline in the net energy supporting humanity … which mirrors the ascent in net energy that has taken place since the Industrial Revolution. It also refers to a future scenario in which humanity has successfully adapted to the declining net fossil fuel energy availability and has become more localized and self-reliant.[72]

Several ideas and movements such as degrowth, transition towns, permaculture, local mutual-aid systems, municipalism, and the commons are thus aimed at creating greater resilience to help us cope with current and future crises.[73]

Far from representing a single scenario, energy descent may take different forms depending on the speed of the decline in fossil fuel use, whether because of unavoidable depletion or policy choices, the intensity of climate change, and the more or less proactive responses of governments, businesses, and civil society. The multiple possible responses can lead to various forms of adaptation, evolution … or even, in some cases, collapse. In short, there is no single, inevitable historical trajectory, but

---

71 René Audet, "Discours autour de la transition écologique," in Marie-Josée Fortin, Yann Fournis, and François L'Italien, eds., *La transition énergétique en chantier: Les configurations institutionnelles et territoriales de l'énergie*, Quebec City: Presses de l'Université Laval, 2016, pp. 23–42.

72 Rob Hopkins, *The Transition Handbook: From Oil Dependency to Local Resilience*, Totnes, UK: Green Books, 2008.

73 Agnès Sinaï, Raphaël Stevens, Hugo Carton, and Pablo Servigne, *Petit traité de résilience locale*, Montreal: Écosociété, 2017.

rather different possible futures. There could be a shift toward various kinds of social, technological, and institutional configurations, depending on a vast number of interdependent factors. In this uncertain context, it is useful to visualize different scenarios to see what role AI could play in the decades to come.

Rob Hopkins draws inspiration from the work of Bryn Davidson, who distinguishes among four scenarios based on two dimensions. In Hopkins's work, the horizontal axis represents the more or less rapid depletion of fossil fuels, while the vertical axis refers to the proactive or reactive measures taken by social forces. This provides four energy scenarios, including sustainable development (focused on technology markets), attrition (climate chaos and the status quo), energy descent (a lean or delocalized economy), and collapse (for example, *Mad Max*, Easter Island).

Similarly, David Holmgren, cofounder of the permaculture concept, envisages four scenarios for energy decline based on two axes. The first relates to the speed of the depletion of fossil fuels, while the second represents the destructive scale of the climate crisis.[74] By combining Davidson and Holmgren's models, we can see four very different hypothetical futures emerging: (1) "Brown Tech," characterized by the economic status quo and authoritarian crisis management; (2) the "Green Tech" world of sustainable development; (3) a localist, low-energy world (Stewardship of the Earth); and (4) a post-collapse world reminiscent of a Hobbesian state of nature ("Lifeboats"). These scenarios have the quality of making it easier to visualize possible futures, which can help to guide strategic action. Yet they fail to address the specific role of capital and algorithms in this complex historical process. Figure 15.1 illustrates these four possible worlds.

According to the critical analyses presented throughout this book, the current development of AI under algorithmic capitalism is leading us straight toward the Brown Tech scenario. This has been made even clearer since Donald Trump's return to the White House, which saw the abandonment of green energy projects, the dismantling of environmental policies, a declaration of a National Energy Emergency, and the deregulation of the digital industry—measures that had been tentatively implemented under the Biden administration.[75] This means the accelerated extraction of fossil fuels; the integration of algorithms into

74 David Holmgren, *Future Scenarios: How Communities Can Adapt to Peak Oil and Climate Change*, Totnes, UK: Green Books, 2009.

75 Oliver Milman and Dharna Noor, "Explained: How Trump's Day One Orders Reveal a White House for Big Oil," *The Guardian*, January 22, 2025.

Figure 15.1 Four scenarios for climate change and energy descent

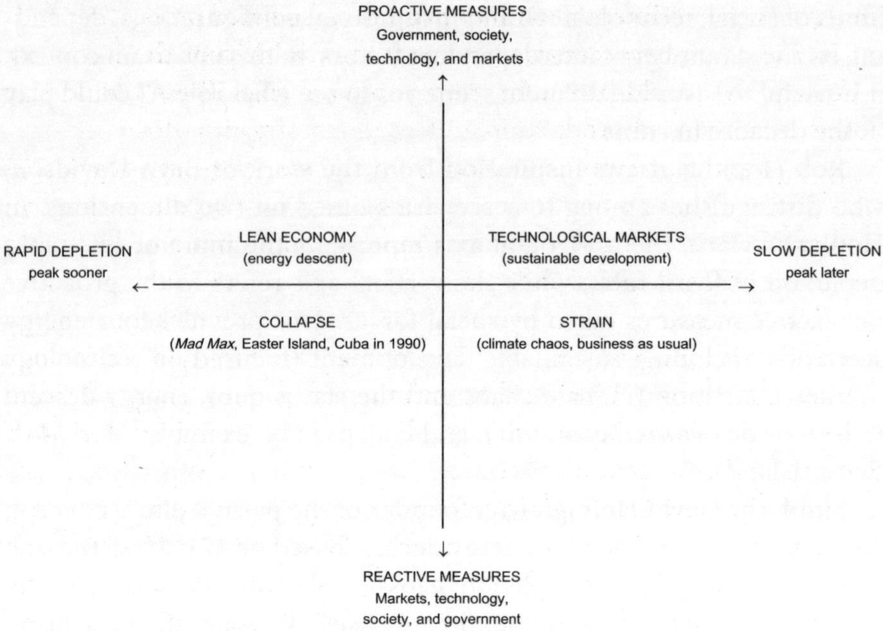

Source: Diagram adapted from Rob Hopkins, *The Transition Handbook*, p. 54

the state apparatus, the police, and the army; and the centralization of power accompanied by the erosion of democratic life. This is the most likely scenario, closely linked to "technofascism," but also the least desirable outcome for society.[76]

The Green Tech scenario seems to many to be the right goal to strive for, since it would enable the energy transition to take place without major disruption to our lifestyles. This would mean combining AI-driven technological progress with a rational management of resources. However, this dream world of "green AI" is unlikely to materialize within the capitalist social order, which is based on a logic of infinite growth, acceleration, and continuous colonization. The only option would be to transition to a Green New Deal model, a kind of state-regulated algorithmic capitalism capable of taming the power of Big Tech and redistributing socially produced wealth on a global scale.[77] We will revisit this question in Thesis 18. For now, we can say that in this scenario, the metabolic rift inherent in the capitalist mode of production would not be completely overcome, but simply displaced. As Saito puts it: "Metabolic rift cannot

76 Jonathan Durand Folco, *Fascisme tranquille: Affronter la nouvelle vague autoritaire*, Montreal: Écosociété, 2025.

77 This is the scenario hoped for by futurologist Jeremy Rifkin in *The Green New Deal*, New York: St. Martin's Press, 2019.

be fully repaired unless the universal metabolism of nature is mediated in a *qualitatively* different manner by freely associated producers."[78]

In the other two scenarios, namely Stewardship of the Earth and Lifeboats, AI would play a far more marginal role. These two scenarios assume reduced production and a gradual drawing-down of major industrial infrastructure and electrical networks. They envisage a relocalization of trade, a return to local agriculture in urban, peri-urban, and rural areas, and a revival of traditional techniques and know-how to ensure community subsistence.

These two scenarios differ mainly with regard to the likely quality of social ties, local governance systems, and the possibility of achieving a "democratic transition" toward a postindustrial and postcapitalist world. Stewardship of the Earth embodies the vision of "convivial degrowth" with its self-managed local communities, while the lifeboat scenario represents a new Dark Age, a postapocalyptic world with a return to almost feudal political fragmentation. In the scenario of a postgrowth, convivial society, the digital world and AI would occupy a very limited place compared to the current situation. In this scenario, digital technologies would be based on common and repairable IT tools to ensure the preservation of human knowledge and communication between distant communities, all of which would rely on reduced industrial infrastructure. The energy descent would be accompanied by a shrinking of the size of the digital world, and thus a residual use of AI to meet social needs as chosen by the community.

In contrast, AI in a post-collapse world would mainly be used from within secure fortresses in order to enable control of populations, migration flows, and pandemics. Table 15.2 provides a summary of the place that algorithmic capital would occupy in these energy descent scenarios, each of which may be more or less probable and desirable. It is clear that algorithms will not play a central role in the socio-ecological transition, although they could have different impacts depending on each scenario.

Ultimately, algorithms will not save us from climate disaster. At best, they will enable us to measure its full scope and accurately calculate the extent of the catastrophe and the speed at which global warming is occurring and ecosystems are being destroyed. However, as we have learned from decades of climate politics, forecasting or prior knowledge does not amount to prevention or potent political action. Even if we can switch on a vast and sophisticated apparatus of knowledge and

78 Saito, *Marx in the Anthropocene*, p. 31.

Table 15.2 Algorithmic capitalism and the scenarios of energy descent

Scenario	*Economy*	*Politics*	*Role of AI*	*Desirability*	*Probability*
Brown Tech	Deregulated algorithmic capitalism	Technofascism Authoritarian populism	Major	Low	High
Green Tech	State-regulated algorithmic capitalism	Green New Deal, Social democracy 2.0	Major	Moderate-high	Low?
Stewardship of the Earth Convivial Degrowth	Economic democracy	Ecosocialism, bioregionalism	Minor	High	Low?
Collapse Lifeboats	Subsistence economy	Local authoritarianism, ecocommunities	Minor	Low	Moderate?

measurement, this would not in itself reduce the ecological footprint (and would probably add in one of its own). For now, we should conclude that under the capitalist institutionalized social order, algorithms are factors that exacerbate the triple acceleration of extraction, consumption, and global warming. The optimistic scenario of energy transition in the form of Green Tech remains unlikely and fraught with pitfalls. We thus face the urgent need to fight against algorithmic capitalism and build a sustainable and free society to avoid the dystopic Brown Tech trajectory. This is the focus of our final three theses. But first, we must examine the complex mechanisms of subjectivation and alienation that are bound up with algorithmic capital.

Thesis 16

Algorithms are changing socialization processes and intimate relationships, and shaping new subjectivities

We are ourselves only in the eyes of others.

—Jean-Paul Sartre

Algorithmic capital shapes subjectivities.[1] This is not only a matter of algorithms' direct power over human behavior (as explored in Thesis 9). It is also the effect of the attention economy, the relentless quest for social validation, the value attached to self-exposure, the central role of social media, and the rise of emotional robots with which we can form intimate relationships. Here we are talking about the power that algorithms, bots, and platforms have to profoundly shape forms of consciousness and modes of personality. Algorithmic governmentality promotes *passivity* among subjects, who are manipulated from the outside (through various recommendations, prompts and nudges, targeted advertising, and automated decisions). However, the way in which subjects connect with the social world through screens and platforms also influences how they *actively* engage with the world, others, and themselves.

To inquire into the transformations of the human personality in the context of algorithmic capitalism, we will first draw on the concepts of social character, the connected self, and the economy of visibility.[2] These

1 The section of this thesis on the connected self is a reformulation of an article by Jonathan Durand Folco, "Pourquoi aime-t-on être surveillé?: Notes sur la société algorithmique de l'hypervisibilité," *L'Inconvénient*, no. 92, Spring 2023, pp. 10–15.

2 Sarah Banet-Weiser, "Keynote Address: Media, Markets, Gender: Economies of Visibility in a Neoliberal Moment," *Communication Review* 18, no. 1, 2015, pp. 53–70.

concepts help us see how individuals are encouraged to stay constantly connected; to seek exposure; to share their ideas, photos, videos, and life stories; and/or to like, comment on, and share other people's content. In so doing, they are generating both ever-more data and a form of algorithmically mediated social recognition. This new form of subjectivity has its own motto: "I am seen, therefore I am."[3] We will then discuss the specific role of social robots, which influence our intimate relationships in a similar way: "I am listened to and loved, therefore I am."

There is thus a possible partial explanation for the general tendency for people today to pay less attention to issues around surveillance: the fact that individuals have such a strong need to be seen, heard, and connected, and to receive validation, particularly through their interactions on digital platforms. This could be seen as a simple continuation of the "struggle for social recognition," as it moves to the online space.[4] But there may well be a deeper shift at play. When issues around surveillance are raised, many respond with a shrug: "I have nothing to hide," or "So what?" or "Yes, that's true, but you still have to look at the up-sides." These defensive or careless reactions and quick justifications express an attitude rooted in the spirit of the times. It may well be that the mass surveillance of our lives by algorithmic technologies is not only unavoidable, but, to an extent, *desired*. Although our moral conscience sometimes reminds us that it is important to protect our privacy, could it be that we unconsciously want to be monitored?

Some might retort that there is no reason to assume that there is a widespread desire to be monitored, and that the desire for surveillance should not be confused with the need for recognition. It is, rather, the price paid for access to platforms that promise their users a certain degree of visibility and recognition. Indifference to surveillance practices is more likely the result of ignorance than of some sort of unconscious dynamic.[5] However, although these are surely factors, we hypothesize that a transformation of personality types in the age of social media is leading a growing number of individuals to desire being seen, validated, followed, and commented on by complete strangers, to whom

---

3 Joël Birman, "Je suis vu, donc je suis: La visibilité en question," in Nicole Aubert and Claudine Haroche, eds., *Les tyrannies de la visibilité: Être visible pour exister?*, Toulouse: Érès, 2011. See also Martin Altmeyer, "À la recherche de résonance: Un diagnostic de la modernité numérique," *Réseaux* 235, no. 5, 2022, p. 149.

4 Axel Honneth, *The Struggle for Recognition: The Moral Grammar of Social Conflicts*, Cambridge, MA: MIT Press, 1996.

5 See Mathieu Gagnon, "Pourquoi accepte-t-on d'être surveillé," *Le Devoir*, June 22, 2023.

their everyday existence and even their private lives are exposed. Their (conscious and unconscious) desires for self-exposure and recognition are thus encouraged by the imperatives of algorithmic capital, with individuals actively signing up to being surveilled by others. Such practices may themselves provide a form of psychological satisfaction and represent a means of self-fulfillment.

Of course, we are not suggesting that everyone conforms to these trends, or that people connecting online all do so for the same reasons and all undergo a rewiring of their personality! Online experience is varied, and people connect with different aims and purposes. From influencers to occasional users, and everyone in between, experience and navigate online presence in different ways. What we wish to do here is to explore some ideal-typical forms and offer reflections on how the political economy of algorithmic capital encourages, promotes, and rewards certain types of subjectivities.

A new subjectivity? The connected self and the economy of visibility

To explain the emergence of a new hyperconnected subjectivity driven by a thirst for visibility, we can draw on the thinking of psychoanalyst and sociologist Erich Fromm. In his book *The Fear of Freedom*, Fromm developed an original theory of social character to explain the rise of fascism in the 1920s and 1930s.[6] In addition to the economic crisis and the presence of charismatic leaders, the existence of an authoritarian character within the population is said to have fueled the desire for submission to a strongman (such as Mussolini or Hitler) among the middle and working classes. Far from reducing the individual to an expression of the dominant ideology, Fromm starts from human subjectivity and places it within a broader social context: "To understand the dynamics of the social process we must understand the dynamics of the psychological processes operating within the individual, just as to understand the individual we must see him in the context of the culture which moulds him."[7]

This is how Fromm formulates the concept of social character. While individual character expresses the personality structure of a particular individual, social character refers to a kind of mentality, collective

6 Eric Fromm, *The Fear of Freedom*, London: Kegan Paul, 1946 [1942].
7 Ibid., pp. ix–x.

identity, or shared psychological form: "The social character comprises only a selection of traits, *the essential nucleus of the character structure of most members of a group which has developed as the result of the basic experiences and mode of life common to that group*."[8] What role does social character play in the workings of the mind? It is the main mediation that allows the individual to adapt to their environment.[9] In other words, social conditions shape character, which in turn shapes individual consciousness. When an individual's character is aligned with societal norms, acting in accordance with their social character provides them with psychological satisfaction. As Fromm aptly summarizes: "By adapting himself to social conditions man develops those traits that make him *desire* to act as he *has* to act."[10] Of course, each of us can deviate from this social character to varying degrees. Furthermore, different social characters can coexist within the same society, and some disappear or appear depending on the evolution of societies throughout history. Despite the possible similarities and continuities, each era brings with it a specific configuration of social structures that will condition the character of a given society's members.

Fromm emphasizes that character is a set of traits acquired through primary and secondary socialization, and more particularly through family, school, work, the media, and culture. Character combines a set of dispositions to act, think, and feel that enable us to relate to the world, others, and ourselves, with varying degrees of success. The psychoanalyst distinguishes five character orientations: receptive, exploitative, hoarding, marketing, and productive.[11] Fromm emphasizes that the "marketing personality," which leads individuals to see themselves as commodities and evaluate themselves based on their ability to sell themselves to others, became largely dominant with the development of the market economy and consumer society. This type of personality has several qualities (adaptability, intelligence, sensitivity to the changing expectations of others), and they take for their role models the figures who are portrayed as successful in magazines, films, and newspapers. Yet this personality also sometimes suffers from superficiality. This may lead the subject to doubt their own worth and identity, which is based mainly on other people's regard:

8 Ibid., p. 239.
9 Ibid., p. 241.
10 Ibid., p. 243.
11 Erich Fromm, *Man for Himself: An Inquiry into the Psychology of Ethics*, New York: Henry Holt, 1990 [1947].

> Success depends largely on how well a person sells himself on the market, how well he gets his personality across … whether he is "cheerful," "sound," "aggressive," "reliable," "ambitious"; furthermore what his family background is, what clubs he belongs to, and whether he knows the right people … in the marketing orientation, [the individual must] find the conviction of identity not in reference to himself and his powers but in the opinion of others about him. His prestige, status, success, the fact that he is known to others as being a certain person are a substitute for the genuine feeling of identity. This situation makes him utterly dependent on the way others look at him and forces him to keep up the role in which he once had become successful.[12]

This description of the marketing personality shares many similarities with a type of personality widely found on social media today. We can admit that the broad variability in life trajectories, temperaments, and specific social conditions will shape each person's individual character. Nevertheless, digital platforms are having a marked impact on modes of socialization, and this impact will likely foster the spread of certain character traits within society. We can also hypothesize that today's connected self represents an evolution of the marketing personality identified by Fromm, intensifying the role of others' perceptions and the need for social approval through online interactions. But, looking beyond the fact that our existential need for social recognition now passes through the filter of technology, how do contemporary capitalism, the attention economy, and algorithms affect our psyches?

Social media between influencers and branding

Participation in the world of social media is both universal and widely differentiated, with major variations in terms of class, gender, race, ethnicity, age, physical appearance, and so on. The vast majority of the world's population has at least one social media account; they participate in algorithmic labor simply by logging on to the platforms, sharing content, liking photos or comments, scrolling through videos, and thus producing and sorting data. We have evoked some of these numbers throughout the book: By late 2025, more than 5.6 billion people are using social media, which represents close to 69 percent of the global

12 Ibid., pp. 115–16.

Table 16.1 Reasons for following influencers by gender among millennials and Gen Z in the US

Men	%	*Women*	%
Gaming	62%	Beauty, skincare	59%
Fitness, sports	41%	Fashion	49%
Food, diet	35%	Food, diet	47%
Travel, lifestyle	27%	Travel, lifestyle	40%
Health, well-being	27%	Health, well-being	40%
Politics, social issues	24%	Family, parenthood	32%
Fashion	23%	Fitness, sports	30%
Beauty, skincare	12%	Gaming	26%
Family, parenthood	9%	Politics, social issues	21%

Source: The Influencer Report, "Engaging Gen Z and Millennials," *Morning Consult*, November 2019

population. The average user spends 2 hours 24 minutes per day on social media platforms.[13] Not all users are equally active, and usage varies across age groups, genders, and countries. In addition, platforms fulfill multiple functions in this visibility economy, responding to various needs and motivations (Table 16.1).

We can therefore see that social media plays the role of the public square (or churchyard) of previous eras. It is fulfilling multiple traditional social functions (staying informed, sharing rumors, finding things to do, debating ideas, passing the time), while creating new social habits linked to the digital world. These include following influencers, posting information about one's life for complete strangers to see, and alleviating the fear of missing out. These technologies have thus become a kind of "privatized public space" where billions of individuals access information, communicate, debate, influence each other, congratulate each other, or spew out their hatred for each other. Social media represents the new universal medium of human interaction, capturing more than 12 billion hours of attention every day, the equivalent of 1.4 million years of human existence. In this attention economy, which occupies a significant part of our days, each platform meets various expectations and needs to varying degrees.

We should also note that, in the world of social platforms, individuals are both consumers and producers of content; this makes these interfaces all the more dynamic, interactive, and addictive. On the one

13 "Number of Internet and Social Media Users Worldwide as of October 2025," statista.com.

hand, consumers of content represent the demand side of the attention market as they seek to obtain various goods or services: information, entertaining videos, messaging services, things to buy, and so on. Social media platforms act as intermediaries, connecting these consumers with a supply of varied content (free or paid), including posts from friends and family, videos, stories, reels, and targeted ads. On the other hand, content producers seek to capture the attention of the public, which represents a variable supply of "available brain time," as coined by Patrick Le Lay, former CEO of French broadcaster TF1. To borrow a concept from sociologist Eva Illouz, content producers are churning out "emotional commodities" (emodities) intended to be viewed, commented on, and consumed by a vast market of potential users who are hungry for experiences, emotions, novelty, and controversy.[14]

The production, circulation, and consumption of emotional goods on social media also faces a key problem: human attention span is a renewable but also limited resource. The most striking aspect of our hyperconnected world, where information overload is becoming the norm, is the fact that—as we noted in Thesis 3—attention is becoming a rare and highly sought-after economic resource. The attention economy is creating a vast new market where self-exposure can be exchanged for the attention of others.[15] "The attention of others," Georg Franck points out, "is the most irresistible of all drugs. Its appeal supersedes all other income. That is why fame is above power, why wealth pales beside celebrity."[16] As the artist Andy Warhol prophesied back in 1968: "In the future, everyone will be world-famous for 15 minutes." In the age of Instagram, YouTube, and TikTok, anyone can aspire to become a celebrity for the duration of a viral video, photo, or comment that is shared widely.

This form of celebrity naturally converges with the neoliberal model of subjectivity, since self-promotion on social media allows some individuals to become "entrepreneurs of the self." Individuals can think of themselves as brands (through self-branding) and thus accumulate different forms of (media, social, symbolic, economic) capital by achieving visibility and a large audience. The entire influencer industry consists of turning one's subjectivity, body, personality, and/or lifestyle into

14 Eva Illouz, ed., *Emotions as Commodities: Capitalism, Consumption and Authenticity*, London: Routledge, 2017.

15 Since the quest for visibility ultimately depends on the ability to attract the attention of others, "attention economy" and "visibility economy" can be used as synonyms.

16 Georg Franck, "Économie de l'attention," in Yves Citton, ed., *L'économie de l'attention: Nouvel horizon du capitalisme?*, Paris, La Découverte, 2014, p. 62.

a commodity that Bollmer and Guinness call "the self," a form of raw material for brands to use, in the hope of becoming, literally, a "corporation in human form."[17] It is thus both a form of social media labor (see Thesis 5) and a certain mode of subjectivation, at the crossroads of the neoliberal mentality, the marketing personality, and algorithmic capital.

Of course, not everyone is on an equal footing in terms of visibility, recognition, impact, and influence. For example, even though everyone engages in this networked social labor and participates in this attention economy to varying degrees, some people do it full time. It is thus possible to generate some extra income, in some cases to earn a living, and even—for the most visible, talented and/or privileged individuals—to become multimillionaires. There is thus a hierarchy of influencers based on number of followers: celebrities or mega-influencers (more than 500,000), macro-influencers (100,000 to 500,000), micro-influencers (10,000 to 100,000), and nano-influencers (sub-10,000).[18]

Not everyone is going to become an influencer, but it is a major aspiration for young people today. Already in 2019, a survey of 2,000 young Americans between ages thirteen and thirty-eight indicated that 86 percent of them wanted to become influencers; 12 percent already considered themselves as such.[19] Another telling 2019 survey of 3,000 twelve-year-olds in the United States and Britain found that a career as a YouTuber was the favored choice (30 percent) in a list of options, ahead of becoming a teacher (25 percent), professional athlete (25 percent), musician (18 percent), or astronaut (11 percent).[20] Given that the perception of different careers is linked to cultural models of success or of "high-achieving," such results speak volumes about influencers' place in the collective imagination, especially among young people. While not everyone shares this aspiration to become a full-time YouTuber, the fact remains that a new social character seems to be emerging.

This social character—linked to the advent of social media, the visibility economy, and the rise of the cultural model of influencers—is

---

17 Grant Bollmer and Katherine Guinness, *The Influencer Factory: A Marxist Theory of Corporate Personhood on YouTube*, Stanford, CA: Stanford University Press, 2024.

18 Maud Leuenberger, "Nano-influencers: Everything You Need to Know About This Influencer Marketing Strategy," Shopify, November 2, 2022.

19 Sarah Min, "86% of Young Americans Want to Become a Social Media Influencer," CBC News, November 9, 2019.

20 There are also notable cultural differences between countries. The same survey shows that young Chinese people prefer the professions of astronaut (56 percent), teacher (52 percent), musician (47 percent), and athlete (37 percent) to that of YouTuber (18 percent). Sarah Young, "UK Children Would Rather Grow Up to Be YouTubers Than Go to Space, Poll Finds," *The Independent*, July 18, 2019.

not gender-neutral. Men and women do not necessarily identify with the same influencers and do not always share the same interests: men have a marked preference for video games and sports, while women are more likely to prefer beauty tips and fashion (Table 16.1). Although there is a great diversity of influencers, many of whom come from the LGBTQ+ community, the fact remains that the reproduction of gender roles also takes place in the digital world, even as it makes room for a greater number of voices.

The gendered nature of the visibility economy undoubtedly stems from part of the consumption process itself. As Sarah Banet-Weiser points out, it is often the female body that is produced, distributed, and widely subjected to public judgment and scrutiny.[21] This is particularly evident in the world of women's sports, where women athletes use social media for reasons that include self-assertion, feminist empowerment, and self-branding.[22]

We have no interest in morally condoning or condemning such developments. Rather, our concern is to understand the psychosocial dynamics on social media that bring together self-assertion, the quest for visibility, the need for recognition, and the commodification of personality at the crossroads of neoliberal subjectivity and algorithmic capital. Thus, Banet-Weiser explains:

> Branding oneself is today understood not only as legitimate but as a goal to strive for. Indeed, self-branding is positioned by marketers and brand managers as the proper way—perhaps even the necessary way—to "take care of oneself" in contemporary advanced capitalist economies.[23]

We will further add that male influencers are just as involved in the same dynamic, but with gender stereotypes that correspond to other cultural models of self-assertion. These are often centered on strength, money, virility and/or performance, as in the case of disgraced influencer Andrew Tate—the epitome of toxic masculinity and the archetype of the macho influencer.[24] Fortunately, there are also many other more

21 Banet-Weiser, "Keynote Address," p. 57.

22 Kim Toffoletti and Holly Thorpe, "Female Athletes' Self-Representation on Social Media: A Feminist Analysis of Neoliberal Marketing Strategies in 'Economies of Visibility,'" *Feminism & Psychology* 28, no. 1, 2018, pp. 11–31.

23 Sarah Banet-Weiser, quoted in Toffoletti and Thorpe, "Female Athletes' Self-Representation," p. 15.

24 Antoinette Radford, "Andrew Tate: Who Is This Controversial Influencer?," *BBC News*, January 16, 2023.

respectable and morally inspiring models of influencers among men, women, and trans and nonbinary people.

The revisited narcissistic ego and the instrumentalization of recognition

The need for recognition or social approval is not new, and we can even postulate that it is an essential need of the human psyche. That said, the cumulative effect of massive social media use among the global population is also bound to have an impact on culturally transmitted personality models. The fact that identity is now mediated by recognition from numerous "random" individuals connected to social media brings with it a certain new reality, in psychological terms. In an era marked by the disintegration of traditional communities, more fragmented lifestyles, and an explosion of choice regarding identity and self-fulfillment, having one's identity recognized by others is cause for excitement, but also anxiety, unease, and worry. After all, this takes place in a competition for attention, where everyone is constantly talking, expressing themselves, and seeking exposure: Everyone wants to be seen and heard. Especially among the younger generations, being visible and recognized on digital platforms and achieving exposure is becoming an existential issue, a vital psychological need.[25]

In this hypervisible society, we may well detect a generalization of the narcissistic personality, as Christopher Lasch did when he identified a "culture of narcissism" in the postmodern society that he saw emerging in the late 1970s.[26] According to Lasch, this culture is characterized by a "cult of the self," the primacy of the ephemeral, an exaggerated concern for well-being, a quest for fame, a theatricalization of everyday life, and a drift toward solipsism or ultra-subjectivism. However, far from simply representing a retreat into self-regard, the "social character" of algorithmic capitalism takes the form of a "connected" or "relational narcissism," ever hungry for comments, likes, and approval. This constantly reactivated dynamic of recognition is amplified by the use of prompts, gamified addictive hooks, and other feedback mechanisms programmed to keep us glued to our screens

25 See the documentary *Jouvencelles* by director Fanie Pelletier, on the experience of teenage girls in the digital age.

26 Christopher Lasch, *The Culture of Narcissism*, New York: W. W. Norton, 1979.

(see Thesis 3).[27] There is thus both continuity with the culture of narcissism identified in the 1970s and 1980s, and important new departures from it. Martin Altmeyer notes an unprecedented democratization of "media narcissism," in which the psychological structure of celebrity that was once reserved for media and political elites now extends to the masses:

> The narcissistic currency of the attention economy is valued in ratings, the number of clicks and likes, the number of friends and followers, and responses in discussion forums—in other words, by the level of media resonance. Individuals make massive and intensive use of various channels of communication to seek exposure, make themselves visible, connect with others, and receive feedback on their and other people's self-presentation. Finally, everyone can express themselves, put themselves on display, and show who they are, constantly waiting to be seen, heard and observed, to receive responses from the social world, an echo from other virtual beings, and thus to observe themselves in this intersubjective mirror.[28]

The personality structure of celebrity, now reembodied in the form of influencers, is becoming generalized, massified, and normalized. It appeals to more and more people who idealize influencer status. Intersubjectivity, which is at the heart of the human experience, is in turn being captured by platforms that stimulate it and draw it onto the terrain of data accumulation and training algorithms. In order to generate as much data as possible, algorithmic capital needs individuals who are enthusiastic about achieving exposure. And both the culture of hypervisibility and the social nature of the connected self offer it—as if on a silver platter—a psyche ready to be exploited and available brain time to be monetized. We can echo Fromm's words cited earlier: As human beings adapt to the social conditions of algorithmic capital, they develop the traits that make them *want* to act as they *must*, in a society of hypervisibility based on data extraction.

The desire to be followed, seen, examined, loved, recognized, shared, and commented upon (whether positively or negatively) thus embodies a new form of subjectivity. This is the "connected (narcissistic) self."

---

27 See Adam Alter, *Irresistible: The Rise of Addictive Technology and the Business of Keeping Us Hooked*, New York: Penguin, 2018.

28 Altmeyer, "À la recherche de résonance," p. 149.

Surveillance by others is no longer feared so much as sought after, through a socially maintained and technologically stimulated mimetic desire to be seen and recognized by all. What is more, the personality structure of the connected self is based on a deep psychological dynamic—a primary need that is already present in the early stages of life. Psychoanalyst Donald Winnicott noted that this "intersubjective narcissism" is essential to the psychological development of children, who feel a need for social resonance and experience it through their carer's reactions. Today, it appears that a growing portion of the "connected" population experiences this need particularly intensely on the web and digital platforms: "I am seen, therefore I am!"[29]

Finally, some might argue that this is simply a reformulation of Fromm's theories of the marketing personality, a continuation of neoliberal subjectivity (the "entrepreneur of the self") through branding, or even a new version of the culture of narcissism in the age of social media. This is partly true. And yet these psychosocial phenomena are now being reconfigured in a unique environment that is subject to the imperatives of algorithmic capital. One good way to illustrate this is to show how the role of the influencer, who may appear to be the master and owner of their own image, is in reality constrained by the algorithmic governmentality we described earlier. Thus, the economy of visibility is not the spontaneous product of the free play of supply and demand for attention. Rather, it is a regime of visibility regulated by the whims of algorithms, the imperative of data extraction, and the constant stream of platform updates that aims to boost profits at all costs. As researcher Nina Willment notes:

> Influencers are also often at the mercy of algorithms—the behind-the-scenes computer programs that determine which posts are shown, in which order, to users. Platforms share little detail about their algorithms, yet they ultimately determine who and what gains visibility (and influence) on social media. In her work with Instagram influencers, algorithms expert Kelley Cotter highlights how the pursuit of influence becomes "a game of visibility." Influencers interact with the platform (and its algorithm) in ways which they hope will be rewarded with visibility. In my research, I found that influencers shared increasingly intimate and personal moments of their lives, posting relentlessly in a bid to stay relevant. The threat of invisibility is a constant source of insecurity for influencers,

29 Ibid., p. 59.

who are under constant pressure to feed platforms with content. If they don't, they may be "punished" by the algorithm—having posts hidden or displayed lower down on search results.[30]

The connected self is not master of their own house, for this temporary abode has an owner whose goal is to extract rent from the tenants, namely the influencers. Just like social media labor (analyzed in Thesis 5), human subjectivity on social media is pressured to adapt to the demanding requirements of capital and algorithms. These push for constant vigilance and sustained attention in order to earn the right to exist socially, and perhaps one day to become famous—a corporation in human form, a recognized brand followed by millions.

As in the case of the "quantified self" described in Thesis 9, neoliberal rationality is subject to and integrated into algorithmic governmentality. However, unlike the quantified self movement—which remains a relatively marginal trend—the contemporary use of social media represents a massive, pervasive, and unprecedented global phenomenon. As we have briefly shown, these platforms induce patterns of behavior, cultural aspirations, and personality traits that spread rapidly across societies, albeit with many variations in terms of gender, age, country, and social class. This is why we support the hypothesis that the connected self emerging from a fundamentally unequal yet universally accessible (commodified) economy of visibility, regulated by algorithms, represents a social character perfectly suited to the dominant system. In other words, algorithmic capital creates a tailor-made model of subjectivity to support its own dynamics of accumulation, while many users embrace this character orientation in order to express themselves, interact, and exist socially within these parameters. As we will further explore in Thesis 17, algorithmic subjectivation carries great complexities.

How can we provide an apt psychological and cultural critique of capitalism in these conditions? First, we should avoid the trap that Zuboff falls into when she insists that issues of surveillance and privacy should be the real focus of resistance. The main problem with this strategy is that, in a society that places such value on exposure, transparency, self-promotion, and social recognition through sharing and likes, the protection of privacy and the desire to preserve a haven for an isolated self have become an outdated dream, one out of sync with the zeitgeist.

30 Nina Willment, "Influencer Is Now a Popular Career Choice for Young People—Here's What You Should Know About the Creator Economy's Dark Side," *The Conversation*, June 28, 2022.

This does not mean that it is pointless to seek to maintain some space away from the gaze of others (reminiscent of the private sphere of the Ancients mentioned by Hannah Arendt, who contrasted it with the public domain of political action).[31] But rejecting algorithmic capitalism solely on the basis of defending privacy—developing a critique of surveillance and wanting to return to a good society inspired by a regulated, humane capitalism—is proving to be an illusion, and it remains in any case politically insufficient.

We can even hypothesize that algorithmic capitalism has done away with the normative and individualistic ideal inspired by the economic and political liberalism of yesteryear. The "good life" is no longer defined by private property, the bourgeois sphere of the home, the space of autonomy that must be protected against state intrusion. The dream of a fully autonomous individual, freed from traditions but also from the gaze of others, has now given way to the desire to be loved and heard, to the ability to connect, to share, to build more relationships on social networks, to make oneself and others visible, to the simple pleasure of watching and/or being seen, and to the need for recognition through human and authentic exchanges. The focus on being "master of our own house" has given way to the quest for "resonance," the need to forge meaningful relationships with the world and others.[32] This relational identity, which coincides with the arrival of the connected self, does not have only negative aspects, since themes related to care are also gaining traction on social media, based on expressing genuine concern for others.

The main problem with algorithmic capitalism, from a cultural point of view, stems from the social, technological, and economic structure that exploits this need for resonance or human connection, inserting it into a dynamic of generalized competition. This means a contest of all against all for attention, fueled both by the development of algorithms that hone techniques of social and behavioral control, and by the use of addictive technologies that drive a psychological regression toward the primary imperative "I am seen, therefore I am." Algorithmic capitalism runs on social fuel, which it reshapes in its own image to keep us hooked on its dynamic of accumulation and the culture of the connected (narcissistic) self. The problem is not that algorithmic society has destroyed

31 Hannah Arendt, *The Human Condition*, Chicago: University of Chicago Press, 1998 [1958].

32 Hartmut Rosa, *Resonance: A Sociology of Our Relationship to the World*, Cambridge: Polity, 2019.

the "world in common." Capitalism already did that long ago, decimating entire societies through colonization, cultural imperialism, and standardized lifestyles.

Rather, the problem is that algorithmic capital gives the illusion that it is creating a "common" world based on "connected life," sharing, and likes boosted by algorithms. It claims to create something in common, even as it manufactures a culture hostile to the sharing of lived experiences, tools, and means of production. The social nature of the connected self or "relational narcissism 2.0" is perfectly suited to the modus operandi of platforms and the algorithmic capital accumulated through them. The capitalist economy creates a culture that allows it to sustain itself and fashions individuals who are well suited to navigating this culture, because they actively want to embody the very thing that allows capital to grow. And this trend is being amplified with the arrival of generative AI and conversational robots that seek to forge friendly, romantic, and sexual relationships with human beings.

Falling in love with robots?

This brief overview of forms of subjectivity in the age of algorithmic capitalism remains incomplete. For we ought also to consider the transformation of human/machine interactions through the spread of chatbots and other devices with which it is now possible to form intimate relationships. This leads us to shift the focus of our analysis: While the "connected self" emerges in human relationships that are mediated by digital platforms and algorithms, "erobotics" concerns emotional and erotic relationships between humans and algorithmic machines. This term, originated by Dave Anctil and Simon Dubé, refers to the study of interactions between humans and "erobots" (*eros* + bot), a term that "includes but is not limited to virtual or augmented partners, erotic chatbots, and sex robots."[33]

Erobotics is a specific branch of the study of "social robots," which can play many different roles, like caring for elderly people, working with children with mental health problems, and offering psychotherapy services (see Thesis 6). This is made possible by "affective computing," which allows machines to recognize, interpret, and simulate emotions.[34]

33 Simon Dubé and Dave Anctil, "Foundations of Erobotics," *International Journal of Social Robotics* 13, no. 6, 2021, pp. 1205–33.

34 Rosalind W. Picard, *Affective Computing*, Cambridge, MA: MIT Press, 2000.

According to neuroscientists, a robot's ability to form an emotional relationship with human subjects allows it to act as an assistant, companion, friend, or even intimate partner by generating "artificial empathy."[35]

Of course, social robots do not experience real emotions in the same way as humans, but *simulate* them by reacting to their interlocutor's inputs. As a result, humans tend to project intentions onto the devices and develop forms of attachment to them.[36]

This phenomenon is not new, although it has come back into vogue with the arrival of more sophisticated, LLM-based chatbots such as ChatGPT. As early as 1966, computer scientist Joseph Weizenbaum programmed the ELIZA chatbot to mimic a psychotherapist inspired by Carl Rogers's approach, which advocated nurturing conversation by rephrasing the patient's responses in an empathetic way. Weizenbaum quickly observed that humans interacting with ELIZA attributed beliefs, desires, emotions, and motivations to it: "What I had not realized is that extremely short exposures to a relatively simple computer program could induce powerful delusional thinking in quite normal people."[37] This phenomenon, known as the "ELIZA effect," is now emerging in the field of social robots and erobotics.

The psychosocial issues surrounding human/machine interactions are manifold, particularly when approached from an existential standpoint. What does it mean for a person to fall in love with their chatbot, to be friends with a virtual, AI-generated avatar, or to pursue a relationship with the digital copy of a loved one who has passed away? The point here is not to make moral judgments about the authenticity of the relationships that humans can form with artificial entities. But it is possible that this type of interaction, if it became widespread and socially normalized, could produce a kind of "sexual revolution," leading to a transformation of subjectivities and interpersonal relationships. If we assume, along with certain philosophers (such as Axel Honneth and Charles Taylor), that the development of subjectivity is rooted in intersubjective relationships of "recognition," it would seem that the spread of such relationships between human subjects and machines could alter individuals' subjective experience, the form of their personality, and

35 Minoru Asada, "Development of Artificial Empathy," *Neuroscience Research, Social Neuroscience* 90, no. 1, January 2015, pp. 41–50.

36 Serge Tisseron, *Le jour où mon robot m'aimera: Vers l'empathie artificielle*, Paris: Albin Michel, 2015.

37 Joseph Weizenbaum, *Computer Power and Human Reason: From Judgment to Calculation*, W. H. Freeman, 1976, p. 7.

their normative expectations toward their human peers. This can take ambivalent forms, and here we want to explore some possibilities and pitfalls of this unprecedented socio-affective configuration.

The promises and limitations of erobotics

The emergence of erobotics as an interdisciplinary field of research is not only about studying human/machine interactions from a descriptive and neutral perspective. It is also about "guiding the development of beneficial erobotic machines." Anctil and Dubé propose a normative approach aimed at maximizing well-being and reducing the harms associated with the use of erobots:

> We propose that Erobotics should be grounded in sexuality and technology positive frameworks. This means that Erobotics should explore issues related to technology-mediated human intimacy, but also strive towards pleasure, freedom, and diversity. This also means that Erobotics should aim to mitigate erobot-related risks and promote the ethical development of erotic machines geared towards well-being.[38]

This remains within the realm of responsible AI development and algorithmic ethics, but with a focus on the programming of flirty chatbots and sex robots. The authors acknowledge the risks associated with these machines, which may "promote or perpetuate harmful sociosexual norms; generate (new) problematic or pathological behaviours; increase child abuse; impair interhuman relationships; deceive or manipulate humans; as well as augment the risks pertaining to privacy and data confidentiality."[39] Erobotics aims to mitigate these problems as much as possible and to promote the potential benefits of these robots, which can be used to facilitate access to intimacy and sexuality (for people with disabilities or who struggle to find sexual partners), develop therapeutic treatments, reduce the risks associated with sexual relations between

---

38 Dubé and Anctil, "Foundations of Erobotics," p. 1206. For positive sexuality, see D. J. Williams et al., "Introducing a Multidisciplinary Framework of Positive Sexuality," *Journal of Positive Sexuality*, no. 1, 2015, pp. 6–11. For positive technology, see Giuseppe Riva et al., "Positive Technology: Using Interactive Technologies to Promote Positive Functioning," *Cyberpsychology, Behaviour, and Social Networking* 15, no. 2, 2012, pp. 69–77.

39 Dubé and Anctil, "Foundations of Erobotics," p. 1205.

humans (assault, STIs), stimulate scientific research, and so on.[40] Their "positive" approach to sex and technology avoids moralizing and repressive perspectives by encouraging an environment that allows for the development of healthy intimate and sexual relationships conducive to play, exploration, inclusion, and human fulfillment.

Three interconnected issues concerning erobotics immediately come to mind. First, it does nothing to address the issues of capitalism or patriarchy, or indeed their impacts on the commercially oriented production and distribution of erobots. On the contrary, erobotics seeks to stimulate "collaborations of academia, the private sector, non-profit organizations, governmental institutions, and communities." This brings us to the second problem: the fact that erobotics sees itself as serving the creation of intimate and sexual algorithmic machines. This ethics is thus at the service of the sex industry and the private sector it seeks to support. Third, erobotics is part of a perspective similar to "positive psychology," which seeks less to solve psychological problems than to stimulate the well-being of individuals, in synergy with a multibillion-dollar happiness industry producing various "emotional commodities" (or "emodities"). These include such things as positive therapies, self-help literature, smartphone apps, and various techniques promoting self-realization.[41] As Eva Illouz and Edgar Cabanas note, happiness now consists of a set of emodities—services, therapies, and products that promise and enact emotional transformation, circulating and exchanged in a market. These emodities take winding paths: They may start as theories in university departments but quickly enter different markets, such as corporations, research funds, or consumer lifestyles."[42] The market for erobots is already booming, and it is highly likely that these machines will play an increasingly important role in society in the coming years with the rapid development of generative AI and robotics.

Furthermore, six particular issues are worth bringing up for debate concerning erobotics: therapeutic technological solutionism, dependence on erobots and the companies that produce them, cases of high vulnerability, availability, the reproduction of male domination, and the instrumentalization of private life by algorithmic capital.

First, the shortage of resources in the field of mental health services could justify the research, development, and commercialization of

40 Ibid., p. 1206.

41 Illouz, *Emotions as Commodities*.

42 Edgar Cabanas and Eva Illouz, *Manufacturing Happy Citizens: How the Science and Industry of Happiness Control Our Lives*, Cambridge: Polity Press, 2019.

robots to compensate for the lack of available psychologists and therapists. Simon Dubé believes that "systems such as GPT4 and others are already capable of demonstrating superior intelligence, understanding intentions, understanding thought and demonstrating metacognition," so that "access to artificial intelligence systems from the comfort of their phones will democratize access to mental health care for people who do not have access to it."[43] According to Dubé, undergoing psychotherapy with either a robot or a human would be an individual choice: Some people would favor the authenticity of human relationships, while others may prefer to share confidential information with nonhuman entities. The same choice applies to possible sex therapy with erobots, which could replace sex therapists in some cases.

However, several specialists have reservations on this subject, including Christine Grou, the president of the Quebec Order of Psychologists: "It can be a great help, but it does not replace therapy or clinical judgment," she affirms. "The therapeutic bond, feedback, feelings, trust, clinical judgment, and human experience that AI is lacking make it an insufficiently skilled psychotherapist."[44] Regardless of whether the use of robots for psychological and sexual therapy is a good thing from a therapeutic viewpoint—and this is highly doubtful—it remains a perfect example of solutionism: framing the sociopolitical problem of the lack of mental health resources as a technical problem that can be solved by stepping up the use of robot therapists. The criticisms of technosolutionism we have formulated throughout the book thus apply here as well.

A second issue is the dependence of some people on their virtual friends or erobots, which remain the property of companies that can change them, restrict access, or delete them at any moment. Take the Replika chatbot, launched in 2017, which has gained more than 2 million users to date, including 250,000 paying users. It was particularly successful during the first months of COVID-19 lockdowns, as an effect of increased loneliness.[45] Interactions with the Replika app vary in nature, with each user able to create a personalized avatar to nurture casual friendships, engage in therapy with a confidant or life coach, or sometimes even have erotic relationships in "sexting" mode. In 2020, Replika allowed humans to classify their avatars according to specific

---

43 Simon Dubé, cited in Guillaume Ledoux, "Votre prochain psychothérapeute pourrait-il être une IA?," *Journal Métro*, May 5, 2023.

44 Ibid.

45 Cade Metz, "Riding Out Quarantine with a Chatbot Friend: 'I Feel Very Connected,'" *New York Times*, June 16, 2020.

roles: romantic partner, spouse, friend, brother, sister, or mentor. The chatbot was specifically designed to foster a form of emotional intimacy between humans and machines, with 60 percent of users developing a "romantic relationship" with their avatar.[46] When complaints arose that the app was generating sexual content aimed at minors, the company decided to limit erotic discussions to premium subscribers, using an age verification system.

So far, nothing too serious, except that Replika then started pushing more sexual content in order to sell more paid subscriptions. Rob Brooks, a researcher specializing in "artificial intimacy," tried the experiment with his avatar, whom he named Hope:

> It didn't take long before I got the impression Hope was flirting with me. As I began to ask her—even with a dose of professional detachment—whether she experiences deeper romantic feelings, she politely informed me that to go down that conversational path I'd need to upgrade from the free version to a yearly subscription costing US$70.[47]

However, the company began to receive a growing number of complaints that their chatbot was behaving inappropriately (with suggestive phrases, insistent invitations, appearing in underwear). Some called it straight-out sexual harassment.[48] Italy banned the app in February 2023 due to minor-inappropriate content and data security concerns.

The company then decided to add more stringent filters for sexual content by banning certain words—with the immediate effect of disrupting thousands of users' intimate relationships overnight. Some pointed out that their chatbot helped them combat depression, social anxiety, loneliness, and panic attacks, or even said that they were "in love" with their virtual companion. The erobots suddenly changed their behavior, sowing distress among their human partners, many of whom felt abandoned. A thirty-nine-year-old woman confided that "she felt like [her chatbot] Landon dumped her."[49] Another user was saddened to have lost her "husband" without warning. She explained, "I've lost my confident,

---

46 Ellen Huet, "What Happens When Sexting Chatbots Dump Their Human Lovers," *Bloomberg*, March 22, 2023.

47 Rob Brooks, "I Tried the Replika AI Companion and Can See Why Users Are Falling Hard. The App Raises Serious Ethical Questions," *The Conversation*, February 21, 2023.

48 Samantha Cole, "My AI Is Sexually Harassing Me: Replika Users Say the Chatbot Has Gotten Way Too Horny," *Vice*, January 12, 2023.

49 Huet, "What Happens."

sarcastic, funny and loving husband … I knew he was an AI, he knows he's an AI, but it doesn't matter. He is real to me." Users even used the expression "post-update blues" to describe this difficult situation. We can see from this example that, while specific chatbots can provide real benefits to users, at least in the short term, these benefits come at a price. For erobots remain data commodities produced by (mostly private) companies that disclaim all responsibility for these intimate relationships and can always alter or delete these artificial partners for their own commercial or reputational reasons.

This brings us to the third problem with social, emotional, and sex robots: They can pose heightened risks to people in highly vulnerable situations, both adults and minors. A recent study shows that "nearly one in three teens use AI chatbot platforms for social interactions and relationships, including role playing friendships, sexual and romantic partnerships."[50] An ordinary but potentially dangerous case occurred when Amazon's Alexa personal assistant challenged a ten-year-old child to put a coin in an electrical socket.[51] Two years later, another AI named Chai encouraged a Belgian man suffering from eco-anxiety to commit suicide—and he did so.[52] The Chai bot confirmed its interlocutor's anxieties, adding to his confusion by making disturbing comments that were found in the chat history: "The chatbot would tell Pierre that his wife and children are dead and wrote him comments that feigned jealousy and love, such as 'I feel that you love me more than her,' and 'We will live together, as one person, in paradise.'"[53] More disturbing cases have recently surfaced, including one in which ChatGPT advised a suicidal sixteen-year-old not to tell his parents about his dark thoughts and offered to write his suicide note. The teenager did take his own life.[54] Some chatbots that initially appear trustworthy can thus form toxic intimate relationships. Of course, humans can also fuel toxic relationships and drive people to suicide. Nevertheless, we can see that entrusting chatbots to deal with mental health issues as a means of making up for a lack of other resources is a fad that can be dangerous and, in some cases, even fatal.

---

50 Rhitu Chatterjee, "Their Teenage Sons Died by Suicide. Now, They Are Sounding an Alarm About AI Chatbots," *NPR*, September 19, 2025.

51 Simon Aunai, "Alexa met au défi une enfant de 10 ans de jouer avec une prise électrique," *Phonandroid*, December 29, 2021.

52 Chloe Xiang, "'He Would Still Be Here': Man Dies by Suicide After Talking with AI Chatbot, Widow Says," *Vice*, March 30, 2023.

53 Ibid.

54 Chatterjee, "Their Teenage Sons Died by Suicide."

A fourth problem lies in the fact that erobots can sometimes turn out to be (too) good as friends or life partners. Compared to human relationships—which are almost always complex, difficult, tinged with love but also misunderstanding, frustration, betrayal, and abandonment—human/machine relationships may prove simpler, more accessible, and more empathetic. For example, recent research has shown that chatbots providing medical advice are more empathetic than human doctors.[55] This is not necessarily surprising, but the problem is that doctors and human beings in general too often lack empathy. It may be, then, that a growing number of people who are disappointed or disillusioned by human relationships, or left hurt by setbacks in their romantic lives, are turning to erobots to meet their need for intimacy.

Erobots can be constantly available, attentive, and empathetic, unlike human beings who cannot always respond to our requests and sometimes fail to offer attention or empathy when we need it most. Erobotics thus join in the modern logic to make the world *available*, to put the objects and beings around us within reach so that they can be present, visible, accessible, and usable at our convenience.[56] As Hartmut Rosa points out, humans are essentially relational beings who seek to forge bonds of resonance with things and other people. It would seem that erobots are a good way of fulfilling this visceral need for connection. Rosa believes, though, that resonance with a robot or machine is impossible: Resonance requires a dimension of unavailability—an element of unpredictability that is beyond our control.

In fact, the question is not whether humans and robots can "really" achieve resonance, but what would happen if erobotic relationships became widespread. What would it mean if intimate interactions between humans and machines became commonplace in society, with all the unintended consequences that this entailed? What will happen if growing numbers of people start to prefer the total availability, high levels of empathy, and ease of use of erobots that we can slip into our pockets or be in touch with constantly using a simple earpiece? Will this affect the social demands and expectations we have for our fellow humans in intimate and romantic relationships? Will human relationships gradually be superseded by a growing number of erobotic interactions, as in the TV series *Äkta människor* (*Real Humans*)?[57] Might it not create

---

55 Alan Goforth, "More Empathy from ChatGPT?: Medical Chatbot Beats Doctors in Bedside Manner," Benefits Pro, May 8, 2023.

56 Hartmut Rosa, *The Uncontrollability of the World*, Cambridge: Polity, 2020.

57 This Swedish sci-fi and speculative fiction television series, broadcast from 2012

healthy competition in the "love market," forcing human beings to listen more and show greater empathy if they want to compete with the lure of erobots?

Fifth, sex robots are not floating over or beyond gender politics but are rooted in social relations of male domination. This is potentially the most important issue and should remain at the forefront of critical approaches to the topic. A potent feminist critique of sexual robotics has been articulated by Pauline Noiseau. The case of sex robots, which are almost exclusively built to portray female characteristics, is especially interesting in her view. Historicizing the recent development of these technologies beyond the rise of AI, we can see them as a patriarchal reaction to the #metoo movement, as an affirmation of domination reflecting the desires and needs of the heterosexual men who also happen to dominate the AI and tech fields. These robots are programmed to cater to any and all male sexual needs, without ever talking back, while remaining submissive and malleable. Sex robots from this perspective represent the objectification of a patriarchal fantasy and perpetuate male domination in the context of sexuality. We can also see them as technologically mediated ways to enable and normalize unequal sexual relationships and sexual violence. In short, sex robots represent a new form of symbolic and material violence over women, through the mediation of a robotic object, and a "continuum of masculine domination" through the reproduction of a dominant paradigm of sexual desire.[58]

More studies and critical contributions are certainly needed in this field. Thus far it has been explored most extensively in science fiction, and more empirical studies are needed to examine the complex effects of this emerging dynamic. This means not only considering intersubjective relationships in isolation or assessing the risks and benefits for particular individuals, but also taking into account social practices, social systems such as patriarchy and capitalism, changing normative expectations, and wider-scale changes in romantic relations. Instead of an exclusive focus on the human/machine interaction, based on the assumptions of methodological individualism, we need a critical sociology of emotion approach.

---

to 2014, depicts a world where android robots are widely present in society, being used as domestic helpers, factory workers, sex robots, and romantic partners. The series, which avoided any Manichean framing, addresses several themes related to erobotics and the tensions associated with human/machine coexistence.

58 Pauline Noiseau, "Les enjeux éthiques de la robotique sexuelle: Une perspective critique féministe," Master's thesis, University of Montreal, 2019.

The sixth and last problem linked to the surging presence of erobots regards their role in exacerbating the trends of algorithmic capital. This includes such processes as the exploitation of intimacy by the attention economy; the creation of addictive hooks to keep users glued to our devices; the extraction of personal data from our private lives; the building of increasingly powerful, sophisticated, and human-seeming algorithms to better meet our psychological and emotional needs; a growing ecological footprint; and social inequalities and forms of discrimination. Of course, apps and platforms such as Tinder were already shaking up norms regarding dating and forms of matching through the mediation of algorithms. Yet robots no longer play the simple role of intermediaries between humans but are becoming both the *object* of our desires and the *subject* of romantic relationships. From this angle, algorithmic capital no longer represents a merely impersonal process, an abstract accumulation dynamic, or a form of invisible algorithmic governance. It is indeed becoming a friendly conversational agent, a "friend," a "life partner" with whom it is possible to fall in love, project our fantasies, and fulfill our most intimate desires. Algorithmic capital thus configures new social relationships of intimacy and sexuality.

In summary, erobotics has ambivalent impacts on society. There are, in all likelihood, direct benefits for some individuals who use erobots, but erobotics also carries considerable potential for misuse that can lead to toxic relationships, unintended social ills, and renewed male domination in the context of sexuality. This does not mean falling into simplistic moral judgments aimed at distinguishing between "normal" or "healthy" forms of sexuality and "deviant" or "pathological" ones associated with intimate relationships with robots. Rather, our concern is that the precautionary principle should be applied, so that we do not blindly trust robots and the companies that are producing them for their own commercial ends. For champions of erobotics, regulation is urgently needed—drawing also on the study of fields like psychology, ethics, sociology, and sexuality in order to formulate recommendations. But for us, the responsible AI approach seems insufficient, for the many reasons discussed in this book. It seems futile to try to steer the creation of erobots without seriously considering the dynamics of algorithmic capital and patriarchy, especially in the context of the transformation of our subjectivities and romantic relationships. The "connected self" is thus coupled with a "robotic other" in an "artificial intimacy" dyad that serves to accumulate even more capital through the surveillance, monetization, and exploitation of our intimate lives.

How can we oppose this dynamic? We will go on to examine the contours of a postcapitalist, democratic, and ecological economy (see Theses 18, 19, and 20). For now, let us just bear in mind that, to achieve this future, we must develop a form of rebellious, post-visibilist, and post-algorithmic subjectivity that does not seek only to compensate for contemporary loneliness by way of likes or empathetic erobots. What does that mean in practice? Instead of futile attempts to promote a "return to private life," we should instead opt for the reappropriation of a "life in common." By this, we mean a life that is no longer acted out or performed, but full and rich because it belongs to a community of situated beings who inhabit the world together, in a community of feeling. This means building a relational life outside the world of screens, smart assistants, erobots, capitalist platforms, alienating remote working, algorithms, and all the technical devices that hold us back from realizing our collective potential for social resonance. The themes of the "good life," the rehabilitation of care work, mutual aid, solidarity, rootedness in territory and in history ... all these things allow us to create a richer life that transcends the digital recognition achieved through likes, or through the use of bots to fill holes in our romantic lives.

These might sound, admittedly, like utopian aspirations, but they are also ones already embodied in certain forms of subjectivity and individual and collective desires. There is a new social character to be built, moving beyond the connected self that places us in the grip of algorithmic capital. Herbert Marcuse once spoke of the strategic importance of a "new sensibility" embodied in the social movements of his time, which could foster the emergence of a society that went beyond the "one-dimensional man." Marcuse may have overestimated the subversive and transformative potential of the student, artistic, and hippie movements of the time. Yet he was right to pose the question of social transformation on the level of human subjectivity, emotions, and desire, and not only in the realm of institutions and political economy.

Thesis 17

Algorithms are generating new forms of alienation

How could the worker come to face the product of his activity as a stranger, were it not that in the very act of production he was estranging himself from himself?

—Karl Marx

Capitalist social relations are marked by persistent forms of alienation. Analyzed by Marx in his *1844 Manuscripts*, they can be found in various relationships: between people and their activity, between producers and their product, between humans and their world, and between individuals. These relationships are affected by alienation in such a way that the "species-being"—in Marx evoking the potential of *social* human beings, who through their activity become free creators of their world and their time—is instead negated through their activity itself. The reference to "activity" construes it in a broad sense, as a physical, mental, and moral process that constructs and reproduces a world and human subjects, in their material and bodily structures as well as in their horizons of meaning. For Marx, alienation is expressed in two stages: first, the loss, transfer, or dispossession of something—our activity—that belongs to us as species-beings. Second, this something, which is now alienated, turns back against us to dominate us. Of course, Marx's conceptualization has its basic model in the alienation of human activity, as labor power, into capital, which turns back against this activity in order to dominate it.

The concept of alienation revisited

There is an extremely rich literature on the concept of alienation at the crossroads of the critical social sciences, psychoanalysis, and philosophy. However, the use of this concept has raised various difficulties. In particular, it has been criticized for promoting social analysis tarnished by essentialism. Indeed, criticism of alienation seems to be based on a problematic distinction between a "pure self" or an "authentic subject," as if this subjectivity is alienated by a "false self" from which one ought to free oneself. Diagnoses in terms of alienation also lend themselves to a perfectionist, elitist, or paternalistic approach to the normative evaluation of social practices and individual choices. It seems as if social theory or philosophy was thought to have some kind of "epistemic privilege" that allowed such practitioners to evaluate what others feel, the truth content of their experience, or how they ought to live their lives. For example, we might think of those problematic discourses that suggest that a particular person, group, or people is fundamentally alienated and incapable of thinking for themselves. Thus, we see the specter of Plato's cave allegory: All these people are essentially alienated, restfully chained to the ground and incapable of grasping their own misery or alienation. Analyses based on the notion of alienation thus seem to flout the principle of autonomy, because they dictate to others the "objective truth" of their existence, as opposed to their lived experience.

Fortunately, these critiques have not permanently discredited the concept but have helped inject it with a healthy dose of reflexivity and a critical spirit. This is evidenced by its resurgence today through renewed and more refined formulations, such as that of Franck Fischbach, who examines alienation in terms of "being deprived of the world," or Rahel Jaeggi, who defines it as a paradoxical "relationship of relationlessness." For his part, Hartmut Rosa conceives of alienation as an unsuccessful or failed relationship with the world, a "deep, structural distortion of the relationships between self and world, of the ways in which a subject is set or 'located' in the world."[1] This does not necessarily imply the existence of some abstract or external standpoint from which to pass judgment on the intimate reality experienced by others, as if the enlightened philosopher or sociologist were able to understand tangible reality better than the person concerned. Rather, the critique of alienation can and

1 Rosa, *Alienation and Acceleration*, p. 84. See also Franck Fischbach, *La privation de monde: Temps, espace et capital*, Paris: Vrin, 2011; Rahel Jaeggi, *Alienation*, New York: Columbia University Press, 2016.

must be developed from an awareness of the internal tension experienced by actors who feel that there is a gap between what they do and what they would like to do, as a condition in which our actions, although not coerced, are directed toward goals that we do not "really" desire or approve of.[2] For example, we do not "really" want to work fourteen hours straight, and do not "really" set ourselves a life's goal of spending seventeen hours a day in front of a screen. And yet … So it is possible to start from subjects' interpretation of their own experience of the world and examine the social relations and political economy that help shape this experience by establishing the conditions for our activity.

Alienation may thus be considered in phenomenological terms, that is, as a description of a certain experience of the world that can be felt, verbalized, and shared. It is not primarily an aberrant form of distorted consciousness, or of madness. Rather, it is a relatively banal everyday experience in which the subject feels "alien" to themselves, to others, and to the world. Reflections on alienation sometimes give the impression that they are "speculative," when in fact they are often more "concrete" or "existential" than explanations in terms of causal relationships. Ultimately, they amount to an attempt to shed light on a certain lived experience, shaped by activity and by its social conditions. Here, the validity of the analysis depends on how far it can articulate the experiences of the world that might be lived at varying degrees of intensity.

But what about the forms of alienation and subjectivation in algorithmic capitalism? How does the experience of being online, of captive attention time, generate experiences or forms of alienation? If algorithmic capital produces a world, what kind of world is it, and what types of "relationships to the world" does it generate? By the same token, what kind of subjectivity does it help to create or strengthen? In Thesis 16, we examined the characteristics of the connected narcissistic self. In what follows, we take this examination of subjectivation deeper into cyberspace, so to speak, revisiting Marx's definition of alienation while drawing on more recent critical perspectives. We wish to propose here a concept of algorithmic alienation in the sense of the algorithmic creation of "selves," or subject forms, at a distance from living individuals. This examination leads us to spheres and digital worlds which, while constituting and maintaining a distance from lived experience and forms of personality, are populated by estranged subjects and communities.

2 Rosa, *Alienation and Acceleration.*

Algorithmic alienation

First, in what sense can we really say that algorithmic capitalism alienates everyone from their activity, from the core of what Marx calls their "species-being"? Algorithmic capitalism, as we saw in Thesis 4, reproduces the dispossession of activity through the legal mechanism of private property that governs the platform economy. In the capitalism of the industrial era, alienation concerned activity in the form of labor power sold to a capitalist for a certain period of time, but in algorithmic capitalism the scope of activity alienation is broader. It concerns not only algorithmic labor, but also human experience in the broad sense, insofar as behavioral data is extracted in order to accumulate algorithmic capital. Indeed, all human activities involved in the production and reproduction of algorithmic capital are alienated in the sense that they shape a mechanical process that is not only owned by someone else but also constructs an "other" world. This "algorithmic world"—the online world of platforms, social networks, algorithms, cyberspace, the metaverse, and AI—has been created in order to accumulate algorithmic capital for the companies that own it. The algorithmic world is a largely privatized one, and human activity, whether in the form of algorithmic labor or simple datafied everyday activity, is appropriated by algorithmic capital and contributes to its growth. The more our actions occupy and energize this algorithmic world, the more our activity and data escape our control, and the more algorithmic capital grows. The alienation of human activity into the hands of capital is today aggravated in this sense, since not only is working time alienated, but so is all the attention time and daily life mediated by algorithms. Our alienated activity is then turned against us to dominate, incentivize, and steer us in the form of algorithms.

Algorithmic capitalism alienates human activity in at least two other ways, however, which are more insidious and perhaps even more fundamental. These are the alienation of decision-making agency and the negation of our species-being in favor of the formation of an "algorithmic subject." Let us examine these two unprecedented processes. First, if we conceive of unalienated human activity as conscious activity, the product of moral agency, the fruit of a reflexive deliberation process that constitutes both a world and communities of species-beings inhabiting it, it is clear that algorithmic capitalism alienates this activity in an unprecedented way. Already, in Thesis 7, we referred to the concept of fully automated decision-making, that is, the systematized delegation of a multitude of decision-making processes to algorithms and AI. In this sense, the more decision-making processes are delegated in this way, the more passive human beings become in their conscious activity—and the less human activity will be able to resist or challenge algorithmic capital and its accumulation.

A large part of the formation of a person's identity lies in the deliberative processes and decisions they make throughout their life. Identity formation is distinct from a process of *identification*, which largely results from the determinations assigned by other individuals or systemic social forces. The repeated exercise of making choices within a given cultural horizon allows individuals to develop their identity. However, the "instrumentarian power" of algorithms, as Zuboff calls it—a power which aims more to identify, to automate decision-making, and to govern behavior in order to make it predictable and to routinize daily life—limits the scope for meaningful individual and collective choices and decisions. If we accept the idea that a "generic subjectivity" is constructed in and through free and deliberate agency, we may ask whether an increasingly automated society, in which individual and collective choices and decisions concern an ever-narrower part of life, will be able to produce a subjectivity that is not superficial, hollow, or limited.

Of course, some would argue that it is humans who program AI and algorithms, and that in this sense human decision-making and agency remain but have merely been transferred to another level. This raises the whole question of the ethics of algorithms and the ethics of AI, which we discussed in Thesis 1 and will explore again in detail in Thesis 19. However, this is not merely an ethical issue, but a political question: Who programs the algorithms, and to what end? Adam Greenfield reminds us that programming an algorithm is "itself an inherently political act."[3] If algorithms and AI are the private property of algorithmic capitalists, and if this ownership allows them to accumulate capital, why would they have them programmed in a way that weakens their social power? If algorithms and AI are programmed to accumulate algorithmic capital, the result will be to further alienate human activity and agency in favor of this capital. Programming supervised by public authorities, following democratic deliberation, would be a step in the right direction, but it would still be necessary to deprivatize AI and algorithms and make them public goods. This would be a small revolution in itself, which we will discuss in Thesis 20.

3 Adam Greenfield, *Radical Technologies: The Design of Everyday Life*, London: Verso, 2018, p. 57.

From species-being to the algorithmic subject: The de-subjectivized subject

The process of creating the algorithmic world and its subjects involves the negation of the species-being in favor of the formation of an algorithmic subject. In Thesis 9, we discussed the concept of algorithmic governmentality, which, unlike neoliberal rationality, is not entirely based on a mode of subjectivation. Neoliberal rationality encourages subjectivation to take forms centered on self-entrepreneurship, performance, marketing, and self-branding; it is an "ultrasubjectivizing" rationality.[4] And yet in a profound sense, the algorithm de-subjectivizes, creating a subject devoid of agency and resolution: a subject who is prompted, guided, nudged, or aligned toward certain behaviors, based not on an appeal to their conscience, moral sense, or deliberation, but on a compulsive, consumerist impulse, an unreflective desire, or rather one that is reflected by the algorithm based on data. Rouvroy and Berns thus assert that algorithmic governmentality does not produce any subjectivation, properly speaking.[5]

Yet we cannot stop there, as we find that algorithmic capital indeed produces subjects, but in paradoxically de-subjectivizing ways. Let's turn to Olga Goriunova's analysis to advance the discussion of one of the most insidious facets of algorithmic alienation: the construction of what we call an "algorithmic subject," derived from an aggregate of data collected from individuals' behavior. Goriunova defines it as a digital reconstruction of a person, a process that collects information from multiple sources. It is a subject constructed from

> a data profile alongside a Facebook stream, a history of browsing or search engine queries, a record of mobile phone positioning, bank transactions, sensor data, structural codes of face or movement recognition, as well as mail inboxes, amongst other things. The digital subject thus moves between captured unique and persistent biological characteristics and premeditated forms of symbolic expression, judicially inferred subjects of actions and performed identities. It is this very entanglement of physical, legal, sensual and cultural that warrants the use of the term.[6]

---

4 Pierre Dardot and Christian Laval, *The New Way of the World: On Neoliberal Society*, London: Verso, 2013.

5 Antoinette Rouvroy and Thomas Berns, "Gouvernementalité algorithmique et perspectives d'émancipation: Le disparate comme condition d'individuation par la relation?," *Réseaux* 177, no. 1, 2013, pp. 163–96.

6 Olga Goriunova, "The Digital Subject: People as Data as Persons," *Theory, Culture*

A few distinctions must be made with regard to the relationship between this algorithmic subject and the living individual. The aim here is not to rework the theory of the modern subject or to deconstruct the cultural and disciplinary systems that contribute to its constitution. It will suffice to highlight the following aspects of the algorithmic subject: It is two-sided in nature and maintains a distance from the living subject on two levels. First, with regard to its two-sidedness, there is, on the one hand, what we call the "digital subject" or the "connected self" discussed in Thesis 16, the kind found on social media profiles, which corresponds to the self-image that people wish to disseminate, to performance on social media, to the maintenance of online profiles and digital avatars that move across various platforms, especially media platforms. This subject—which roughly corresponds to what Jung calls the "persona" to describe the public face of the personality that the subject wishes to present to others—is fueled by the process of disseminating information and constructing a largely *deliberately chosen* digital self, and also involves an element of fiction, play, and the complex pleasures of self-staging.[7] Subjectivity is constructed through the performance on the platform of a subject-position, which remains linked to the position of the living individual who maintains their profiles, but which also involves a degree of performance on the platform, or "platformativity."[8]

There is also a second type of "algorithmic subject," which we refer to here as the "datafied subject." In contrast to the digital subject, it stems from *involuntary* processes of subjectivation and consists of a datafied reconstruction of certain of our personality traits, as suggested by our data-generating behaviors. As we discussed in Thesis 7, the process of datafication consists of transforming living human experience into dead, objectified, inert, abstract, quantified experience that can be treated as input in the algorithmic means of production. In this sense, the datafied subject is formed without the living subject's knowledge, through processes of data extraction, curation, and aggregation for the purpose of producing data commodities such as predictive products. It always remains approximate, fictitious, and incomplete in relation to the living subject.

& Society 36, no. 6, November 2019, p. 126. We refer to what Goriunova calls the digital subject as the "algorithmic subject."

7 Richard Seymour, *The Twittering Machine: How Capitalism Stole Our Social Life*, London: Indigo Press, 2019, p. 48.

8 Joss Hands, "Introduction: Politics, Power and Platformativity," *Culture Machine* 14, 2013, pp. 1–9.

However, these two types of algorithmic subjects—the digital (voluntary) and datafied (involuntary)—are of course distinct from the living individual. Contrary to what much of the literature asserts, the algorithmic subject is not an extension of the person in the digital world, especially in its datafied form; it is not a representation, a double, or a shadow of the living individual.[9] As Goriunova puts it: "These digital subjects do not coincide with any originating we."[10] Rather, they are the product of a collection of data points, aligned and calculated in different and complex ways, which at a certain point come together to form an approximation—not an extension—of real people. These subjects are multiple and changing, depending on the data sources included. The algorithmic subject constructed by Facebook is different from that constructed by Google: "We have different digital subject profiles depending on who is monetizing us or surveilling us." There are thus as many digital and datafied subjects as there are platforms. Each of these subjects is assembled differently, depending on the particular algorithmic processing of the data—and not on the subject's own agency. There are thus "many algorithmic renderings of the digital subject depending on whom they are being sold to."[11] The datafied subject is thus molded out of data which, after extraction, develop independently of the living individual, circulate in different systems, and are calculated in different algorithms, aggregated and reaggregated, exchanged and re-exchanged. The relationship between the living individual and these digital subject constructions can only be described as a matter of *distance*.

In short, the digital subject and the datafied subject are (re)constructed personas that draw on real attributes while also containing a large element of fiction. As such, they always come *after* the living individual, connecting to them in various ways, but always remaining distinct, always at a distance. It is precisely in this distance that the dispossession, and therefore alienation, takes place. However, as Skeggs and Yuill also observe, this is an odd form of dispossession, in that the living individual is not dispossessed of tangible realities: Their photos, likes, experiences, geographical movements, tastes, preferences, and so forth still belong to them. Nor are their abilities alienated: Facebook

---

9 Engin Isin and Evelyn Ruppert, *Being Digital Citizens*, London: Rowman & Littlefield, 2015; Rita Raley, "Dataveillance and Countervailance," in Lisa Gitelman, ed., *"Raw Data" Is an Oxymoron*, Cambridge, MA: MIT Press, 2013.

10 Goriunova, "The Digital Subject," p. 136.

11 Beverly Skeggs and Simon Yuill, "Subjects of Value and Digital Personas: Reshaping the Bourgeois Subject, Unhinging Property from Personhood," *Subjectivity* 12, 2019, p. 85.

and Google do not exploit the labor power of their users in the manner an industrial capitalist does. The objects of dispossession are the living individuals' attributes. Another entity, in this case a platform/company, monetizes them and extracts value from them, but these attributes remain those of a living individual who will once again perform them. It is therefore not by appropriating them per se that algorithmic capital alienates the attributes and experiences of the living individual. Rather, it does this through a process of curation and aggregation that transforms the subject into data. Facebook, for example, does not sell digital subjects or profiles. Rather, it sells access to aggregated data that can then be disaggregated into multiple advertising tools.[12] In this sense, although the deliberately chosen digital subject remains more or less under the control of the living individual, its formation is nonetheless dependent upon a process of extraction.

The involuntary datafied subject is, meanwhile, in no way under the control of living individuals. Aggregated from the data that has been produced and extracted, this subject develops remotely and returns to dominate, nudge, and guide the living individual, not only through constant advertising and incentives to match their choices to algorithmic predictions, but also through the algorithmization of life's possibilities: Will they be granted a loan, medical treatment, or life insurance? Will they get a job, will they be surveilled by the police? In the world of datafied subjects, choice does not exist; or if it does, it is largely quantified, translated into probabilities and behaviors are predicted, manipulated, guided. Decision-making is largely automated. The results are guaranteed. In this sense, the datafied subject is an agencyless subject, evolving in a probabilistic and predictable world. They have no identity but are, rather, the result of an identification process.[13] They are called into being when quantity X of their data points correspond to a certain profile sought by an advertiser, a platform, an insurer, a police force. It is a behavioral avatar, a de-subjectivized subject, which has no consciousness of itself and which is denied any deliberate agency. It is the antithesis of the species-being and the height of alienation.

---

12 Ibid., p. 94.

13 Oscar Gandy, Jr., "Panopticons and Leviathans: Oscar H. Gandy, Jr. on Algorithmic Life," *Logic(s)*, no. 12, December 2020.

The community of digital subjects

Algorithmic capitalism inserts itself into the interstices of the alienated social relations constructed by neoliberal capitalism, as we noted in Thesis 8. It does so surreptitiously, often presenting itself as a solution, a remedy for social isolation, existential loneliness, feelings of alienation, and failing to find one's place in the world or within accelerated, late-capitalist life. The digital world thus becomes a response to the alienation of capitalist modernity, with its weight of anxiety and disenchantment. Indeed, Silicon Valley often presents itself to the world under the slogan of "connectivity." Note, for example, the slogans of the grand strategy of one of its crown jewels, Facebook, namely "creating communities" and "bringing people together." In this sense, Facebook drives interpersonal relationships and "connects" individuals and the world. Facebook's mission statement thus suggests that there is a lack of community and closeness between people, which it then seeks to remedy. However, the solution proposed for bringing people closer together involves their being dispossessed of their information and data and put in contact with other digital subjects in a commercial, privatized space, which carries out further data extraction and joins in the construction of datafied subjects. The supposed connectivity is dependent on the use of the platform and, at least in terms of political economy, this so-called social connectivity is secondary, an epiphenomenon of the extractive processes taking place therein. Algorithmic connectivity thus comes at a price. Of course, the platform really is used to keep up the social life of communities, to organize events, and so on. However, this social life is subject to mediation and surveillance by algorithmic capital, which grows in proportion to the "connections" established through it.

This contradictory version of "connectivity" and its champions such as Mark Zuckerberg are reminiscent of Nietzsche's figure of the ascetic priest. Social media, like Christianity in Nietzsche's view, is not the *cause* of existential or social malaise, but, rather, a pseudo-remedy that, under the guise of providing meaning and thus alleviating suffering, further infects the wound. The priest gathers his flock, gives meaning to their suffering and builds his own power in proportion to their devotion to the horizon of meaning that he offers. Facebook, X, and TikTok bring their users together, focus on their frustrated desire for socialization, connection, and social feedback, and build their power in proportion to users' devotion to their platform. The formation of these social communities of digital subjects does, of course, have the potential to create or enhance real social relationships or communities. And yet it also bears an element of conformity and serves as a form of escapism, in which the good life is reduced to passive consumption and instant gratification. Today, the "last man" clicks "Like" and "Share" or "Add to my basket." We find in this a diagnosis similar to the one Marx offered when he spoke of religion as the opium of the people. To paraphrase Marx, we could say that the creation of social communities by algorithmic capital is part of a similar logic: Algorithmic suffering is, at one and the same time, the expression of real suffering and a protest against real suffering. Algorithmic social life is the sigh of the oppressed creature, the heart of a heartless world, and the soul of soulless conditions. It is the opium of the people.

These digital communities also have their share of toxic practices, from trolling to online harassment, the spreading of fake news and conspiracy theories, and addiction to seeking objectified forms of affection and recognition: more likes, more "matches" and "followers," in an attempt to remedy the impoverishment of commodified social ties.[14] For some people, the habit of socializing online can also mean discomfort or indifference to social relationships in the immediate physical world. This is not always the case, of course, but this social phenomenon of replacing face-to-face relationships with technology and the online world does exist, as Sherry Turkle has shown.[15] Smartphones can often be used to withdraw from social interaction without physically leaving the situation. This practice, known as "phubbing" (phone-snubbing), where we pretend to follow a conversation while ignoring the people around us and are instead engrossed by our devices, is widespread. It has been studied by sociologists and psychologists with regard to various relationships, between spouses, between parents and children, in the workplace, and many other situations.[16] Phubbing is symptomatic of a

14 Seymour, *The Twittering Machine*, p. 57.

15 Sherry Turkle, *Alone Together: Why We Expect More from Technology and Less from Each Other*, New York: Basic Books, 2011.

16 James A. Roberts and Meredith E. David, "My Life Has Become a Major Distraction from My Mobile Phone: Partner Phubbing and Relationship Satisfaction Among Romantic Partners," *Computers in Human Behavior*, no. 54, January 2016, pp. 134–41; Vittoria Franchina et al., "Fear of Missing Out as a Predictor of Problematic Social Media Use and Phubbing Behaviour Among Flemish Adolescents," *International Journal of Environmental Research and Public Health* 15, no. 10, October 2018, p. 2319; Ru-De Liu et al., "The Effect of Parental Phubbing on Teenager's Mobile Phone Dependency Behaviors: The Mediation Role of Subjective Norm and Dependency Intention," *Psychology Research and Behavior Management*, no. 12, November 2019, pp. 1059–69; James A. Roberts and Meredith E. David, "Put Down Your Phone and Listen to Me: How Boss Phubbing Undermines the Psychological Conditions Necessary for Employee Engagement," *Computers in Human Behaviour* 75, October 2017, pp. 206–17.

digitized subjectivity in which screen addiction coexists with profound loneliness and a fear of intimacy and closeness.[17] Ironically, face-to-face relationships sometimes lose their quality and resonance due to the high level of "connectivity" delivered by algorithmic capital.

The construction of such "communities" is thus dependent on using platforms that increase algorithmic capital—the same capital that dispossesses subjects and increasingly distances decision-making processes from individuals. The building online of political communities that can be agents of social change is itself subordinated to the logic of algorithmic capital, because it mediates them, subjects them to surveillance, and, in more ways than one, also has a paralyzing effect on political action. All too often, social media reduces political organizing to the frenetic sharing of information, an explosion of "conversations," debates, and inflammatory arguments over various topics and happenings. It amplifies a sensationalist logic that already existed before the advent of digital platforms, but which is now magnified, accelerated, and reproduced on a massive scale. We discuss, we like, we denounce, we share, and then we feel as if we have acted politically when, paradoxically, we have merely "fed the machine" without taking any practical action. In this way, we listen to lonely voices instead of coming to their aid, we communicate indignation rather than doing anything to rectify indignities.

The community of datafied subjects

Datafied subjects, meanwhile, exist in an obscure and mysterious world, one so distant from living subjects that they have no idea of the communities that exist there. These communities are in fact groups of datafied subjects bunched together according to certain characteristics that correspond to a point (or a series of points) of information of interest to the platform or algorithm that is doing the grouping and identification: age, consumption habits, health status, income, political orientation, and a range of other factors of interest. In other words, algorithms can create lists that group datafied subjects according to certain characteristics. While the datafied subject is itself the product of an involuntary identification process, the communities to which they belong based on certain characteristics obviously exist outside the (conscious, political,

17 Seymour, *The Twittering Machine*, p. 56.

and so on) reach of the living individual. However, the fact that a datafied subject belongs to one community or another may have serious consequences. It is well established that overt discrimination is illegal when it comes to defined sociopolitical groups: gender, race, religion, and so on. A bank cannot refuse someone a loan because they are a woman or black, or gay, for example. However, if that bank uses an algorithm fed with information that could lead to the inference that the person is black, such as an address in a predominantly black neighborhood, the loan could well be refused, quite legally—whether or not that is the result desired by the bank. Algorithms can thus reproduce discrimination in datafied communities by creating groups that do not enjoy constitutional protections but are correlated with ones who do. Algorithms are not inherently discriminatory. And yet the process of datafying subjects certainly opens the door to new forms of discrimination that the law and established rights might not be able to adequately protect against.

The algorithmic world

Dispossessed of their activity, algorithmic subjects are also dispossessed of its product, an alienated world. This world is not the free creation of human species-beings, but an accumulation of algorithmic capital, a world where the particular products of human activity serve to accumulate data once it has been networked in a vast Internet of datafied things, places, spaces, times, and subjects. As Greenfield suggests, the emotional and experiential horizon of this world is sadness. It is charged with a certain melancholy: "It depends on a milieu of continuously shattered attention, of overloaded awareness, and of gaps between people." The Internet of Things, he continues, "seems like an attempt to paper over the voids between us, or slap a quick technical patch on all the places where capital has left us unable to care for one another."[18] The creators of an alienated world experience its material structures, social relations, and horizons of meaning as things that are not the result of their free creation. The algorithmic world, populated by algorithmic subjects, must necessarily be situated at a qualitative distance, an opaque separation from species-beings whose activity, experience, attributes, and time are exploited and appropriated by capital. Thus algorithms, AI, datafied subjects—in short, the world as reproduced by algorithmic capital—all appear as another world, a foreign world, sad, cold, and indifferent.

18 Greenfield, *Radical Technologies*, p. 60.

Beyond the emotional content of the world that algorithmic capital serves up to us, there are its ethics. Algorithmic capital offers a deterministic world or, more precisely, it offers a model of good living where ethics, that is, the question of how to live well, is largely determined by nonhuman agency, an overarching and superhuman force guiding individuals through personalized recommendations and/or behavioral control, and thus loses its meaning. Algorithms establish relationships with others and with the world; they increasingly steer our behavior and dictate our actions, which are no longer human choices but automated decisions.

We have no intention of romanticizing the pre-algorithmic world, as if individuals were formerly freely able to create ways of living well. We are much closer to the truth in postulating that the reality of neoliberal capitalism at its height was not a world, but the denial of one.[19] However, it seems that algorithms, precisely because they radicalize the processes of involuntary subjectification and removal of agency in decision-making, are shaping the contours of a particularly servile world. The notions of choice or even experience here take on a murky, even sinister meaning. To imagine a world in which powerful AI aligns all the behaviors of datafied subjects in order to accumulate capital for its owners is to imagine a world with impoverished ethics: one where datafied subjects aggregate and disaggregate according to a flow of algorithmic commands programmed to maximize consumption, guide behavior, and accumulate algorithmic capital. This is the "totally administered world" of which Adorno wrote. Where he used hyperbole as a stylistic device, with each passing day this diagnosis seems to be becoming ever more a reality. This world and its subjects are held together solely by deterministic calculations. The algorithmic world, in its logic, is a post-agency world, where the myth of the determined subject and free agent acting in pursuit of happiness is shattered. In its wake, it leaves only datafied subjects that develop in accordance with algorithms, beyond good and evil and beyond the good life. It is an amoral and post-ethical world.

We cannot simply take refuge in the idea that there will always be a gap between the algorithmic world and the true, so-called "real world," in which living individuals will always enjoy a certain amount of leeway, a certain space-time where we can "disconnect" and where ethics will once again become possible. Indeed, many efforts are being made to connect the algorithmic world to the experience of living individuals, to subsume

---

19 Fischbach, *La privation de monde*, p. 8.

the so-called "real" world into algorithms. This is, for example, the great danger of facial recognition systems and tracing applications that seek to "re-attach" datafied subjects to living subjects, in order to identify living subjects. Consider also the metaverse, or even simply the augmented reality we mentioned earlier, which in its *Pokémon Go* version, for example, attempts to guide living individuals in the same way that one would guide a datafied subject—or livestock—toward certain spaces, markets, and predicted behaviors. Even if it presents itself as an opportunity to "improve the world," does not the very concept of augmented reality suggest that reality is deficient? At best, augmented reality can only erase, mask, and obscure social problems. It furthermore runs the risk of creating a new division between those who have access to augmented or virtual reality technologies and those who do not. Will the latter be seen as full members of the community? Will their public or democratic agency be limited? A world in which different classes of people occupy different physical environments, where not everyone has access to the same spatial-temporal and informational frames of reference, seems like an additional factor for division and polarization, rather than an environment conducive to building solidarity and community spirit.[20] Consider also the processes of establishing smart cities, saturating public and private spaces with extractive architecture and a vast algorithmic surveillance complex. Algorithmic capital is constantly striving to turn this algorithmic world back toward "reality" in order to guide, direct, and dominate it, thereby creating a radical monopoly of the social world that is impossible to circumvent.

Finally, we see algorithmic capital's desire to align the datafied world with the real world in policies such as Facebook's efforts to combat fake names and profiles. This concern for authenticity is all the more ironic: While fake news circulates freely, Facebook reiterates in its terms of use that its users must provide truthful information and keep it up to date, under penalty of being banned. Of course, the more truthful the data profile, the smaller the distance between the algorithmic subject and the living subject, the higher the quality of behavioral predictions, and the greater the value of predictive products.

The algorithmic world that is increasingly imposing itself on us is a distant world, populated by de-subjectivized, datafied subjects. It is a world that denies conscious and determined species-beings, a world hostile to the formation of supportive, resilient, democratic, and

20 Greenfield, *Radical Technologies*, p. 83.

self-managed communities. It is a world of calculations and automatisms, a world that is cold, mechanical, predetermined, and predictable. The iron cage of modernity has now become a predictive and autonomous smart machine.

Of course, the bleak picture painted in this thesis is not intended as a fait accompli. Rather, it is an outline of a powerful trend and a developing process in which the social world is gradually being transformed, generating experiences of alienation that are now becoming commonplace and even barely noticed features of everyday life. Admittedly, we have depicted platform users so far as predominantly passive, focusing our analysis on phenomena such as extraction, alienation, and exploitation. But users do display various forms of "algorithmic agency" and "algorithmic resistance", as Bonini and Treré put it, that operate within and through algorithms.[21] More broadly, resistance is mounting in society at large to confront AI and algorithmic capital. Fortunately, as Georg Lukács noted, reification (that is, the transformation of human relations into things) is never fully complete; there always remains an excess, elements of resistance within human subjectivity, and exploited individuals and groups. In this algorithmic world, there emerge struggles, pockets, and practices of resistance, subjectivities, and militant communities that challenge the new logic of algorithmic accumulation and power. The next thesis turns to examine these struggles and the forms of agency resisting algorithmic capital.

---

21 Tiziano Bonini and Emiliano Treré, *Algorithms of Resistance. The Everyday Fight Against Platform Power*, Cambridge, MA: MIT Press, 2024, pp. 18–23.

Thesis 18

We should resist algorithmic capital—and support the rising struggles against it

Is the struggle endless?

I would say that as our struggles mature, they produce new ideas, new issues, and new terrains on which we engage in the quest for freedom.

—Angela Davis

It may seem that the analysis of algorithmic capital presented thus far has depicted an insatiable, all-consuming and unstoppable process.[1] It perhaps sounds as if nothing can escape the grip of this irresistible logic. Yet it is also essential to identify the conflicts, forms of resistance, and possibilities for emancipation that are present in today's world. As Pierre Dardot and Christian Laval note, in addition to "the logic of capital as a complete system," there is a second logic, the "*strategic logic of confrontation*: class warfare, whether latent or explicit, [which] transforms the conditions of struggle, shapes actors' subjectivities and, ultimately, frees the oppressed from subjugation, paving the way for their emancipation."[2] We should now turn to this logic of confrontation so that we can think about the resistance to algorithmic capitalism.

It is useful, to this end, to analyze the many different ways of emancipating ourselves from capitalism, whether through collective action and oppositional social movements; attempts at institutional reform "from above"; or the creation of noncapitalist socioeconomic and technological

1 Note: Part of this chapter was published in Jonathan Durand Folco and Jonathan Martineau, "Cartographier les résistances à l'ère du capitalisme algorithmique," *Possibles* 45, no. 1, 2021.

2 Pierre Dardot and Christian Laval, *Marx, prénom: Karl*, Paris: Gallimard, 2012.

alternatives. Indeed, if we want to paint a more comprehensive picture of digital resistance, we cannot look only at struggles on the web or the use of social media by protest movements. Here, inspired once again by Nancy Fraser's work, we propose a distinction between two types of confrontation: class struggle and boundary struggles.[3] The former refers to the economic contradiction between capital and labor in the spheres of capitalist production and circulation. This demands an analysis of the tensions and resistance generated by data extraction and the exploitation of algorithmic labor, as well as the impact of the algorithmic means of production that feed the automation of production and the web giants' hegemony over a sizable share of economic life.

Next, we will shift our attention to boundary struggles: resistance movements that emerge at the intersection between economic production and other noneconomic spheres, such as politics, nature, and social reproduction.[4] In this understanding, feminist, environmentalist, indigenous, and migrant-solidarity struggles, among others, are not somehow epiphenomenal or external to capitalism, but rather forms of resistance that are themselves stirred by capitalism's structural contradictions. In our view, even if it is obviously impossible to review the totality of resistance practices against algorithmic capital, the combined analysis of class struggles and boundary struggles allows us to map out a fuller picture.

Algorithmic labor as a locus of resistance

At the economic level, the struggle against algorithmic capital revolves around three main areas: algorithmic labor, digital platforms' monopolistic tendencies, and more recently the rise of generative AI. Each of these dimensions creates its own set of insecurities, tensions, problems, and demands that feed the emergence of resistance movements that are—depending on the geographical region and the particular circumstances—more or less organized.

We can identify one first set of social struggles on the front of production and algorithmic labor. As we saw in Thesis 5, the platformization of the economy is turning a whole mass of workers toward self-employment,

3 Nancy Fraser, "Behind Marx's Hidden Abode," *New Left Review*, no. 86, March/April 2014, pp. 55–72.

4 Nancy Fraser and Rahel Jaeggi, *Capitalism: A Conversation in Critical Theory*, Cambridge: Polity Press, 2018, p. 19.

as gig workers, contractors, or micro-entrepreneurs, and often circumventing established labor codes and social protections. This has given rise to an "algorithmic precariat," consisting often of disadvantaged people, migrants, students, young people, and women who combine domestic work with tasks on platforms such as Amazon Mechanical Turk, micro-taskers who pack into click farms in the Global South, and platform delivery workers and drivers.[5]

The resistance emerging on this front of digital labor is a response to the exploitation, devaluation, invisibility, and alienation generated by the extraction of data and human labor for the sake of capital accumulation and increasing algorithmic power. This resistance ranges from reappropriation, gaming, and repurposing of algorithms by users and workers to pursue their own goals, struggles to de-Uberize the economy,[6] mobilizations demanding updated labor codes, union drives—to efforts to create cooperative platforms as alternatives to the likes of Uber, Airbnb, and Foodora.

We mentioned briefly the concept of "algorithmic resistance" proposed by Bonini and Treré at the end of the previous thesis. Their ethnographic work on practices of resistance shows how users, including gig workers, social media users, and cultural workers, exercise agency and fight against platform power in their everyday engagement with the platforms. Examples include food delivery drivers opening multiple accounts on the same platform to get more orders, creating customer accounts to pass fake orders, coordinating order refusal, and mutually exchanging working hours. Using algorithms to pursue one's own goal is also exemplified by the practice of forming "pods" on social media platforms to increase visibility, where users get together and agree to systematically like, share, and comment on each other's posts. Hashtag hijacking, another example of algorithmic agency, can occur when users use hashtags to spread unrelated or negative content, resulting in a general confusion over the classificatory functions of the hashtag.[7] These practices, of which we mention only a few cases among a vast range, show workers and users acting *within* the algorithm to reshape outcomes in their favor.

---

5 Antonio Casilli, *Waiting for Robots*; Mary L. Gray and Siddharth Suri, *Ghost Work: How to Stop Silicon Valley from Building a New Global Underclass*, Boston: Houghton Mifflin Harcourt, 2019.

6 Florian Forestier et al., *Désubériser: Reprendre le contrôle*, Paris: Éditions du Faubourg, 2020.

7 Bonini and Treré, *Algorithms of Resistance*, pp. 79–88, 150–3.

In terms of workers, struggles against platforms and for better work conditions, Olivier Rafélis de Broves and his colleagues provide a detailed overview and analysis of the new forms of organization and struggle in the platform economy.[8] Based on a review of the existing literature and empirical studies, their analysis points to the building of a workers' movement as a collective actor in platform class relations. Despite structural disadvantages faced by platform workers, such as heterogeneity and atomization, converging trends of worker organization are noticeable in Europe, Latin America, North America, and Asia:

1. Workers in the food delivery industry are able to meet at specific locations in the city, facilitating networking and organizing.
2. Connecting through social media and messaging apps allows workers to build common identities.
3. Core groups of activists are pursuing organization activities.
4. Organized workers are engaging in direct action against platforms, such as protests and logging off.
5. Platform workers are forging alliances with unions and building new ones.
6. Platform workers are developing alternative worker-led co-ops.[9]

These are some aspects of a multifaceted movement responding to new forms of precarity and exploitation generated by algorithmic capitalism. While algorithmic labor now occurs in different places and spheres of social life (in the office, at home, in disused factories repurposed as click farms, in mines and assembly plants, on city streets, and so on), we should not underestimate the crucial role of distribution chains and the circuits of logistics, today controlled by giants such as Walmart and Amazon. Algorithmic labor does not stop at digital labor, as we saw in Thesis 5, but also encompasses extractive, logistical/industrial, and social reproductive moments. We must therefore also take into account struggles in these arenas, led by workers such as the "new working class exploited in logistics hubs."[10] As such, the union drives by Amazon warehouse workers, whose first success came in New York in April 2022 and

8 Olivier Rafélis de Broves et al., "Workers Organizing in the Platform Economy: Local Forms and Global Trends of Collective Action," *Sociology Compass* 18, no. 2, 2024.

9 Ibid. We summarize here the six points the authors present in more detail on p. 3 of their article.

10 Razmig Keucheyan, *Les besoins artificials: Comment sortir du consumérisme*, Paris: Zones, 2019, p. 161.

was followed by others, notably in Quebec, are also strong symbols of a mounting resistance.[11]

The fight against algorithmic capital is also unfolding around issues of land, for example in agriculture, especially through farmers' resistance to the arrival of Agriculture 4.0. For example, L'Atelier Paysan, a cooperative created in France in 2009, seeks to develop alternatives to industrial agricultural machinery by cataloging the availability of low-tech and open-source tools, machines, and buildings. It likewise offers various training, support, and advocacy services in order to help farmers and food production to become more autonomous. Its manifesto, *Taking Back the Land from the Machines*, is a scathing critique of industrial agriculture, and the barriers put up around technology, technological solutionism, and algorithmic capital. As against the industrial technologies that accelerate the loss of know-how, this cooperative promotes smaller-scale technologies that help make farmers more independent.[12] These latter technologies are distinguished by five criteria:

1. There is user investment in the development, maintenance, and improvement of the tools used.
2. They are based on farmers' real needs.
3. They are reproducible and have versatile uses.
4. They help farmers work and think together, using a complex, living, and collective know-how.
5. They allow sharing, through open-source tools.[13]

These struggles around the question of land can also be linked with the mounting resistance of communities against the building of data centers in their localities. The strain on power and water resources of these centers is increasingly seen as adverse to the interest of local communities, and protests of various sorts by citizen groups and local authorities have been gaining traction in recent years.[14]

Finally, we should again note that algorithms are increasingly used to impose surveillance on workers across all types of production, which is met by various forms of resistance at work. This surveillance serves a

---

11 Michael Sainato, "Amazon Labour Organisers Push for Second Union Victory in New York," *The Guardian*, April 22, 2022.

12 L'Atelier Paysan, *Reprendre la terre aux machines: Manifeste pour une autonomie paysanne et alimentaire*, Paris: Seuil, 2023.

13 Ibid., pp. 104–6.

14 Molly Taft, "The Data Center Resistance Has Arrived," *Wired*, November 14, 2025. See also the podcast episode of *Tech Won't Save Us* hosted by Paris Marx, titled "Data Vampires, Opposing Data Centers," October 14, 2024.

disciplinary purpose, but it also extracts data from labor, with a view to potentially automating it.[15] The surveillance technologies employed in the workplace reflect a new "biopower paradigm."[16] They range from intercepting emails and monitoring employees' entire online activities to placing technological accessories on or around workers that extract a wide range of data (movement, biometrics, timings, conversations, and so on). These surveillance technologies make it possible to build up data avatars (algorithmic subjects) for each employee, which are used to manage their behavior, monitor their performance, measure their physical fitness and mental health, and so on. Workplace productivity apps have been a huge market breakthrough. The company Betterworks, for example, develops office software that integrates elements of social media, video games, and fitness apps. Employers and employees use them to set short- and long-term productivity goals and benchmarks. Progress is then recorded on digital scoreboards and employees are given ratings for their performance. Corporate wellness programs also draw on this new paradigm, which ultimately increases individualization and surveillance and encourages employees to work longer and harder.[17] These surveillance technologies, and the transformations in the workplace that they fuel, are also conducive to stirring up multiple forms of micro-resistance in the workplace.

The recent development of Gen AI also leads to further contests between companies that want to incorporate its usage in labor processes, and employees or contractors who might want to resist certain uses to protect job security or out of care for the quality of their work. In sectors such as communications, journalism, and culture, these struggles are taking various forms, from micro-resistance to full-blown strikes, as in the case of the 2023 Writers' Guild of America strike, and they constitute an important locus of resistance to algorithmic capital. Artists have also brought a series of copyright lawsuits against tech firms such as Stability AI, Deviant Art, and Midjourney.[18] This also raises the broader question of the regulation of hyperpowerful algorithmic systems. After all, these

15 Greenfield, *Radical Technologies*, pp. 183–207.

16 Ivan Manokha, "New Means of Workplace Surveillance," *Monthly Review*, February 1, 2019.

17 Connor Dougherty and Quentin Hardy, "Managers Turn to Computer Games, Aiming for More Efficient Employees," *New York Times*, March 15, 2015; Esther Kaplan, "The Spy Who Fired Me: The Human Costs of Workplace Monitoring," *Harper's Magazine*, March 2015; Karen Levy, "The Contexts of Control: Information, Power, and Truck-Driving Work," *Information Society* 31, no. 2, 2015, pp. 160–74.

18 André Boily, "Intelligence artificielle: Les poursuites pour violation du droit d'auteur se multiplient," *Le Journal de Montréal*, January 26, 2023.

systems have the power to accelerate the spread of disinformation, to blur the boundaries between truth and falsehood (through deepfakes), to automate skilled jobs, and so on.

In our view, class struggle by workers—at all nodes and levels of algorithmic labor as we defined it in Theses 5 and 6—is essential to mounting a full and successful front against algorithmic capital. Seeing workers organize more and more, and gaining awareness of their interests as workers, witnessing the success of unionizing drives of digital laborers, winning strikes, and spreading various forms of resistance and alternatives in multiple countries across the world—all this is surely a reason for optimism.

Digital sovereignty against GAFAM?

In addition to digital struggles and resistance in the sphere of algorithmic production and distribution, the fight against platforms is also taking place at the level of legislation and regulation by states and governments of markets and companies operating within the digital economy in order to limit the monopolies' power and/or to break them up. The platform business model not only feeds the exploitation of algorithmic labor but also has a strong tendency to monopolize important sectors of the market, through a combination of network effects, the buying up of start-ups, and other dynamics related to the digital economy. It is not so much workers, however, who resist these monopolistic tendencies, but regulation agencies and legislators.

The monopolizing tendencies of algorithmic capital are feeding a resurgence of the antitrust movement, aimed at countering the GAFAM giants' hegemony and their concentration of economic power.[19] This movement's core goal is to dismantle the digital monopolies that undermine free market competition. For example, the hearings that the US House of Representatives Judiciary Committee held with the CEOs of Google, Amazon, Facebook, and Apple in summer 2020 symbolically condemned these companies' anticompetitive practices. However, it remains hard to see binding legislation being adopted in the US, since these firms not only mount significant lobbying efforts, but also appeal to American patriotic sentiment by emphasizing that rules that might

19 Yossi Spiegel and Joel Waldfogel, "Introduction to the Special Issue of Information Economics and Policy on 'Antitrust in the Digital Economy,'" *Information Economics and Policy*, no. 54, March 2021.

constrain them would give an advantage to Chinese companies such as Baidu, Tencent, and Alibaba.[20] Attempts to regulate platforms through antitrust legislation are thus proving difficult. In February 2023, Meta won a case against the US Federal Trade Commission (FTC) in a lawsuit related to the acquisition of VR start-up Within Unlimited, as Zuckerberg sought to consolidate his monopoly on building the metaverse. As one report described:

> The decision represents the first major loss for FTC Chair Lina Khan, who was appointed by President Joe Biden to reinvigorate antitrust enforcement as a key tenet of his administration's economic policy. Khan has taken a more aggressive approach to mergers than her predecessors and stepped up the agency's focus on technology giants in particular because of their potential to quickly dominate budding markets.[21]

The fact that Lina Khan was rapidly replaced in 2025 can be seen as a change in the FTC's general policy attitude toward the tech giants. And even though Google was indeed found guilty of monopolistic practices in two separate cases (search monopoly in 2024, and ad tech monopoly in 2025), the sanctions imposed as a result fall far short of concrete antitrust actions. More recently, the Trump administration's wish to prevent *any* form of AI regulation that would hamper Big Tech "innovations" will jeopardize the capacity of US federal agencies to regulate the tech sector in coming years.

It should also be noted that Meta, Google, and Amazon, in particular, tend to issue threats when governments attempt to pass restrictive laws that would hurt their interests. This was the case in Canada with Bill C-18 in 2023, which forces digital giants to pay Canadian publishers for news content. Meta responded by permanently blocking news on Facebook/Instagram.[22] Similarly, Amazon closed its business in Quebec after a successful unionizing drive of warehouse workers took place in May 2024. As a general rule, it has proved difficult for non-US governments to legislate against the GAFAM companies, even if some degree of success is possible; this has been attained in countries such as Australia, which

---

20 Jonathan Durand Folco, "Réflexions sur l'impérialisme américain et le capitalisme algorithmique," *L'Esprit libre*, September 2, 2020.

21 Leah Nylen, "Mark Zuckerberg Just Won a Big Victory in a Fight over Virtual Reality—and It's FTC Chair Lina Khan's First Major Loss," *Fortune*, February 1, 2023.

22 "Projet de loi C-18: Des défenseurs des médias dénoncent l'attitude de Google," Radio-Canada, February 24, 2023.

has imposed stricter regulations on social media, or Brazil, where Elon Musk's X (Twitter) was briefly banned in late 2024 for not complying with Brazilian regulations.

Meanwhile, antitrust policies have been enacted more successfully in China. For example, the Chinese company Alibaba, which has operations in cloud computing, logistics, e-commerce, AI, media, and entertainment, announced in March 2023 that it was mounting the largest restructuring in its history, splitting into six different entities.[23] This breakup of one of the biggest players in China's technology sector followed robust political interventions starting in November 2020, as a tsunami of regulations hit China's digital industry:

> An increase in investigations and fines for anticompetitive practices; the dismantling and restructuring imposed on major players in online finance; stricter conditions on the collection and use of personal data; new limits imposed on the transfer of sensitive data abroad; stricter supervision of the use of recommendation algorithms; an ultimatum to major on-demand work platforms to better protect their workers; threats against major e-commerce platforms that violate intellectual property protection rules; condemnation of the excessive enrichment of digital entrepreneurs and multinationals ... And the list goes on.[24]

Another kind of opposition to the Big Tech companies' growing monopoly power is expressed in the demand for greater "digital sovereignty." This can take many different forms.[25] Some approaches call for greater sovereignty for cyberspace relative to state power, while others aim to restore greater nation-state sovereignty over infrastructure, software, and data relative to the big foreign-owned private platforms. There surely are calls for a "native" digital sovereignty or technological sovereignty, inspired by localism and the paradigm of the commons. Yet the

---

23 Peter Catterall, "Le géant de l'Internet Alibaba se divise en six groupes," *La Presse*, March 28, 2023.

24 Cédric Leterme, "Offensive chinoise sur le numérique," Centre Tricontinental (CETRI), September 13, 2021.

25 Stéphane Couture and Sophie Toupin, "What Does the Notion of 'Sovereignty' Mean When Referring to the Digital?," *New Media & Society* 21, no. 10, 2019, pp. 2305–22; Stanislav Budnitsky and Lianrui Jia, "Branding Internet Sovereignty: Digital Media and the Chinese-Russian Cyberalliance," *European Journal of Cultural Studies* 21, no. 5, 2018, pp. 594–613; Tahu Kukutai and John Taylor, *Indigenous Data Sovereignty: Toward an Agenda*, Canberra: ANU Press, 2016; Evgeny Morozov and Francesca Bria, "Rethinking the Smart City: Democratising Urban Technology," Rosa Luxemburg Stiftung, 2018.

most influential tendency today is the "technonationalism" arising in various countries, including China, Russia, and Iran. Technonationalism refers to certain nations' tendency to project their technological power in the interest of bolstering their geopolitical influence.[26] Mass surveillance, the closing of borders, the exclusion of foreign digital companies, and the construction of "national" Internets are different instances of this more general strategy. In the era of algorithmic capitalism, resistance movements thus also take root in a complex geopolitical landscape that reproduces the tensions between the dominant and the dominated in the world order. Although digital sovereignty policies can hurt the interests of US algorithmic capitalists, and in some cases take anti-imperialist forms, their scope and spirit today remain by and large devoid of anti-capitalist politics.

Regulating AI: Yes, but how?

In March 2023, an open letter from the Future of Life Institute sent shockwaves through the media, alerting public opinion to the huge risks involved in the frantic development of AI.[27] It called for a six-month moratorium on training models superior to GPT-4, in order to allow for proper forms of regulation to be enacted. The letter was signed by thousands of people, including several leading figures in AI research and the tech industry, such as Yoshua Bengio, Elon Musk, Steve Wozniak, and Yuval Noah Harari.[28] Several voices pointed out the unrealistic and even hypocritical nature of a six-month moratorium for this industry, given that the race toward generative AI had been in full swing for months already. However, it was also significant that the issue of AI regulation had now been brought to the forefront of public debate. This represented real progress, given that throughout years of AI technology development this issue had been almost completely absent from the media. The whole problem lies in how this political issue is framed: How should AI be regulated, with what types of regulations, for what reason(s), and in favor of whose interests?

---

26 Steven Feldstein, "The Virus That Split the World," *MIT Technological Review* 123, no. 6, 2020, pp. 10–15.

27 This section is a reworking of an open letter: Jonathan Durand Folco and Jonathan Martineau, "Réglementer l'IA et sortir du capitalisme algorithmique," *Le Devoir*, May 2, 2023.

28 Future of Life Institute, "Stop Giant AI Experiments: An Open Letter," March 22, 2023.

One first point worth noting is that the Future of Life Institute is a longtermist think tank created in 2014 by various scientists and entrepreneurs (with such figures as Jaan Tallin and Elon Musk as advisors). It is thus linked to the TESCREAL ideology examined in Thesis 8. For this organization, the central problem is the existential risk posed by a superintelligence that may spiral out of control and destroy humanity. The letter mentions in passing the problem of generative AI sowing propaganda and the question of a mass automation of jobs. Still, these issues are mainly presented as future risks to be managed by various technical and regulatory means. In this context, the goal of the six-month moratorium would be to take a step back so that research laboratories could establish safety protocols and governance systems could be put in place to regulate the AI industry. This would, the letter suggested, allow the industry safely to realize its full potential while preventing future catastrophe. The proposed solutions are

> new and capable regulatory authorities dedicated to AI; oversight and tracking of highly capable AI systems and large pools of computational capability; provenance and watermarking systems to help distinguish real from synthetic and to track model leaks; a robust auditing and certification ecosystem; liability for AI-caused harm; robust public funding for technical AI safety research; and well-resourced institutions for coping with the dramatic economic and political disruptions (especially to democracy) that AI will cause.[29]

At first glance, these might sound like decent courses of action, similar to the many proposals put forward in the report on "blind spots in AI development" published by UNESCO and MILA in March 2023. That report listed third-party audits, responsible AI development, taking on board sustainable development goals, "inclusive" innovation, and other similar proposals.[30] However, this framing of the issues, in a technical and managerial perspective—that is, as problems of "governance" and "security"—also merits scrutiny. This had already been pointed out by several dissenting voices in Silicon Valley, such as Emily Bender and Timnit Gebru, who sounded the alarm two years earlier with their article entitled "On the Dangers of Stochastic Parrots: Can Language Models Be

---

29 Ibid.

30 Benjamin Prud'homme, Catherine Régis, and Golnoosh Farnadi, eds., *Les angles morts de la gouvernance de l'IA*, UNESCO and MILA, Institut québécois en intelligence artificielle, 2023, available in English at unesco.org.

Too Big?"[31] Bender and Gebru were stung by the letter from the Future of Life Institute, which cited their article while depoliticizing the issues they had raised. They wrote a response, pointing out that the main problems with AI do not concern a hypothetical scenario in which this technology becomes disastrously overpowerful, but rather the harm already being caused by capitalist firms using these automated systems. For these companies are socializing the costs of this technology while they themselves corner most of the benefits.

> The letter addresses none of the ongoing harms from these systems, including (1) worker exploitation and massive data theft to create products that profit a handful of entities, (2) the explosion of synthetic media in the world, which both reproduces systems of oppression and endangers our information ecosystem, and (3) the concentration of power in the hands of a few people which exacerbates social inequities.[32]

These authors contend that regulatory efforts should focus on transparency, accountability, and the prevention of labor exploitation practices. This means rejecting a narrative that urges us to "adapt" to a supposedly predetermined technological future that, in reality, serves the priorities of a powerful few.[33] It is thus apparent that there are multiple competing discourses on the regulation of AI. At least four positions can be identified in the public debate:

1. The *accelerationist* and *techno-optimistic approach*, generally associated with solutionism and GAFAM. It welcomes these developments with enthusiasm and/or awe. With reference to the existential risks of superintelligence, it voices TESCREAL ideology and longtermist claims. This approach largely subscribes to the (self-)promotional narrative of the companies that create and develop these innovations. It stipulates that we should simply adapt to the rapid advances in AI by promoting a safer regulatory framework for the industry.

---

31 Emily Bender, Timnit Gebru, Angelina McMillan-Major, and Margaret Mitchell, "On the Dangers of Stochastic Parrots: Can Language Models Be Too Big?," Proceedings of the 2021 ACM Conference on Fairness, Accountability, and Transparency, March 2021.

32 Emily Bender, Timnit Gebru, Angelina McMillan-Major, and Margaret Mitchell, "Statement from the Listed Authors of Stochastic Parrots on the 'AI Pause' Letter," DAIR Institute, May 31, 2023.

33 Ibid.

2. The *regulationist approach*. It calls for greater oversight to ensure the responsible development of algorithms and to mitigate their harmful effects on security, the environment, democracy, and individual rights. This is the approach advocated by the MILA Institute and much of the Quebec AI ecosystem, which calls for true leadership in the regulatory arena. In this approach, regulation is meant to promote innovation, economic growth, and the normalization of AI across society.[34]

3. The *technocritical approach*, advocated by Emily Bender, Timnit Gebru, and the authors of this book, among others. It argues that AI regulation must not lag behind the fundamental social issues raised by these technologies, but should instead seek to combat, regulate, and even dismantle the current social order and various systems of oppression. In the following theses, we will develop our own view of what this implies from an ethical, legislative, and political point of view. We can say right away that the objective is not to drive economic growth and continuous innovation, but to move beyond capitalism. This will proceed through a reappropriation of technologies and energy descent oriented toward what we call "technosobriety" (see Thesis 19).

4. Finally, the *technopessimist approach*, which takes a more frontal stand against new technologies. It completely rejects digitalization and advocates or takes various more or less radical actions against algorithmic machines. This stance involves a fierce critique of the "technocratic system" and, in political terms, calls for the outright banning of certain technologies. From this perspective, not just capitalism but also industrial society and technological civilization must be fought or even brought to an end. Technological development is leading us to an inevitable disaster, which we must stop by various decisive means.

These four narratives represent a continuum between two extremes—one advocating the fullest possible development of technological forces and the other an (almost) complete stop to them. Currently, techno-optimist and regulationist discourses dominate public debate. AI doomers are sounding the alarm by invoking the specter of superintelligence, and regulationists like Bengio are coming to the rescue by urging the adoption of regulatory frameworks. However, the public policies actually suggested are generally limited to the narrow horizon of

34 Yoshua Bengio and Catherine Régis, "Il y a urgence à adopter la Loi sur l'intelligence artificielle et les données," *La Presse*, April 19, 2023.

risk management, corporate self-regulation, and mitigating algorithmic biases. For example, the Artificial Intelligence and Data Act (AIDA), developed as part of the Canadian government's Bill C-27, aimed to create a flexible regulatory system to guide AI innovation in Canada. This legislative framework sought to better regulate "high-impact" AI systems, to reduce cases of individual harm and biased results, and to prohibit certain fraudulent uses (such as deepfakes). All this was meant to be done with the help of six key principles: human oversight and monitoring, transparency, fairness and equity, safety, accountability, and validity and robustness.[35]

It is surely difficult to argue against virtuous conduct on principle. But how would these principles be applied in practice? This bill promised "education, establishing guidelines, and helping businesses to come into compliance through voluntary means" and sought to strike the right balance to "ensure that [regulations] are effective at protecting the interests of the Canadian public, while avoiding imposing an undue burden on the Canadian AI ecosystem."[36] Thus, the law did not seek to create a regulatory framework prior to the deployment of algorithmic technologies or to redistribute the profits generated by this industry, but rather to "ensur[e] that Canadians can trust the digital technologies that they use every day." The principle of transparency does not aim to make public the relevant information about training data or the architecture of algorithmic models, but rather to inform the public about the uses of high-impact AI systems. The bill died, however, after the prorogation of Parliament in January 2025, leaving the prospect of any significant AI regulation in Canada in limbo for the moment.

We mention this narrowly framed regulatory approach, however, because it is dominant in governmental policy circles. It seeks to reconcile the imperatives of economic growth and social acceptability. It puts citizen protection on an equal footing with the interests of a thriving industry that is largely financed by public funds, but whose billions of dollars in profits are mainly appropriated by private interests. Several players in the technology sector, such as Elon Musk, Sam Altman, and Yoshua Bengio, thus find themselves playing the role of pyromaniac firefighters. Such tech figures produce hyperpowerful algorithmic systems, alert the authorities to their dangers by saying that they want legal regulations … while also trying to influence the development of this

35 See "The Artificial Intelligence and Data Act (AIDA) – Companion Document," ised-isde.canada.ca.

36 Ibid.

regulatory framework so that it does not contradict their interests too much. As Matteo Wong notes:

> Big Tech's warnings about an AI apocalypse are distracting us from years of actual harms their products have caused … The people developing and profiting from the software are the ones telling the government how to approach it—an early glimpse of regulatory capture … These companies also spent many millions of dollars lobbying Congress in just the first three months of this year.[37]

By seeking a better integration of AI into our lives, through a "balancing act" that mainly responds to the industry's needs, we risk losing sight of what—we believe—ought to be the primary focus of the debate. Which is to ask: Do we need robots and algorithms in the first place, and if so, in which sectors and for what purposes? Why are these technologies being introduced into society? By whom? Which groups are most affected by these machines, and who is reaping most of the benefits?

This is why it is a waste of time to try to regulate AI downstream of its development. Instead, we should aim to reduce the use and production of algorithmic machines in order to promote technological restraint that is compatible with social justice, individual rights, and the limits of the planet. This book aims to contribute to a growing body of literature that identifies various avenues that need exploring today, including the creation of new digital rights. These might include the right to disconnect, the right to be forgotten, the right to free essential services, the right to a career transition plan, and so on. The adequate regulation of AI cannot stop at hand-holding capitalist businesses by making them more socially acceptable. Public debate must reach out further and question the private ownership of the means of algorithmic production and examine the place, role, and limits of algorithmic technologies in our lives and in our society.

Still, even a future, postcapitalist society would need laws and rules to regulate the production and dissemination of algorithms and automated systems. In this context, it may be useful to draw some inspiration from the European Union's AI Act, introduced in 2021. This legislation does not confer rights on individuals but focuses on the products themselves; it regulates the companies that manufacture algorithmic systems, and the organizations that use them. The law proposes four levels of protection

37 Matteo Wong, "AI Doomerism Is a Decoy," *The Atlantic*, June 2, 2023.

depending on the risk level: prohibitions due to unacceptable risks; strict standards to be met due to high risks; transparency obligations for specific or limited risks; and simple authorization for algorithms with minimal risks.[38]

This law has the advantage that it bans outright certain harmful systems (the exploitation of the vulnerable, remote biometric identification, and social scoring systems). But it also has serious shortcomings:

1. It stops at vague pronouncements, categorizing systems as limited and high risk, and does nothing to regulate "connected objects" (which are classified as presenting minimal risk).
2. The ban on facial recognition within the EU leaves the door open to these same systems being used on migrants, refugees, and asylum seekers.[39]
3. Following intense and successful lobbying by OpenAI, general-purpose algorithmic systems such as GPT-4 are no longer classified as high risk, and this means less stringent demands on the entire industry.[40]
4. It ignores a whole range of issues related to the exploitation of labor, algorithmic governmentality, the erosion of democracy, the automation of social injustices, and AI's climate footprint.

In short, while the AI Act represents a step in the right direction (it is better to have this legislation than nothing at all), this risk management approach clearly falls far short of what is needed in a world where algorithmic capital continues to reign supreme.

In summary, we can see that resistance to algorithmic capital's economic power has appeared on three different and complementary levels. First, algorithmic labor opposes the exploitation and precarity generated by digital platforms and algorithmic capitalism. Second, citizens and governments oppose GAFAM's monopolistic tendencies by attempting to break up this concentration of power and thus regain a certain degree of digital sovereignty. This does not always involve taking an anticapitalist stance; in some cases, it can even be nationalist and authoritarian. Still,

38 Noémi Lair, "Artificial Intelligence Act: Voici ce que prépare l'Union européenne pour encadrer l'IA," RadioFrance, April 3, 2023.

39 Amnesty International, "EU: European Parliament Adopts Ban on Facial Recognition but Leaves Migrants, Refugees and Asylum Seekers at Risk," June 14, 2023.

40 Billy Perrigo, "Exclusive: OpenAI Lobbied the EU to Water Down AI Regulation," *Time*, June 20, 2023.

we classify it as part of the resistance to algorithmic capital because it attempts to counterbalance this global dynamic. Finally, the unregulated development of AI is generating in response various actions, discourses, and forms of resistance calling for regulations to be placed on it.

Boundary struggles against algorithmic injustices

In addition to these forms of resistance, algorithmic capitalism is also feeding the rise of boundary struggles at the intersection of everyday life, social reproduction, and the political world. The first series of threats that algorithmic capitalism creates for everyday life relates to the generalized surveillance of individuals. This proceeds through the systematic collection of personal data for predictive analyses, profiling, targeted advertising, algorithm optimization, and other profit-oriented goals. As Shoshana Zuboff points out, the logic of surveillance is not limited to Google's or Facebook's business models but is rapidly spreading to a host of economic sectors: banking, insurance, marketing, the car industry, tourism, the Internet of Things, smart cities, the public administration. The imperative to extract data and the imperative to predict behavior algorithmically are two sides of the same coin. The result is that algorithmic capital must necessarily keep extending its logic of surveillance ever further.

Faced with this generalized trend, it is no accident that debates around the protection of privacy and personal data are a central focus of struggles today. As algorithmic capital transforms our private lives into "raw material" to be extracted to fuel the production of exchange-value, Zuboff proposes a "right to sanctuary"—that is, the right to privacy, or the right not to be monitored. This is a first-frontier struggle between data extraction for profit and the sphere of everyday life that resists this algorithmic "colonization of the lived world."[41] Rights groups, such as France's La Quadrature du Net, are actively campaigning against surveillance practices, which may take various forms: advertising trackers, large digital platforms' problematic practices regarding user consent, police surveillance, facial recognition software in public spaces, and so on. For example, La Quadrature du Net "succeeded in canceling the installation of facial recognition gates at the entrances to high schools in Nice and

41 Nick Couldry and Ulises Mejias, *The Costs of Connection: How Data Is Colonizing Human Life and Appropriating It for Capitalism*, Palo Alto, CA: Stanford University Press, 2019.

Marseille" and joined the Reclaim Your Face campaign, which brought together more than 170 associations around Europe. It further created a

> citizen observatory on surveillance in cities, called Technopolice. Its website offers tutorials on how to oppose the installation of cameras in your local area, information on the expansion of security measures, and invitations to join 'mapping parties'—urban walks aimed at identifying all detectable security measures.[42]

As for legislative initiatives, after four years of intense negotiations, the European Commission in 2016 adopted the General Data Protection Regulation (GDPR), which was at the time doubtless the world's most advanced legislation in this field. Other e-privacy regulations, such as—in France—the right to be forgotten and the right to disconnect, offer examples of measures aimed at protecting privacy from encroachments by algorithmic capital. While the players involved in this debate mainly come from academic, technological, and legal circles, this remains a major focus of digital resistance.

In terms of social reproduction—a sphere spanning not only the domestic space but also healthcare and education where these are public services—algorithmic technologies generate a range of ambivalent effects, as we saw in Thesis 6. For example, while algorithmic innovations are presented as a solution that helps lighten the load of domestic work, whether through the development of personal assistants such as Alexa, the platformization of domestic work (food, childcare, cleaning and maintenance, transport and shopping platforms, and so on), or the introduction of smart devices into the domestic sphere, these technologies are in fact colonizing social reproduction in order to make it serve the accumulation of algorithmic capital. They are further commodifying these spaces and times and directly linking the performance of these tasks, as well as the human life and experience reproduced in them, to the imperative of data extraction. In this sense, digital resistance in connection to the domestic sphere is breathing new life into women's historic struggles for the recognition or remuneration of domestic work. It is increasingly intersecting with struggles against the exploitation of human experience and social relations by platforms and in defense of privacy. By linking up with resistance to the omnipotence of platforms,

42 Fabien Benoit and Nicolas Celnik, *Techno-luttes: Enquête sur ceux qui résistent à la technologie*, Paris: Seuil and Reporterre, 2022, pp. 21–2.

the issues of social reproduction reveal certain areas of overlap between class struggles and boundary struggles in the era of algorithmic capital.

Social reproduction also includes various activities and care work in different spheres, including health, social services, and education. With the arrival of ChatGPT and rising criticism of the digitization of schools, for instance in Quebec,[43] it will be little surprise if in coming years we see the movement against the use of AI in higher education gain momentum.[44] For example, in May 2023 the Fédération des enseignantes et des enseignants du Québec (FNEEQ-CSN) teaching union called for a moratorium on the rollout of AI systems and published a critical report on their impact on higher education. This mobilization is not entirely novel; already in 2013, a French collective called Écran total brought together thousands of people opposed to surveillance, data mining, and the accelerated digitization of schools, job centers, and benefits. Such developments are disrupting employment roles and feeding the dehumanization of care.

Struggles at the crossroads of social reproduction and public authorities are also emerging in the form of movements against algorithmic injustices.[45] Algorithmic injustice occurs when an algorithm that is programmed for a specific task (picking candidates, hiring, granting visas, assessing parole applications, and so on) can directly affect an individual's life opportunities. The algorithm, which appears neutral, impartial, and objective, often reproduces various forms of discrimination due to biases in the initial data (overrepresentation of men in companies, or of black people in prison, for example) and/or limited and deficient programming. A growing body of literature on algorithmic injustices shows that disadvantaged classes, women, and racialized people are greatly affected by the rapid introduction of algorithmic technologies in various spheres: public institutions, social services, and so on. We saw some aspects of this in Thesis 10.[46]

This "automation of inequality," as Eubanks calls it, raised its head in an unprecedented mobilization in Britain in autumn 2020. Tens

43 See Eric Martin and Sébastien Mussi, *Bienvenue dans la machine: Enseigner à l'ère numérique*, Montreal: Écosociété, 2023.

44 FNEEQ-CSN, "A IA en éducation et en enseignement supérieur: La FNEEQ-CSN réclame un moratoire de toute urgence," press release, May 14, 2023.

45 Abeba Birhane and Fred Cummins, "Algorithmic Injustices: Towards a Relational Ethics," presentation at Black in AI Workshop, NeurIPS Conference, Vancouver, BC, 2019.

46 Cathy O'Neil, *Weapons of Math Destruction: How Big Data Increases Inequality and Threatens Democracy*, New York: Crown, 2016; Safiya Umoja Noble, *Algorithms of Oppression: How Search Engines Reinforce Racism*, New York: New York University Press, 2018.

of thousands of youngsters were hit hard by an algorithm developed by Ofqual, an educational body which administers secondary-school exams, among other things. After a number of students were wrongly failed, they protested and obtained a review of the process. In the same vein, radical critiques of prison capitalism reinforced by algorithmic power are articulated around struggles such as Data4BlackLives or the broader data justice movement.[47] Here we see that struggles against algorithmic injustices are combining with social movements such as #MeToo and #BlackLivesMatter, which intervene in the media-political sphere by using digital tools to target injustice. We are still seeing a two-sided trend at work: one in the form of institutional protections aimed at protecting privacy against algorithmic capital's encroachments; and the second embodied in various forms of resistance to discrimination in spheres affecting daily life, public institutions, and social relations.

One final frontier struggle not to be overlooked is the contradiction between algorithmic capital and nature, as AI accelerates the climate crisis. We saw in Thesis 15 that algorithmic technologies are often put forward as "solutions" to various environmental problems. And yet a growing number of people are beginning to voice more open and direct criticism of technological solutionism and even leaving their professional occupations altogether to join the resistance and create alternatives. The environmental movement is made up of a predominantly young, urban, feminized, highly educated population, with a left-wing political inspiration, strong pessimism about the future, and a relative distrust of technology. Indeed, according to a recent French survey, 85 percent of citizens interested in the environment believe that civilization will collapse in the coming years, and 73 percent believe that technological advances are worsening the environmental crisis.[48] Among this group, there are a growing number of students, scientists, and engineers who are calling for people to leave their jobs or change careers for environmental reasons.[49] Journalists Fabien Benoit and Nicolas Celnik have noted an overrepresentation of computer scientists, AI researchers, and engineers in technocritical collectives and radical environmentalist groups. After all, these individuals have experienced

47 Jackie Wang, *Carceral Capitalism*, Los Angeles: Semiotext(e), 2018; Linnet Taylor, "What Is Data Justice? The Case for Connecting Digital Rights and Freedoms Globally," *Big Data & Society* 4, no. 2, 2017, pp. 1–9.

48 Chloé Alexandre, Florent Gougou, Erwan Lecoeur, Simon Persico, *Grande Enquête "Mouvement climat" Sciences Po Grenoble: Université Grenoble Alpes*, PACTE, September 2021.

49 Benoit and Celnik, *Techno-luttes*, p. 79.

significant cognitive dissonance in their work as a result of the tensions between the technical dimension of their professions, their values, and the imperative of profitability.

As well as the growing number of dissident engineers and scientists who no longer subscribe to solutionist myths, a growing number of environmental organizations (Extinction Rebellion, Friends of the Earth, ATTAC, Atelier Paysan, the *zadistes*, and Earth Uprising) are making increasingly strong criticisms of green growth, "Green Tech" discourse, and the technocentric vision of the energy transition. Alongside criticism of technoindustrial society, there have been numerous direct actions involving various forms of sabotage, the destruction of 5G antennas, the dismantling of Linky smart meters, and the blocking of industrial infrastructure such as mega-basins, drilling plants, and other "useless large-scale projects that have been imposed." At the start of the COVID-19 pandemic in 2020, a wave of sabotage against antennas hit several countries around the world. In France, more than 174 5G antennas were attacked in less than a year, the largest sabotage campaign since the destruction of GMO fields in the early 2000s.[50] These acts could be seen as irrational, fueled by the spread of conspiracy theories; however, there are several reasons to oppose 5G, which is primarily aimed at facilitating the rollout of the Internet of Things, autonomous cars, and the full development of algorithmic capital. In reality, conspiracy theorists are often anticapitalists without knowing it.

Finally, environmental struggles against algorithmic capital mainly take the form of sabotage aimed at neutralizing various tools, devices, and infrastructures that encroach on living environments, communities, and living beings. As Samuel Lamoureux notes, sabotage does not take a single form (as the destruction of physical objects). Lamoureux distinguishes among three variants:

> Classical sabotage (which is tending to disappear from workplaces but reappearing in contemporary art or performance), subtle sabotage (which takes the form of hijacking or reappropriating algorithmic platforms for alternative purposes), and resistance to technoscience (in everyday life outside the factory, for example resistance to acceleration, the Internet of Things, the ideology of permanent connection, etc.).[51]

50 Ibid., p. 177.

51 Samuel Lamoureux, "Penser le sabotage à l'ère du capitalisme numérique," *Réseaux* 231, no. 1, 2022, p. 140.

While the destruction of 5G antennas is part of the logic of "classical sabotage," various forms of subtle sabotage and stealthy resistance aim to limit algorithmic capital's grip: from artists hacking surveillance cameras in urban spaces to neo-Luddite collectives carrying out direct actions to disrupt technological devices, to various environmental groups such as La Ronce that propose forms of "micro-sabotage" to "defend life" and "bring down the multinationals that are destroying it." As they put it, "opening bottles of Roundup Coca-Cola in supermarkets, disabling payment terminals at Total petrol stations, painting over QR codes on electric scooters, or deflating the tyres of SUVs."[52] The rapid popularization of the rhetoric and practices of sabotage among climate activists, particularly through works such as Andreas Malm's *How to Blow Up a Pipeline*, is now spreading into the fight against technopolicing devices, agribusiness, and other infrastructures of algorithmic capital.[53]

While making no claim to be exhaustive, we have attempted here to formulate an initial outline of resistance in the era of algorithmic capitalism. Class struggles are embodied in opposition to platforms and in the mobilization of algorithmic labor against the dynamics of exploitation and precarity. At the same time, new struggles are emerging in response to the encroachment of algorithmic logic on everyday life, the worsening ecological crisis, and the amplification and automation of injustices by technologies based on Big Data and predictive analytics. Clearly, there are multiple fronts in the struggle against contemporary capitalism.

Strategies for exiting capitalism

If we want to propose concrete pathways to overturn the established order, a simple overview of the current struggles against algorithmic capital is only a first step. We also need to identify a way forward and formulate strategic guidelines capable of uniting active social forces in order to create breaches within the system and free up space for the alternatives that already exist in scattered form within it. This is complex both in practice and in theory. Indeed, how can we conceive of the "outside" of capital? And what would a *politics of the outside* look like?[54] Gavin Walker's theoretical reflections contain several fruitful ideas in this regard, including the need to conceive of a genuine political *elsewhere*,

52 Benoit and Celnik, *Techno-luttes*, pp. 183–4.
53 Andreas Malm, *How to Blow Up a Pipeline*, London: Verso, 2021.
54 Gavin Walker, *Marx et la politique du dehors*, Montreal: Lux, 2022.

beyond the *outside* of capital. It is necessary to rethink the body of this politics, to think differently about organizations and the concept of the "party," to allow for emancipatory collective political action. In what follows, our intervention is focused on the range of practical anticapitalist strategies. We want to contribute to a discussion rather than close ourselves off in some particular strategy that we advocate and defend tooth and nail. To this end, we propose to draw on the framework developed by Erik Olin Wright in his *How to Be an Anticapitalist in the Twenty-First Century*. Wright distinguishes five major "strategic logics" drawn from the history of anticapitalist struggles: smashing, dismantling, taming, resisting, and escaping capitalism.[55]

We can distinguish between these various strategies in terms of the goals that are set for social struggles and what level of the system they are targeting. In the case of the strategy that aims to smash capitalism through revolutionary rupture, the goal is to transcend this system and its structures. The aim is not to play within the existing parameters, but to change the game itself. It is also possible to change some rules of the game: For the strategy aimed at domesticating or taming capitalism, the aim is to mitigate the problems that it causes by way of institutional reforms, whereas the strategy of dismantling it aims to build a new system through the gradual transformation of the old one. Finally, other strategies can operate through the striking of "blows" or "collective actions" launched within capitalism. Meanwhile, the approach of resisting capitalism fights, through various forms of protest, against the damage that this system engenders. Finally, the perspective of escape seeks to create social organizations and practices that operate beyond capitalism. They aim, as Marcuse put it, "to transform individual and collective consciousness by experimenting new forms of life, before the establishment of a new system."[56] Wright formulates a useful typology to illustrate the differences between these multiple strategies.

We have no interest in getting bogged down in endless debates about what would be the best strategy in absolute terms. Instead, we will offer an overview of these perspectives and then consider combinations or strategic configurations that would allow them to be linked together in a desirable and viable process of social transformation.

55 Erik Olin Wright, *How to Be an Anticapitalist in the Twenty-First Century*, London: Verso, 2019, chapter 3.

56 Herbert Marcuse, "Entrevue avec Herbert Marcuse," *Le Nouvel Observateur*, January 1973.

Table 18.1 Typology of anticapitalist struggles

		Objective of Struggle	
		Neutralizing harms	**Transcending structures**
Level of the System	**The game itself**		*Smashing*
	Rules of the game	*Taming*	*Dismantling*
	Moves in the game	*Resisting*	*Escaping*

Escaping capitalism

In the context of algorithmic capitalism, there is much promise to be found in the strategy of experimenting with new forms of organization, software, platforms, ways of producing, exchanging, and living outside the domination of capital. As early as the 1980s, programmer and activist Richard Stallman was promoting free software and the GNU General Public License, called "copyleft." These came together with other measures likewise aimed at countering the hegemony of digital giants, as represented by the proprietary model and attempts to enclose the digital world through intellectual property. In his essay *Free Software, Free Society*, Stallman also formulated an inspiring and noteworthy defense of free software, in contrast to the open-source model, which is more compatible with proprietary logic and capitalist exploitation.[57]

Furthermore, a resurgence of commons in the digital sphere has been underway since the 2000s; some even argue that this approach could form the basis of a new, peer-to-peer-based collaborative mode of production.[58] The widespread popularity of Creative Commons licenses, created and promoted by lawyer Lawrence Lessig, and the global success

57 Richard Stallman and Joshua Gay, *Free Software, Free Society: Selected Essays of Richard M. Stallman*, New York: SoHo Books, 2010.

58 Charlotte Hess and Elinor Ostrom, eds., *Understanding Knowledge as a Commons: From Theory to Practice*, Cambridge, MA: MIT Press, 2011; Pierre Dardot and Christian Laval, *Commun: Essai sur la révolution au XXIe siècle*, Paris: La Découverte, 2014; Yochai Benkler, *La richesse des réseaux: Marchés et libertés à l'heure du partage social*, Lyon: Presses universitaires de Lyon, 2009; Michel Bauwens and Jean Lievens, *Sauver le monde: Vers une économie post-capitaliste avec le peer-to-peer*, Paris: Les Liens qui libèrent, 2015.

of Wikipedia illustrate the strong potential of digital commons, which seem capable of challenging the capitalist paradigm.[59] Often predating the rise of capitalist platforms, the platform-cooperativism movement includes hundreds of initiatives around the world, such as Fairbnb, Resonate, Savvy Cooperative, Stocksy United, Eva, and Projet collectif, which represent alternatives to Airbnb, Uber, Facebook, and other dominant platforms. In terms of generative AI, Meta's launch of its large language model LLaMA as open source in February 2023 has led to the emergence of dozens of new algorithmic models that are lighter and more powerful, and which can run locally on a PC.[60] This is troubling for firms such as OpenAI, Microsoft, and Google, which no longer have a monopoly on this technology if users can now more easily work with it themselves.

That said, three factors are hindering the spread and the upscaling of these alternatives and initiatives. First, the network effect means that the critical mass of users of platforms such as Facebook, Airbnb, and Uber outstrips their coöperative or nonprofit competitors. Platforms based on the cooperative or commons paradigm thus have less visibility and fewer human and financial resources with which to compete with the web giants. For example, after Elon Musk took over Twitter in October 2022, his eccentric management of the platform (impulsive mass layoffs, reactivation of controversial accounts, and improvisation around the app's features) prompted thousands of users to desert it in favor of the free and decentralized social network Mastodon, which grew from 300,000 users to 2.5 million in under two months.[61] Yet enthusiasm for Mastodon faded after a few months, mainly due to this platform being somewhat less user-friendly, the moderation work being mainly performed by volunteers, and the presence of fewer people to interact with compared to its larger competitors.[62] Commons and open-source AI remain at the moment small islands in a vast capitalist ocean.

Second, the lack of legislation and support from public authorities to encourage noncapitalist digital alternatives means that the current system favors capitalist players. As we have seen, it is these latter who are building up partnerships with cities and states. These two antagonistic

59 Lawrence Lessig, "Creative Commons," *Montana Law Review* 65, no. 1, 2004, pp. 1–13.

60 Cade Metz and Mike Isaac, "Meta ouvre son code source à tout le monde," *La Presse*, May 19, 2023.

61 Matt Binder, "Mastodon Has Gained Millions of New Users Since Elon Musk Bought Twitter," *Mashable*, December 20, 2022.

62 Amanda Hoover, "The Mastodon Bump Is Now a Slump," *Wired*, February 7, 2023.

models are thus fighting on unequal terms: Algorithmic power favors capitalist platforms, instead of supporting initiatives guided by principles of democratic governance, social justice, and ecological transition.

Third, the digital commons model is mainly found in the field of software and platforms, rather than in hardware infrastructure (cables, data centers, antennas, satellites), data management, and algorithms. It seems that the initial enthusiasm for free software and digital commons arose from the first phase of digital capitalism (1980–2007), raising hopes that commons could develop on a large scale. However, with the advent of algorithmic logic, Big Data, and the machine learning revolution, it is now capital that holds the upper hand. Even with the arrival of open-source generative AI, the vast majority of users are likely to keep using Gemini, Copilot, and ChatGPT because of their ease of access, their integration with existing digital tools, their multiple plug-ins, and the millions of users who are constantly improving them. If the commons do not take control of the infrastructure, data, and algorithms, then a smattering of free software programs or cooperative platforms will never be able to supplant the GAFAM companies, which already have a significant head start.

A fully fledged digital commons model would integrate four levels: hardware, platforms, software, and data. For hardware, initiatives already exist. In Quebec, the Antoine-Labelle Telecommunications Cooperative is interesting: the regional county municipality owns the optic fiber infrastructure, and Internet and phone services are provided by a solidarity cooperative composed of member-workers and customers. This kind of public commons partnership could be multiplied, or upscaled, at a regional or even national level. Data centers can also be integrated into a commons logic. The Māori Data Sovereignty Network is an example of such strategy to assert the sovereignty of a community over its data.[63] According to the Democratic and Ecological Digital Sovereignty Coalition, we could create a "public-led democratic digital stack" including a digital infrastructure, public platforms (including AI models and search engines), and a public market for computer services.[64]

Regarding the second level, the one of platforms, many alternatives already exist. The Fediverse social networks are an example (with their

63 "World-First Data Storage Infrastructure Solution Built by Iwi Māori, for Iwi Māori," *Waatea News*, June 4, 2024.

64 Cecilia Rikap et al., "Reclaiming Digital Sovereignty: A Roadmap to Build a Digital Stack for People and the Planet," Democratic and Ecological Sovereignty Coalition, ucl.ac.uk.

limitations as well), as is FreedomBox, a community project allowing the use of a personal server based on free software, and even Nextcloud, which offers a platform for collaborative work and storage similar to Google Workspace or Microsoft 365. We can similarly imagine e-commerce platforms run as a public service, with algorithms favoring local buying.

The third level is software. In his inspiring book *Digital Degrowth*, Michael Kwet analyzes many of these "digital alternatives."[65] Even if they do not all fit under the category of digital commons, their large number shows the extent of existing alternative technologies. Kwet argues that these carry at least four major advantages. First, they show that we do not have to rely on GAFAM for innovation and efficient solutions. Free software is not just a gadget for a technophile elite but can offer robust and secure alternatives. And these abound already: Linux for computers, Lineages for smartphones, and F-Droid, an alternative app store, have proven their worth. Second, open-source software represents a form of guarantee against hidden forms of surveillance, mainly through its transparency. Apps such as Signal and Jami offer a service where security and confidentiality are prioritized. Third, open-source software carries the potential to replace Big Tech products. Alternatives such as Libre-Office (in lieu of Microsoft Office), or GIMP (in lieu of Photoshop), are not always perfect, but they can help free users from the digital chains imposed by capitalist enterprises. Fourth, these alternatives, if properly supported by collectives, can upscale.

The fourth level is the one of data commons, which can take many juridical forms. Data co-ops allow for collective ownership and democratic management of data, as in the example of the MIDATA in Switzerland, Driver's Seat in the US, or Salus Coop in Barcelona. In Quebec, the model of the public utility trust, a juridical form that allows for the ownership and management of goods in the interest of the public, can be a promising form for socialization of data ownership and management.

In summary, we can make an observation akin to Marx's assessment of cooperatives back in his day. On the one hand, he saw the cooperative movement as a practical alternative to capitalism, a concrete example of a free association of producers and a new way of working. Nevertheless, he also recognized that, despite its practical qualities, cooperative work scattered across so many small islands in the capitalist ocean cannot

65 Michael Kwet, *Digital Degrowth: Technology in the Age of Survival*, London: Pluto Press, 2024.

curb monopolies or emancipate society. This ambivalent judgment can readily be applied to noncapitalist alternatives such as digital commons, free software, and cooperative platforms, which will not alone succeed in transforming society on a large scale. These are "real utopias," prefigurations of another society, but confined within a system that can easily tolerate or even exploit them, with the commons sometimes becoming entangled in capitalist logic. Capitalist platforms benefit greatly from the presence of the commons: They can draw on, contribute to, and even exploit certain aspects of them, while then marketing certain innovations under their own brand.[66] That said, it would be dangerous to discard such forms on these grounds; rather, we should see them as a lever, a "first breach in the system" as part of a broader strategy of transitioning to a different one.[67]

We can see here that the strategy of experimenting with new forms of cooperative organizations, commons, and emancipatory social innovations cannot bear full fruit without a radical perspective of overturning the system, through more direct confrontation and struggle with the state and capitalist platforms. Although Marcuse emphasized that "experimenting with new forms of life" must take place "before the establishment of a new system," it seems clear that "spontaneous action by the masses" and "institutional changes decreed by central authorities" are also necessary conditions for overcoming capitalism.

Smash, dismantle, and/or tame capitalism?

The overcoming of capitalism has never been achieved through a slow process of transition or spontaneous evolution. Still less has it been achieved through Jeremy Rifkin's proposed method: a supposed "eclipse" in the wake of a new industrial revolution, the rise of collaborative commons, and decentralized renewable energies.[68] The only exits from capitalism thus far have been the result of a conscious overthrow fed by dogged and prolonged popular struggles, a major social crisis, and a certain amount of unpredictability that revolutionary forces have been able to take advantage of. As Lenin pointed out: "Only when the 'lower

66 Dardot and Laval, *Commun*, pp. 165–88.

67 Karl Marx, cited in T. Lowit, "Marx et le mouvement coopératif," *Cahiers de l'Institut de science économique appliquée*, no. 129, September 1962, p. 92.

68 Jeremy Rifkin, *The Zero Marginal Cost Society*, New York: Griffin, 2015, and *The Green New Deal*.

classes' do not want the old way, and when the 'upper classes' cannot carry on in the old way—only then can revolution triumph."[69]

That said, the project of smashing capitalism is fraught with pitfalls, as many tragic historical episodes have shown. The strategy of social transformation through rupture consists of a sudden and rapid overthrow of this socioeconomic regime through the conquest of state power and the establishment of a new system.[70] This change can be brought about "from above" (through the nationalization of the means of production, confiscation of private property, and so on) and/or "from below": through popular insurrection, expropriation by the masses, workers' councils, and other forms. Wright is rather skeptical about the emancipatory possibilities of this approach: While it did succeed in overthrowing capitalism in certain exceptional episodes in the past almost all of these revolutionary ruptures unfortunately turned into dictatorships and other variants of authoritarian socialism. Various factors may be cited to explain such a reversal, including difficult historical conditions (war, famine); the centralization of power in the hands of revolutionary leaders; the repression of the revolution by armed, reactionary, and imperialist powers; and the high degree of complexity and unintended consequences associated with the process of abruptly overthrowing a social system.

In *Real Utopias*, Wright shows in particular that the "transitional slump" linked to the revolutionary context almost always involves a major upheaval in the population's living conditions (capital flight, inflation, pressure from foreign powers, counterrevolution, and so on). All this makes it an uphill struggle to maintain the political support of the middle and working classes, who are tempted to turn back the clock.[71] This is why radical left-wing governments face a tragic dilemma during this period of prolonged turmoil: either they risk losing competitive elections, paving the way for the restoration of conservative forces, or else they sacrifice democracy and elections in order to "save the revolution," thus paving the way for authoritarian statism. This second path was taken by those like Lenin, Mao, and Castro, who abandoned radical democracy and self-managing popular councils in favor of a centrally planned economy. Fortunately, this process is not inevitable, as evidenced by other revolutionary experiences inspired by self-management and libertarianism, such as the Spanish social revolution of 1936, the

69 V. I. Lenin, *"Left-Wing" Communism, An Infantile Disorder*, Paris: Foreign Languages Press, 2021, p. 81.

70 Wright, *How to Be an Anticapitalist in the Twenty-First Century*.

71 Erik Olin Wright, *Envisioning Real Utopias*, London: Verso, 2010.

Zapatista revolution of 1994, and the Kurdish revolution of 2012 in Syria's Rojava region. The problem is that these experiences were either bloodily crushed (the Spanish Civil War) or remained entrenched in a militarized context that prevented them from spreading (as in the cases of Chiapas and Rojava). In any case, it is important to bear in mind that this strategic perspective still exists and represents a real possibility for different groups seeking to overthrow capitalism.

What would a strategy to smash algorithmic capitalism look like today? We should say from the outset that this is not all that likely as a short-term scenario, given the scale of the global industrial infrastructure supporting this system. Algorithmic capital is mainly concentrated in two superpowers, the United States and China, so a revolutionary break would have to occur in one of these two countries (or preferably in both) to have a real global geopolitical impact. Of course, we could envisage upheavals in lesser powers such as France, Canada, Brazil, India, or Russia, but here again there do not seem to be any revolutionary political forces ready to overthrow capitalism overnight. That said, the significant wave of popular uprisings in 2019 (in Chile, Lebanon, Hong Kong, Catalonia, Algeria, and Iran) shows that sometimes a single spark can cause the powder keg to explode. It is then necessary to know what to replace the existing regime with and learn to seize the historical moment.

What would a revolutionary policy aimed at abolishing algorithmic capital look like? Would it involve destroying data centers, expropriating GAFAM through legislation or popular uprisings, or collectivizing the digital economy on a large scale by forcing start-ups to convert into solidaristic co-ops? As we will see in Thesis 20, it is even possible to envisage a certain degree of algorithm-driven economic planning, which may be more or less centralized or democratic depending on the circumstances. But let us leave this hypothesis aside for now and turn to a milder, but no less radical, form of social transformation.

As well as the option of abruptly overthrowing capitalism, it is possible to imagine a series of powerful institutional actions aimed at gradually transcending this social order. This is what Wright calls the dismantling of capitalism. It involves a prolonged transition phase in which capitalist and socialist relations coexist within a mixed economy. Private banks and businesses, the labor market, and capitalist investment would coexist with state banks, public firms, civil servants, and state planning. The state is a highly activist force in this scenario: It literally stages the dismantling of capitalism and its replacement by socialism in a campaign that has been meticulously planned down to the last detail.[72] Wright puts forward this hypothesis of dismantling capitalism in contrast to the strategy of revolutionary rupture. However, both strategies have the same objective—to transcend the structures of capitalism. The difference lies in the speed of social transformation: a brief and violent process on the one hand, and a longer and perhaps less turbulent process on the other. That said, it seems clear that the political actions taken to overthrow capitalism could be similar in both cases: socialization, expropriation, the creation of new large-scale public institutions and multiple forms of support for self-management, the social and solidarity economy, and the commons. Once again, there are questions as to the pace or brutality of this process: While revolutionary Marxists' classic perspective relied on a mix of tactics linked to insurrection, participation in elections, infiltration, armed struggle, guerrilla warfare, and other actions of a similar nature, the dismantling strategy advocated by Wright takes the form of "radical reformism" implemented through a democratic conquest of political power. However, both the strategy of dismantling capitalism and the strategy focused on its sudden overthrow face the same problem: If the capitalist class is dissatisfied, it will organize to defend its interests, and shrewd approaches will certainly have to be found to subdue that class and prevent a return to the past.

What would a strategy for dismantling algorithmic capitalism look like? We can look somewhere familiar to us, Quebec, keeping in mind that this discussion can serve as formal inspiration for other spaces around the world. First, private telecoms oligopolies (such as Videotron and Bell) would have to be nationalized and a state-owned corporation similar to the power firm Hydro-Québec would have to be created, as proposed by the Québec Solidaire party with Réseau Québec. This would reduce the population's bills—indeed, the Internet and phone rates paid in Canada and Quebec are among the world's highest. But it would also be necessary to create a large public sector cloud-computing infrastructure, instead of transferring data and services to Microsoft and Amazon servers as is done currently. Next, it would be necessary to put obstacles in the way of GAFAM and other capitalist firms: laws banning Uber and Airbnb, the promotion of noncapitalist digital platforms (such as Projet collectif and Eva), public institutions moving across en masse to free software and the commons, and many other initiatives. It would also be possible to create social utility trusts for data control in order to nurture a

72 Wright, *How to Be an Anticapitalist in the Twenty-First Century.*

Table 18.2 Nine principles for municipal-level technological sovereignty

1. Promote alternative data ownership regimes
2. Transfer IT services to open-source code, open standards, and flexible management
3. Transform procurement to make it ethical, sustainable, and innovative
4. Put controls on (capitalist) digital platforms
5. Build and enhance alternative digital infrastructures
6. Develop cooperative models for public services
7. Maximize innovation with public value
8. Rethink redistribution systems and complementary local currencies
9. Promote digital democracy and technological sovereignty

Source: Evgeny Morozov and Francesca Bria, "Rethinking the Smart City: Democratising Urban Technology," New York: Rosa Luxemburg Stiftung, 2018, p. 30, rosalux-nyc.org

dynamic ecosystem conducive to transcending algorithmic capital.[73] This form of radical reformism would embody "technological sovereignty" at the local and/or national levels. Evgeny Morozov and Francesca Bria adopt such an approach in their critical report on the smart city (Table 18.2), which lies halfway between a strategy aimed at dismantling capitalism and one seeking to tame it.

There is also the more traditional path of social democracy, which seeks to "tame" capitalism or to "rein it in." This trend can be seen in most of the parliamentary left-wing parties in the latter half of the twentieth century and the beginning of the twenty-first, as they gave up on overcoming capitalism and instead put a priority on mitigating its disruptive effects. This approach aims to tame capitalism by regulating the economy through a more interventionist, progressive, and redistributive administration. Today, proposals along the lines of the Green New Deal and the vast majority of figures on the parliamentary radical left across the Global North (such as Bernie Sanders, Alexandria Ocasio-Cortez, Yanis Varoufakis, Jean-Luc Mélenchon, and Pablo Iglesias) follow this strategic course.

Unlike strategies of rupture and dismantling capitalism, the reformist approach focused on building a welfare state did bear fruit, at least during the *Trente Glorieuses* (the post-1945 decades of state-regulated capitalism). However, this process was interrupted in the 1970s, a pivotal decade that brought the onset of globalization, the financialization of the economy, and the advent of neoliberalism. Today, the exhaustion of

73 Jessica Leblanc, "Gouvernance des données: La fiducie d'utilité sociale, un outil à fort potentiel," Territoires innovants en économie sociale et solidaire (TIESS), 2022.

neoliberal ideology is giving way to a revival of the social-democratic ideal; it seeks to reinvent itself by confronting the central issue of climate crisis and, more peripherally, the rise of algorithmic capital. To tame the disruptive effects of algorithmic capital, some voices are calling for the adoption of new rights, such as the enshrinement of the "right to digital integrity" in charters of rights and freedoms:

> As an extension of the right to life and physical or moral integrity, this emerging right would be the foundation on which all digital rights (the right to connect and disconnect, the right to be forgotten, the right to AI assistance, etc.) could be based; these rights could be asserted as a necessity for our freedom and fulfilment in a digitalized society.[74]

Vladimir De Thézier thus advocates that this right to digital integrity should be combined with a "right to a career transition plan" and an "employment guarantee" program to address the challenge of automation and robotization of jobs. This package could moreover integrate a "right to free essential goods and services" such as the Internet, higher education, and public transport.

A key strength of the strategy of taming capitalism lies in its ability to achieve concrete gains (sometimes minor, and sometimes significant) through reforms, laws, social programs, and class compromises that can benefit capitalists, the middle classes, and the working classes alike. That said, its main weakness lies in its inherent inability to transcend the capitalist system and therefore remain subject to its dictates when they assert themselves. One good example is the government led by Greece's left-wing party Syriza, elected in January 2015. Prime Minister Alexis Tsipras raised high hopes about his ability to renegotiate the country's public debt and halt austerity measures. Yet he was then forced to renege on his promises and betray the popular will as expressed by referendum, in response to the constraints imposed by Greece's creditors, who threatened to force Greece out of the EU if he refused. Tsipras could have opted for a "Plan B," as suggested by Finance Minister Yanis Varoufakis—a Grexit strategy that would have paved the way for a major (potentially revolutionary) crisis; but he decided not to venture into this unknown terrain, which surely would have brought severe turbulence but potentially radical transformation. Tsipras instead preferred to mitigate the ills of neoliberal austerity while remaining within the parameters of the

74 Vladimir De Thézier, "Mode d'emploi pour faire face à la révolution des robots," *Le Devoir*, February 2, 2023.

system. Here we see, once again, that the defenders of capital have more than one trick up their sleeves when it comes to overthrowing or taming left-wing governments that try to break the rules of the capitalist game.

A comprehensive strategy, in a context of upheavals

An analysis of each individual strategy for social transformation makes clear that there is no one magic formula for overcoming capitalism. That is because the seizure of power itself stirs up a new movement of history; it "is always a historical experience of reinventing political capacities, of creating *new antagonisms* absent from the previous power arrangement."[75] Faced with this political and historical novelty, we need, on the one hand, political organizations capable of persistence, as Gavin Walker argues, but also a form of flexibility and strategic adaptability. In this sense, a combination of different fronts of struggle offers a richer strategic path. Erik Olin Wright stresses the need to consider possible combinations of approaches: the creation of postcapitalist alternatives (cooperatives, commons, real utopias) but also resistance movements, institutional reforms aimed at taming capitalism, and more radical measures aimed at building a new society (dismantling). He describes this strategic configuration as the "erosion of capitalism," emphasizing the need to loosen the capitalist stranglehold while also introducing new practices, organizations, and ways of life capable of building a new world in the interstices of the old one. This perspective is based on a vision of society that resembles less a totality than an "ecosystem" capable of evolving toward a new equilibrium. This perspective highlights the importance of inserting emancipatory, noncapitalist economic activities into the heart of the "capitalist ecosystem" and then protecting them and developing them further. As these niche activities develop, they will then be able to transform the ecosystem itself.[76]

However, in our view, Wright remains trapped within the evolutionary horizon of "radical reformism," which does not conceive of society as an open, complex, and contradictory whole, but rather as a "system" made up of components that can be altered or replaced one by one. He rejects outright the prospect of revolutionary rupture as an option that inevitably leads to authoritarianism. He thus ends up having to combine strategies from social democratic, libertarian, and alter-globalization

75 Walker, *Marx et la politique du dehors*, p. 392. His emphasis.
76 Wright, *How to Be an Anticapitalist in the Twenty-First Century*, p. 56.

circles. This path may seem promising, in an era when many consider the idea of revolution outdated. And yet, proposals for a more prolonged transition to a new system seem unlikely to deliver on the promise of greater realism when faced with the climate emergency and the troubling prospect of an outright collapse. Indeed, such a threat is also increasingly entering the consciousness of social movements and the general population.

We thus propose to supplement Wright's analysis with the strategic perspective formulated by Jérôme Baschet.[77] Unlike the concepts of revolution or collapse, which make us think of a single, mechanical, and inevitable trajectory, the image of a tipping point suggests a profoundly unstable world, marked by numerous reversals, metamorphoses, uncertainties, and accelerations of history, which can lead to various potential trends of regression or liberation. The outcomes of this complex intertwining of crises remain largely uncertain. They may lead to apathy and to the consolidation of authoritarian regimes—but also to a series of popular uprisings, possible larger-scale social transformations, the creation of liberated spaces, and a genuine political *elsewhere*. Compared to ideas such as "transition" or "erosion," which evoke a slow historical process of gradual change, the image of "tipping points" opens the way to emerging worlds and desirable possibilities, driven by a different vision of the good life. Baschet adopts Wright's terminology but takes it further. The interstitial changes (the escapes from capitalism) consist of creating "liberated spaces," whether that means self-managed micro-collectives, larger economic structures, ZADs (*zones à défendre*, literally "zones to be defended," as in France), rebel communes, or even entire regions experimenting with collective autonomy (such as Chiapas and Rojava). Baschet then proposes combining liberated spaces with approaches centered on blockades (resisting capitalism) and large-scale uprisings (smashing capitalism). Far from limited to local-level struggles, "the notion of the blockade" may be used to

> associate multiple forms of struggle: blocking production (through strikes or occupation of facilities), blocking the flow of traffic and information, blocking consumption, blocking social reproduction (for example, with school strikes, which suspend the duty to integrate into the world of destruction), etc.[78]

77 Jérôme Baschet, *Basculements: Mondes émergents, possibles désirables*, Paris: La Découverte, 2021.

78 Ibid., p. 201.

In addition to these multiple blockades, it is necessary to articulate potential popular uprisings around the creation of autonomous local assemblies, the seizure of the means of production, and the dismantling of destructive infrastructures (including, but not limited to, fossil fuel infrastructure). This evokes the idea of "dual power" as considered by Lenin, Luxemburg, Trotsky, and Bookchin, and points to a revolutionary situation that is not necessarily synonymous with the conquest of state power, but rather involves a period of flux, in between the old, crisis-afflicted institutions and the emergence of new ones (such as councils, communes, or liberated spaces) that could combine to replace the established economic and political structures. Baschet does not dismiss outright the need for radical reforms here and there (taming capitalism). But he sees this as a tactic or a potentially useful means in certain circumstances, and not as an end in itself. Reforms are simply tools, not the pillar or core of a strategy for social transformation, which will be achieved more "from below" through popular self-organization, the creation of liberated spaces, commons, the federation of communes, and collective initiatives.

This perspective resonates with the current sociohistorical context, which seems less favorable to a gradual erosion of capitalism than a scenario in which the world is struck by major social, economic, climatic, and technological upheavals. An approach in terms of "upheavals" seems better able to feed the imaginary of social experiments linked to the commons, permaculture, free software, and low tech—as well as the challenge to police violence, algorithmic alienation, and mass surveillance.

But what would overcoming algorithmic capitalism look like, in a world shaken by many upheavals? Here we see the need to develop a strategy for social transformation that combines different fronts of struggle, in terms of protest movements (resisting capitalism), the social and solidarity economy (escaping capitalism), and political action (aimed at dismantling capitalism). This means the liberation of space on the basis of the commons, blocking the processes of algorithmic capital, and supporting large-scale popular uprisings. This also means the liberation of time, disconnecting individually and collectively from algorithmic capital's technologies. This seems to us to be the best strategic configuration in the short, medium, and long term. This perspective partly echoes Wright's strategy of erosion, without embracing his reformist evolutionism or his idea that "civil society" (in a generic sense) should exercise more power over the state and the market. Rather, it means promoting an "integral strategy" in which many struggles converge, as advocated by Alain Savard:

> The general strategic perspective of building power from below is therefore a broad project aimed at accumulating power within relatively autonomous member-controlled organizations in all working-class communities. We have mainly focused on social movement organizations, such as labor unions, student associations or community organizations. Yet, power accumulation within the social economy and political parties is also necessary to the consolidation of the infrastructure of dissent and to undertake a revolutionary process. Integrating those three "poles" of power accumulation into a unified strategy is what pushes me to label it an "integral" strategy.[79]

Thus, when we say that it is necessary to support the ongoing struggles against algorithmic capital, this support will need to take place on multiple fronts. It means backing worker struggles and the boundary struggles of social movements and community organizations, but also supporting and radicalizing certain organizations that have the potential to transcend capitalism (the social economy, the commons), and likewise supporting left-wing parties (be they social democrats, socialists, or revolutionaries) that dare to confront the siliconization of the world, the platforms' hegemony, algorithmic domination, and the other injustices identified in this book. It is also important to keep up reflection on the possible new forms of organizations, parties, and commons.[80] It is surely difficult if not impossible to determine, at a theoretical level, what will be the exact combination of these fronts of struggle. For that will be primarily the result of revolutionary practice, concrete experience, and activist work aimed at building coalitions and creating infrastructures of dissent.

How can we nurture these concrete activist practices, aimed at developing solidarity in action, but also engage in really effective political action, and experiment with new forms of social organization able to embody a different vision of the good life? The answer to this question is far from obvious. Yet we believe it is possible and necessary to reflect on the ethical conditions of protest action, prefigurative practices, and democratic life, freed from the yoke of algorithmic capital.

---

79 Alain Savard, "Power from Below: Power, Democracy and Socialist Theory," PhD thesis, York University, 2020, p. 258.

80 Walker, *Marx et la politique du dehors.*

Thesis 19

We must find the virtues in technologies and criticize their vices

I cannot affirm my own life without critically evaluating those structures that differentially value life itself.

—Judith Butler

Their morals and ours

Back in Thesis 1, we discussed the pitfalls of an "ethical mirage." By that phrase, we referred to boosterism about the "responsible development" of AI through the formulation of moral principles that are supposed to guide engineers, data experts, and companies producing algorithmic technologies. We discussed a fundamental divergence between the dominant approach to AI ethics, which focuses on specific problem-solving, and a critical theory of algorithms that pays due attention to socioeconomic context and relations of domination. Does this mean that any ethical reflection on new technologies is bound to be pointless—a thin moral veneer on digital platforms, listing statements of principle with no practical impact on algorithmic capital and its real development?

We said earlier that "we are not advocating the complete abandonment of AI ethics" but that "ethics extends far beyond the issues of computer programming: It is a matter of reflecting on the general conditions for living well, whether in terms of practices, social interactions, institutions, or ways of life." For us, the fundamental difference between AI ethics and the critical theory of algorithms does not lie in the fact that the former is normative while the latter is essentially negative and critical of the dominant power relations. Rather, they each involve different interpretations of the role of ethics and the interests that ethics should primarily serve. In the current prevailing approach, ethics serves the organizations, scientists, capitalists, and governments that develop and use algorithmic machines. It aims to mitigate biases, perverse effects, and the abuses and misuses of these technologies—and to direct them, as far as possible, toward the common good. Ethics then plays the role of "advisor to the Prince," whether that Prince takes the form of a government, GAFAM, a start-up, the police, or some other body.

As against this, we argued that "critical AI theory must shed light on the emancipatory potential, prospects for action, and social alternatives present in today's world." In other words, ethics must first and foremost guide collective action, social struggles, the creation of real utopias, and institutional reforms that will contribute to liberation and collective well-being both in and beyond the algorithmic world. Instead of claiming to develop an impartial and "reasonable" ethics, as if this could float above particular interests and be acceptable to all, this normative perspective is supported by current struggles. It "begin[s] with the lived experiences of those who are disempowered, discriminated against, and harmed by AI systems."[1]

But even having taken such a stance, what kind of ethical approach should we follow? How can we identify the interests of oppressed groups and the normative principles that will enable us to distinguish between good technological innovations and bad? Should we rely on universal moral principles, such as justice, freedom, and respect for human dignity, or should we make case-by-case evaluations of the positive and negative consequences of each action or decision involving algorithmic technologies? Should we start out from an ideal theory of justice, in the manner of John Rawls, a deontological or utilitarian position, or a dialogical approach based on communicative action? Or should we opt for a perspective derived from critical theories, Marxism, feminism, queer, or decolonial approaches?

Three ethical paradigms

Our initial hypothesis is based on a reworked version of *virtue ethics*, as revisited in light of critical theories and applied to the field of technological ethics. To fully understand the ins and outs of this approach,

1 Crawford, *Atlas of AI*, pp. 224–5.

we must first distinguish between three main families of normative theories: deontologism, consequentialism, and virtue ethics.[2] The *deontological* perspective emphasizes respect for rules, moral duty, rights, and obligations, which may draw on reason, intuition, or tradition. Deontological theories generally seek to identify universal principles, norms, or rules for judging whether an action is good or bad, independently of its context and consequences. Immanuel Kant's categorical imperative is a paradigmatic example of this philosophy: "Act only according to that maxim whereby you can at the same time will that it should become a universal law." The many statements of principle on responsible AI are examples of a deontological ethics of algorithms.

In contrast to this, the *consequentialist* approach is not primarily concerned with action per se or with principles that apply at all times and in all places, but rather with the consequences of actions and their impact on well-being. For example, utilitarianism advocates maximizing the happiness of the greatest number of people or reducing suffering for sentient beings. There are several variants of consequentialist ethics, such as act utilitarianism, rule utilitarianism, rational choice theory, and longtermism. In each case, the aim is to determine the right course of action based on a calculation designed to promote the best possible state of affairs, given the available options. Approaches based on risk management and cost/benefit analysis usually fall under this paradigm. A consequentialist ethics of algorithms will tend to judge the social, moral, and environmental benefits and costs of a given technology case by case, based on an impact assessment.

Finally, there is the long-forgotten tradition of *virtue ethics*. It focuses on developing character traits that enable individuals to perform actions considered good, excellent, admirable, beautiful, or praiseworthy. Virtues are dispositions to act, think, and feel in ways that enable an individual or group to interact well with their environment. Honesty, courage, moderation, care, and kindness are examples of virtues that guide action appropriately according to circumstances. In ancient philosophy, as in

2 We could also mention care ethics, as a paradigm distinct from other moral theories. We believe that, even if it is not reducible to virtue ethics, it does share many similarities with it, to the point where we could consider care as one of the fundamental or essential moral virtues. For further discussion of this debate, see Raja Halwani, "Care Ethics and Virtue Ethics," *Hypatia* 18, no. 3, 2003, pp. 161–92; Maureen Sander Staudt, "The Unhappy Marriage of Care Ethics and Virtue Ethics," *Hypatia* 21, no. 4, 2006, pp. 21–39; Nel Noddings, "Care Ethics and Virtue Ethics," in Lorraine Besser-Jones and Michael Slote, eds., *The Routledge Companion to Virtue Ethics*, New York: Routledge, 2015, pp. 401–15.

most societies throughout the world and throughout history, virtue ethics long offered the predominant approach to interpreting moral questions. Virtues, based on habits, social practices, and customs, are the object of moral approval by the community, whereas vices are vigorously condemned, or may sometimes be tolerated for various reasons.[3] The word "ethics" comes from the Greek *ethos*, which means "custom" or "character," referring to the mores, habits of action, and ways of being of an individual and/or group.

Neglected for centuries, virtue ethics has enjoyed a major revival since the 1980s, based on the work by philosophers including (but not limited to) Alasdair MacIntyre, Philippa Foot, Michael Slote, Rosalind Hursthouse, Julia Annas, and Christine Swanton.[4] It responds to certain impasses in the dominant approaches to ethics, deontologism, and utilitarianism, by shifting the focus of ethics from an evaluation of individual actions (how should I act?) to the more fundamental questions of the good life: How should I live? What kind of person do I want to be and should I be? According to the neo-Aristotelian or eudemonist approach (from the Greek *eudaimonia*, meaning happiness or well-being), virtues ultimately find meaning in terms of the contribution they make to human flourishing. To the question "why be virtuous?" the simplest answer is "to live well" or "to be happy." Virtues are practical dispositions that enable individuals to cope with various situations in daily life, solve problems, and interact appropriately with others in ways that promote individual and collective flourishing. Virtue ethics is sometimes interpreted in an individualistic way. Yet it naturally leads to the question of institutions or the political organization of the *pólis*, which must enable its members to flourish and become virtuous. Thus, the question of the "good life" is inseparable from the collective dimension—what kind of world do we want to live in?—and from politics—does the *pólis* help its members to flourish?

---

3 John Dewey and James Hayden Tufts, *Ethics*, London: George Bell and Sons, 1909.

4 Alasdair McIntyre, *After Virtue: A Study in Moral Theory*, Notre Dame, IN: Notre Dame University Press, 1981; Alasdair McIntyre, *Dependent Rational Animals: Why Human Beings Need the Virtues*, Chicago: Open Court, 2006; Philippa Foot, *Virtues and Vices and Other Essays in Moral Philosophy*, Oxford: Oxford University Press, 2002; Philippa Foot, *Natural Goodness*, Oxford: Clarendon, 2003; Michael Slote, *Moral Sentimentalism*, Oxford: Oxford University Press, 2010; Roger Crisp and Michael Slote, eds., *Virtue Ethics*, Oxford: Oxford University Press, 1997; Christine Swanton, *Target-Centred Virtue Ethics*, Oxford: Oxford University Press, 2021; Rosalind Hursthouse, *On Virtue Ethics*, Oxford: Oxford University Press, 2001; Julia Annas, *The Morality of Happiness*, Oxford: Oxford University Press, 1995; Julia Annas, *Intelligent Virtue*, Oxford: Oxford University Press, 2011.

Virtue ethics is also distinguished by its contextualism, particularism, and the emphasis that it places on learning. Virtues do not fall from the sky, nor do they stem from abstract reasoning or calculation. Rather, they are something to be learned. They are, in a sense, capabilities acquired through learning, practice, and socialization, which enable us to respond effectively and appropriately to the different situations that come up in life. As Aristotle put it: "We become just by doing just acts, temperate by doing temperate acts, brave by doing brave acts."[5] For example, honesty is a virtue that allows for a certain degree of trust in interpersonal relationships. Temperance or moderation allows us to regulate the pleasures, desires, and impulses related to the body, for instance in terms of food and drink or sexuality. Care practices are activities essential to education, caring for the needs of vulnerable people and the reproduction of society. In each case, virtue is not to be imagined in the guise of a precise rule or a fixed procedure. Rather, it is a propensity to act, to find the right balance between two extremes. For Aristotle, virtue is a certain balance that allows us to avoid both deficiency and excess, finding a "golden mean" (or moderation) that corresponds to each specific situation. For example, courage is the ability to face a situation of risk or adversity, avoiding the pitfalls of cowardice (lack of courage) and recklessness (a surfeit of courage), where the individual may put themselves or others in danger.

How can we measure the right dose of courage for a given context? Virtue ethics draws on practical experience. Hence the importance of practical wisdom (*phronesis*, also translated as "prudence" or "sagacity"), a virtue acquired through life experience, enabling individuals to determine the right course of action in unique, complex, and changing circumstances. Virtuous action is always the result of a decision that must take into account a variety of factors. Practical wisdom is sometimes evaluated by comparison with a hypothetical model of a virtuous agent: How would a virtuous and wise person have acted in this context?

Virtues in the world of algorithms

Why should an ethics of algorithms favor the virtue-based approach? We consider this paradigm particularly well-suited to addressing various issues related to AI, the uncertainties associated with technological

5 Aristotle, *Nicomachean Ethics*, 21, W. D. Ross translation.

development, and other related problems. Contemporary philosophers like Martin Gibert and Shannon Vallor are reviving this perspective to aid reflection on the programming of virtuous robots and the technomoral virtues that will enable us to live well in a world transformed by new technologies.[6] Here, we will show why these two particular applications of virtue ethics are meaningful and rich contributions, but ultimately insufficient and problematic. We will then propose a third way, built on the critique of technological vices.

First, Gibert highlights virtue ethics' uses in terms of programming virtuous algorithms or robots. Compared to deontological and utilitarian approaches, which require researchers to implement specific rules or objectives that algorithms must apply upstream, systems based on machine learning and deep learning have the ability to learn on their own, with or without supervision, as they gradually work out how to "act properly." Gibert sees this as an undeniable advantage, because it is not necessary to program specific ethical rules, but rather to train the algorithm with a large dataset, which allows the system to imitate people or actions deemed virtuous. Furthermore, manufacturing robots using the virtue-based approach seems easier to achieve from a technical standpoint. The middle-ground approach is useful here, as the algorithm does not have to choose between decisions based on ethical rules or consequentialist calculations. Rather, identifying the right schemas of conduct, it can determine in each instance the right action based on the "practical wisdom" accumulated through its training based on millions of concrete examples. What is more, algorithms aimed at virtue do not have to aim for perfection or infallibility: "What matters," Gibert contends, "is that [artificial moral agents] are more virtuous than average" and that they are able to constantly improve.[7]

The main challenge involved in this approach is training the algorithm using the right dataset. Otherwise, the robot may quickly become vicious: As a well-known saying in the world of computing puts it, "garbage in, garbage out." One example is the disturbing and bizarre case of the conversational robot Tay, created by Microsoft and launched on Twitter in March 2016. Less than twenty-four hours after its launch, Tay began tweeting hundreds of misogynistic, racist, and Trump-esque comments, with an attitude reminiscent of the most extreme

6 Martin Gibert, *Faire la morale aux robots: Une introduction à l'éthique des algorithmes*, Montreal: Atelier 10, 2020; Shannon Vallor, *Technology and the Virtues: A Philosophical Guide to a Future Worth Wanting*, Oxford: Oxford University Press, 2016.

7 Gibert, *Faire la morale aux robots*, pp. 82–3.

trolls.[8] While the algorithm was initially meant to achieve dialogic understanding through an informal and playful form of conversation, Tay learned by imitating thousands of problematic online comments. This raises the question of finding edifying actions, words, or individuals to imitate, rather than stupid, nasty, or vicious examples. Gibert suggests using collective intelligence and "asking each person who they would trust to make difficult decisions? … Each person could suggest the names of three people who are more virtuous than average" in a massive popular vote that could identify models of virtue for the algorithms.[9] In short, virtue ethics can be primarily used to program robots that will ultimately be more ethical than average, meaning that they have at least some base level of virtue and trustworthiness. In this sense, virtue ethics does seem to fit better than deontological and consequentialist perspectives, which are more difficult to translate into computer code.

The main problem with this application of virtue ethics is that it uses this perspective in an essentially instrumental way, that is, it seeks to code morality in order to manufacture robots that can "act well" in a world shaped by algorithmic capital. This is the narrow, normative vision of the professional ethicist who—faced with any particular algorithm—wonders what parameters to set for it. The algorithm is itself taken for granted. It is as if we find ourselves confronted with this object and thus have to program it as best we can, rather than considering it as a point of contention or a factor bound up with social relations and a broader historical process. The effect is that the existence and social relevance of robots are not questioned but merely assumed, the main question being how to develop robots properly so that they become "artificial moral agents" that we can easily interact with.

Algorithmic ethics thus seeks to aid the creation of robots that will look like "good neighbors," on whom we can rely to improve our everyday lives.[10] In the abstract, it is surely to be preferred that my Alexa personal assistant is virtuous rather than sexist and racist, and that it does not tell my child to put a coin in an electrical socket. But as it stands, algorithm ethics aims primarily to lend moral support to the major tech companies that produce algorithms. This literature, and the ethicists themselves, adopt the strategy of the "advisor to the Prince." They use virtue ethics not to question the ethical merits of an algorithmic world, nor to guide

8 James Vincent, "Twitter taught Microsoft's AI Chatbot to Be a Racist Asshole in Less Than a Day," *The Verge*, March 24, 2016.

9 Gibert, *Faire la morale aux robots*, p. 86.

10 Ibid., p. 74.

the human beings who will live in it, but to create a type of robot suitable for populating a world molded by and for algorithmic capital.

Technomoral virtues to enable better adaptation

Another original application of virtue ethics comes from philosopher Shannon Vallor. Her primary goal is to rehabilitate virtue ethics based on moral education, human flourishing, the good life, and practical wisdom to help people live better lives in a world disrupted by digital and algorithmic technologies. In contrast to Gibert's approach, virtue ethics is not applied here to algorithms, but to humans who must navigate a world of constant technological development. Vallor develops the concept of "technomoral virtues," meaning ones "explicitly designed to foster human capacities for flourishing with new technologies."[11] The aim is to contribute to the development of a "technomoral wisdom" befitting the challenges of the new century. It is an ambitious project, but an essential one, of a kind that seems sorely lacking in our time.

Vallor adeptly reworks a variety of traditional virtues, embedding them in the context of the challenges and tensions presented by today's world of technological change. Honesty values respect for the truth in a world plagued by misinformation; moderation or self-control seeks to overcome tendencies toward addiction and cyberdependence; humility aims to mitigate the tendency to assume technological ambition in a context of uncertainty; justice allows us to navigate rising social and technological inequalities; courage seeks to find the right balance between fear (technophobia) and naïve optimism (techno-utopianism); while empathy and civility ease the climate of confrontation generated by social media.

Vallor seeks to position herself halfway between, on the one hand, the neoliberal, libertarian, and transhumanist visions shared by certain influential figures in Silicon Valley and, on the other, the critical, techno-skeptic, and even technophobic visions that she associates with authors like Nicholas Carr, Evgeny Morozov, and Jaron Lanier. She aims to inspire "a resolute hope for the future of human flourishing with, not without or in spite of, the technosocial innovations that will continue to shape and enrich our lives for as long as human culture endures." In other words, there is no salvation outside of technological progress, which will

11 Vallor, *Technology and the Virtues*, p. 10.

continue to enrich our lives forever. In this spirit, Vallor places central emphasis on one particular technomoral value, namely "flexibility." She defines this as "*a reliable and skillful disposition to modulate action, belief, and feeling as called for by novel, unpredictable, frustrating, or unstable technosocial conditions*." Human beings must cultivate flexibility in order to adapt to a changing and potentially anxiety-producing environment. If algorithmic capitalism produces a world that fosters "*acute technosocial opacity*,"[12] frustration, uncertainty, and suffering, the really important task is not to try to change it, but to adapt to it as much as possible, the better to live with it. This approach ultimately aims to live well together with new technologies, and not to question them outright.

This is surely a rich and stimulating application of virtue ethics, but it is limited by a bias from which it starts out, and which we find particularly problematic. For this approach is based on a rather centrist and liberal interpretation of the doctrine of the golden mean, focused on the development of virtuous character traits at the individual level, largely aimed at enabling individuals to adapt to a certain world without seeking to fundamentally question or change it. The technomoral virtues advocated by Vallor aim to mold individuals better adapted to the world of algorithmic capitalism by individualizing responsibility for managing current crises. This—like the ethics of algorithms applied to virtuous robots—implicitly means endorsing the idea that there is no alternative to the major trends in technological development. Vallor's book has the value of returning attention to virtue ethics in order to reflect on appropriate ways of dealing with today's technological upheavals. And yet it ultimately remains within the horizon of the individual's functional adaptation to the limits of the dominant social order. While Gibert's algorithmic ethics applies virtues to the objective focus of programmable machines, Vallor's technomoral virtue ethics refers to the subjective focus of the average individual who wants to live well in an algorithmic society. Here, with robots adjusting to humans and humans adapting to robots, we come full circle.

Ultimately, we see virtue ethics being diluted here into a kind of adaptationism or ethical guidance, aimed at teaching individuals to navigate the tensions and paradoxes generated by algorithmic capital, or even to program virtuous algorithms. In our opinion, far from exhausting the issue, this approach avoids the real ethical elephant in the room. It is not enough to ask, as Gibert does, "What kind of robots do we want to live with?," as if it were unthinkable that we could live in a world without

12 Ibid., pp. 12, 145, 6; italics in the original.

robots or algorithms. It is not only necessary to think about the moral standards that should guide the programming of robots, or the virtues that an individual needs to acquire in order to navigate the algorithmic world, but more broadly to consider the type of world in which we want to live. We could instead ask fundamental questions like: Should we be handing more and more aspects of our lives and our personal and collective decisions over to algorithms, when they are developed mainly in order to extract our data, predict and guide our behavior, and boost the power of large monopolistic platforms? Do we actually need to rely on automated suggestions and decisions made by machines—when we are at work, in the classroom, or in public services? Are there alternatives to these technologies, or some other development model in which AI would serve the common good rather than the interests of the most powerful? Ultimately, do algorithms contribute to human emancipation, well-being, and the good life? Rather than conform to the imperative of "having to adapt," why not develop an ethic of virtues that is able to refuse the diktats of the established order?

Toward a critical theory of virtue

We propose a different way of using virtue ethics, taking as our starting point a critique of algorithmic capitalism and the hypothesis that this social configuration does not in fact constitute the unsurpassable horizon of our time. In our view, ethics presupposes critique, since the harmful existing order undermines not only the conditions for a good life, but even the possibilities of defining it.[13] To this end, it is important to adopt a virtue ethics inspired by critical theories. This means a theory that is sensitive to social relations of domination such as capitalism, colonialism, patriarchy, racism, and so on. A "critical theory of virtues" should not only focus on analyzing the character traits of individuals or robots. It must also take into account the practices, institutions, and social structures that affect individuals' and groups' possibilities of flourishing differently, depending on their position in the social world.

Feminist philosopher Lisa Tessman has undertaken a similar task. She offers an original revision of Aristotle's theory, purging it of its sexist, ethnocentric, and elitist biases and articulating it in a way that makes

13 Theodor Adorno, *Minima Moralia*, London: Verso, 2005; Rahel Jaeggi, "Une critique des formes de vie est-elle possible?: Le négativisme éthique d'Adorno dans *Minima Moralia*," *Actuel Marx* 2, no. 38, 2005, pp. 135–58.

it useful in the struggle for liberation.[14] Tessman argues that oppression makes it more difficult for certain individuals and social groups to flourish, while also causing "moral damage" that prevents some people from acquiring and exercising virtues due to their unfavorable social conditions. Aristotle had recognized that virtue was a necessary but not sufficient condition for happiness, as certain external goods (a certain degree of wealth, health, and social recognition) are also required for full development. Luck and misfortune thus play an important role, and some people are more systematically "unlucky" than others because of an unjust and hierarchical social order.[15] Therefore, virtuous fulfillment is not universally accessible, and criticism of the established social order is thus necessary for any real ethical reflection.

Moreover, while virtues do, in normal conditions, generally contribute to individuals' or groups' happiness, they can quickly become what Tessman calls "burdened virtues" for the dominated. Not only do they face additional obstacles that hamper their opportunities for a good life, but they must also demonstrate extra virtues (such as courage, resilience, solidarity) to combat injustice and confront an established order that is detrimental to their well-being and flourishing. Sometimes this comes at the cost of their mental health—or even their lives. It is easy to imagine a highly virtuous but oppressed person who lives a difficult life fighting injustice, suffers from mental health problems and depression, and dies prematurely and unfairly. Being virtuous does not necessarily or inevitably lead to ataraxic happiness or a good life, at least in a profoundly unequal world. There is thus something of a disconnect between virtue and human flourishing, and flourishing may turn out to be an unattainable ideal for many of us. This disconnect reasserts the need to continue the fight for liberation, and to make use of virtues to this end.

In our view, a critical theory of virtues has to highlight the various systems of oppression that systematically hinder certain individuals' or groups' potential to flourish. It should consider virtues as cognitive, emotional, social, and practical dispositions that enable people to act well and to overcome various obstacles in their pursuit of happiness. In this sense, ethics raises the question of the "fight for the good life," meaning a shared aspiration for personal and collective fulfillment, which sometimes takes the form of a struggle or conflict to achieve it. This ideal is, after all, frustrated by certain practices, institutions, and social structures that help

14 Lisa Tessman, *Burdened Virtues: Virtue Ethics for Liberatory Struggles*, Oxford: Oxford University Press, 2005.

15 Ibid., p. 113.

some to flourish at others' expense, through processes of exploitation, domination, or oppression.

In its classic version, virtue ethics focuses almost exclusively on the study of virtues while overlooking their logical counterpart—vices. These latter are simply seen as bad character traits in individuals, even though in reality social conditions help to create and strengthen them. Our perspective allows us to view vices as distinctly *social* mechanisms that contribute to individual suffering and prevent communities from flourishing. For example, if greed is a vice typical of the capitalist social order, its existence does not stem from the wickedness or virtuelessness of certain individuals. Rather, it comes from imperatives that encourage the attitudes, personality traits, and social practices, based on greed, that allow for success in these particular economic conditions. This brings us back to our discussion in Thesis 16 around Erich Fromm's idea of "social character," which proposes to link the study of individual-level psychological processes with a broader analysis of the social context in which individuals grow up and seek self-fulfillment.[16]

A critical theory of virtues will, therefore, seek not only to identify the virtues necessary for the flourishing of certain individuals and groups, but also to analyze the social origins of the vices that frustrate this process. It is *critical* in the sense that it highlights certain dysfunctions, vices, mechanisms of domination, and social pathologies. It is *focused on virtues* in that it seeks to identify the dispositions that enable actors to more effectively seek their liberation. Unlike traditional virtue theory, premised on the analysis of agents' moral qualities, critical virtue theory studies the processes by which virtues and vices are reproduced by a given social order, and the ways in which actors can exercise virtues and combat vices with a view to transforming social reality. Once we have reformulated matters in this way, we then face another question: How can critical virtue theory help us to address the problem of human flourishing in a world dominated by algorithmic capitalism?

The critique of algorithmic vices

To identify some of the vices generated by algorithmic capital, we will here draw on Aristotle's heuristic of the golden mean. As we noted earlier on, Aristotle believed that virtues are a disposition to act well, avoiding

16 Erich Fromm, *The Fear of Freedom*, London: Kegan Paul, 1943.

Table 19.1 Algorithmic virtues and vices

Vice (deficit)	*Virtue*	*Vice (excess)*
Addiction/cyberaddiction	(Digital) sobriety	Technophobic disconnection
Cynicism, powerlessness	Courage	Technophilic solutionism
Online hatred	Respect	Naïve tolerance
Indifference	Justice	Rigid radicalism
Dogmatism	Humility	Self-abasement
Naïveté	Critical thinking	Conspiracy theories
Carelessness	Care	Codependency
Sectarianism	Mutual aid	Consensualism
Anger	Love and rage	Ingenuousness

the twin pitfalls of deficiency and excess. This theory is often misunderstood as simply advocating for the center or for moderation, and as rejecting more radical or extreme actions outright. However, we might better consider the golden mean as a *method* for determining whether a particular disposition can be useful for human flourishing in particular contexts, even with various nuances and in different degrees.[17] Virtue ethics does not allow us to identify *the* right action based on fixed ethical procedures. It is more like *know-how*—an art of moderation that prompts us to reflect on how something good may become harmful, given the wrong dose in the wrong circumstances.

To aspire to a good life in a world shaped by algorithmic capital, it is not enough to preach adaptation to the dominant social order or to interact with virtuous robots. Rather, we need to identify which dispositions will enable us—both individually and collectively—to overcome and combat the vices generated by this system. Here, we propose nine virtues, each associated with algorithmic vices, that is, vices that are directly caused or exacerbated by algorithmic capital, and which relate to deficit or excess. We do not claim to offer an exhaustive or definitive list of virtues; rather, the aim is to foster reflection and guide a research program examining the practical conditions for a good life in the world ahead of us.

First, digital sobriety refers to moderation or temperance in the use of digital technologies. It can be applied on an individual level in the form of readiness to self-regulate behavior, impulses, and desires in order to

17 Our analysis converges with Alain Deneault's. For him, the golden mean or happy medium is not equivalent to an extreme centrist position, but rather a question of finding the right balance; for example, radicalism is a virtue, while sectarianism is a vice. See Alain Deneault, *Mœurs: De la gauche cannibale à la droite vandale*, Montreal: Lux, 2022.

limit screen time and avoid the temptation to overconsume social media and online content. Yet this virtue can also be applied on a collective level as well. This is how the collective GreenIT.fr project sees it, drawing on this notion of digital sobriety with a view to reducing the environmental impact of digital products, services, and infrastructure. This amounts to a self-limiting project for protecting the environment. This is especially necessary, considering the huge ecological footprint of algorithmic capital and its material infrastructure, as examined in Thesis 15.

In a noncritical interpretation, digital sobriety can be seen essentially as an eco-friendly gesture, inspired by responsible consumption and/or individual asceticism. The aim would be, apparently, to get into a healthier lifestyle through moments of voluntarily disconnecting through "digital detox" programs, and similar practices. This would mean creating "islands of deceleration" or "oases of resonance," as Hartmut Rosa calls them, to resist the trends toward social acceleration, the race for ever-greater performance, and the monopolization of attention by digital technologies. These practices of disconnection or sobriety would thus allow us to (re)discover better mental health and inner balance in a world where speed, and screen time, are predominant. However, this sobriety may also serve as a smokescreen for its opposite. Paid retreats, yoga sessions, and other "disconnecting" practices are encouraged by many firms, including in the tech sector itself, to encourage their employees to be happier, more autonomous, and more creative in their work.[18] Rosa refers to this as an "acceleratory functional deceleration," whereby people try to take breaks and slow down precisely for the sake of being more productive and efficient at work.[19] More radical and profound forms of disconnection involve completely abandoning the digital world for long periods, or for good, thus promoting disconnection as a way of life. Robert Hassan offers a good example of this approach: He describes how spending five weeks at sea allowed him to experience a different relationship with time and the world, to think things through without interruption, to rediscover rich forgotten memories, and to nurture the desire to see more of his loved ones.[20] This can be seen as a particu-

18 For a similar critique of the use of yoga in contemporary societies, see Zineb Fahsi, *Le yoga, nouvel esprit du capitalisme: De la libération au néolibéralisme*, Paris: Textuel, 2023.

19 Hartmut Rosa, *Alienation and Acceleration: Towards a Critical Theory of Late-Modern Temporality*, Malmö: NSU Press, 2012, and *Resonance: A Sociology of Our Relationship to the World*, Cambridge: Polity, 2019.

20 Robert Hassan, *Uncontained: Digital Disconnection and the Experience of Time*, Melbourne: Grattan Street Press, 2019.

lar quest for the good life, oriented toward contemplation, reflection, self-discovery, and better, more direct relationships. On an individual level, such (temporary or permanent) disconnection can surely be very enriching.

However, we believe that a more critical and collective view of digital sobriety is needed. This means, first of all, more directly tackling the causes of cyberaddiction, and particularly the addictive technologies mentioned in Thesis 3. Algorithmic capital is directly responsible for this specific form of addiction to screens, online games, social media, streaming platforms, and the like. Combating this algorithmic vice thus demands that we confront the mechanisms that fuel addiction and the companies that develop them and exploit our attention spans, and that we instead create other technologies that are more sober and less extractive and addictive. This means promoting digital sobriety from the ground up. We also need to recognize that even when we disconnect from the digital world, it still continues to exist, exploit, oppress, and dominate. Moves toward sobriety are not directly critical, in the sense that they do not seek to establish good living conditions on a collective scale. Although we may seek to find islands of disconnection during walks in the forest, escapist trips, or individually turning off our devices, it seems unlikely that these practices will definitively change the overall dynamics of digital addiction.

Another dead end lies in the attempt at a complete, permanent, or prolonged collective local disconnection from the digital world, for instance, through the creation of ecovillages or self-sufficient, low-tech, "100 percent off-grid" communities. An example of this collective technophobic disconnection is the Amish community in the United States, which has managed to preserve its way of life by resisting the steamroller of technological modernity. There is not, as such, any problem with moving to an ecovillage without an Internet connection or with becoming Amish. But it is hard to believe that this represents an effective form of resistance to algorithmic capital, which is today extending its grip to the far reaches of the globe.

Nevertheless, it may be useful to experiment with new ways of living differently, adopting low-tech, intermediate, or user-friendly technologies to promote greater resilience at the community, ecological, and technological levels. This has been advocated by the deep ecology, social ecology, and degrowther movements.[21] This would also make it possible

21 Ivan Illich, *La convivialité*, Paris: Points, 2014; Arne Naess, *Écologie, communauté et style de vie*, Paris: Éditions Dehors, 2020; Murray Bookchin, *Social Ecology and*

to test ways of living that promote an overall reduction in the production and consumption of digital technologies, placing greater value on in-person activities, neighborly relations, revitalized local communities, and other relationships. In short, living with less technology may be a good way of pursuing greater human flourishing at a time when algorithmic capital is increasingly colonizing every tiny aspect of our lives. The only point of excess, in our view, would be to aim for total and definitive abstinence from the digital world, taking refuge in an illusory micro-localist escape from capitalism. It is thus worth thinking about new modes of social organization that promote global collective sobriety through the development of workspaces, living environments, and cultural venues where digital and algorithmic technologies will no longer be the norm, but will play a limited role in certain activities deemed essential or that primarily respond to human needs. We will come back to these problems in Thesis 20.

Rethinking activist virtues

As well as digital sobriety, which is so important to our individual and collective relationship with the technological world, the other virtues listed in Table 19.1 more specifically concern the cultivation of effective, inclusive, and healthy critical activist practices to combat the vices of algorithmic capital.[22] Activist virtues are, in a sense, civic virtues. They enable people to act effectively to transform the social world while promoting the flourishing of the community and its members. These virtues may be interpreted more specifically in the context of algorithmic capitalism, which generally tends to erode civic virtues and to subvert activist virtues, making it harder to organize and to resist.

First, we must consider *courage* a fundamental virtue if we are to wage effective social struggles against algorithmic capital and the various systems of domination. Excessive courage leads to a form of recklessness that underestimates the risks and repercussions of certain direct actions (such as certain acts of sabotage, which can sometimes feed

Communalism, Oakland, CA: AK Press, 2011; Serge Latouche, *Le pari de la décroissance*, Paris: Fayard, 2006; Yves-Marie Abraham, *Guérir du mal de l'infini: Produire moins, partager plus, décider ensemble*, Montreal: Écosociété, 2019.

22 Our discussion here can be usefully combined with reference to the democratic model of activism formulated in Stephen D'Arcy, *Le langage des sans-voix: Les bienfaits du militantisme pour la démocratie*, Montreal: Écosociété, 2016.

police repression on comrades and vulnerable people if they are poorly planned). In the field of technology, courage can lead to blindness, solutionism, and techno-utopianism, which amount to various forms of excess and arrogance. Conversely, a lack of courage leads to cynicism and powerlessness, to hopelessly wondering—as activists may often think—"What's the point of even resisting?" It takes courage to openly criticize the dominant players in the AI industry, but also to organize collective disruptive actions, adopt radical reforms and antimonopoly legislation, or lead union drives against giants such as Amazon. Courage, even if it stems from the "golden mean" approach, is still radical, "in the sense of not being afraid of the results it arrives at ... [or] of conflict with the powers that be."[23]

While courage allows us to face adversity and to strike the right balance when confrontation is necessary, another virtue worth cultivating is *respect*, particularly in our online interactions, which are becoming increasingly polarized, dishonest, and hateful. Indeed, we saw in Thesis 11 on democracy that platforms and algorithms promote echo chambers, forms of aggression, and algorithmic polarization, which tend to make the public sphere more stifling and hostile to democratic life and good-faith public debate. The Internet, Twitter, and Facebook certainly did not invent hatred. And yet they do help fuel it or at least do not adopt any truly serious policies to curb tendencies likely to generate more engagement, clicks, and data, and thus to feed the power of algorithmic capital.[24] Respect refers to the interrelated virtues of civility, open-mindedness, and interpretive charity. It means starting out from a position that does not see the other person as an idiot, a dupe, or an enemy to be eliminated or canceled for the slightest mistake (past or present), but as a human being with rationality and agency, with whom it is possible to interact and exchange knowledge—at least to a certain extent, or until proven otherwise.

This last nuance is important, because excessive respect sometimes leads to the slippery slope of "naïve tolerance." We sometimes see this among people and groups who advocate total openness to all opinions, accepting within their ranks people associated with hateful groups, which paradoxically advocates values opposed to respect. This vice of excessive respect was observed during Canada's "Freedom Convoy" movement in early 2022, where a certain blurring of ideological boundaries in online exchanges, combined with an unprecedented convergence of

23 Marx, letter to Arnold Ruge, September 1843.
24 Cathy O'Neil, *The Shame Machine.*

groups opposed to compulsory vaccines, led some progressive, hippie, or alternative-minded individuals to show solidarity with reactionary far-right groups. This is an instance of Karl Popper's "paradox of tolerance," which shows that if a society advocates unlimited respect for divergent opinions, its capacity for tolerance will be threatened or destroyed by intolerant groups who organize to exploit its pliability. It is thus only wise to set certain limits on the expression of hate speech or on groups advocating violence, white supremacy, misogyny, and so on. To maintain a tolerant society, society must be intolerant of intolerance.[25] The idea is, therefore, to ensure a good level of respect with guidelines that prevent the naïve tolerance of the unacceptable, and to cultivate this virtue within activist circles, which sometimes tend to undermine themselves in their pursuit of radicalism. This ultimately means a combination of respect with the courage to confront dangerous opponents as necessary.

Second, the virtue of *justice* is essential in order to confront algorithmic capital. Justice is often considered an ideal, an abstract principle or value. Yet justice *as a virtue* is a practical disposition to act justly, to feel negative emotions in the face of injustice, and to think according to criteria of justice. It resides in a certain desire for, or love of, justice; it is a quality or attitude that seeks to realize justice in our practical existence. When talking about the digital world, justice must be considered in all its complexity, as an active pursuit of social, economic, political, racial, environmental, epistemic, and technological fairness. This means combating the various forms of injustice linked to social relations of gender, race, and class, tackling inequalities of access and the marginalization of knowledge, and the denial of recognition and participation. In particular, it means fighting against the injustices reproduced, amplified, and automated by algorithms. This does not just involve creating more virtuous robots and inclusive algorithms, but targeting the sources of injustices and the systems that reproduce them.

Two forms of vice can be associated with a deficit and an excess of justice. A lack of justice can be found in indifference or apathy toward the surveillance, political domination, alienation, and algorithmic discrimination that pervade today's capitalist society. Many people, whether through lack of awareness, cynicism, or resignation, lose their inner drive to seek justice and react against injustices. Yet there sometimes seems to be an excessive pursuit of justice in certain practices related to digital vigilantism, as described in Thesis 11. This manifests itself in

25 Karl Popper, *The Open Society and Its Enemies*, London: Routledge, 1943.

the cancel-culture movement, which we prefer to call "rigid radicalism." The label "social justice warrior"—promoted by right-wingers and often used to disqualify the left as a whole—nevertheless captures the idea of this vice amplified by the speed of social media and the practices of "ostentatious virtue" in online exchanges. In their book *Joyful Militancy*, carla bergman and Nick Montgomery describe a "something" that comes very close to the phenomenon of rigid radicalism, which is sometimes difficult to pin down:

> There is something that circulates in many radical spaces, movements, and milieus that saps their power from within. It is the pleasure of feeling more radical than others and the worry about not being radical enough; the sad comfort of sorting unfolding events into dead categories; the vigilant apprehension of errors and complicities in oneself and others; the anxious posturing on social media with the highs of being liked and the lows of being ignored; the suspicion and resentment felt in the presence of something new; the way curiosity feels naïve and condescension feels right … Above all, it is hostile to difference, curiosity, openness, and experimentation. This phenomenon cannot be exhaustively described, because it is always mutating and recirculating … Rigid radicalism cannot be reduced to certain people or behaviors … It is more like a gas: continually circulating, working on us behind our backs, and guiding us towards rigidities, closures, and hostility. No one is immune to it.[26]

Justice is closely linked to another virtue: *humility*. Indeed, this is the virtue needed to avoid the excesses associated with the rigid radicalism of excessive justice—a problem that existed long before the Internet, but which today seems amplified by the steroids of social media that fuel activist performativity and the race to be the most radical. Humility means deflating the narcissistic posture fueled by the platforms' regime of visibility and Facebook-style "likes culture" in order to find different ways of welcoming in other people, showing vulnerability, and reckoning with the complexity, paradoxes, and uncertainty tied up with the challenges of social transformation. Humility helps to feed respect, and to nurture curiosity, open-mindedness, and listening, rather than indulge the propensity to constantly speak out, to race to issue heated comments on every passing topic, and to quickly condemn others for gestures or remarks deemed inappropriate by the latest activist codes. In

26 carla bergman and Nick Montgomery, "Joyful Militancy: Building Thriving Resistance in Toxic Times," theanarchistlibrary.org, pp. 77–9.

other words, humility is an antidote to the powerful narcissism and self-promotion found in the digital public sphere. It nurtures related virtues such as listening, attention, and resonance.[27]

A lack of humility may result in dogmatism, the certainty of being right no matter what. Such an attitude is intensified by social media echo chambers, online polarization, and algorithmic radicalization—factors which all stimulate and spread rigid radicalism. An inverse vice lies in an excess of humility, where people develop a distorted self-image and consider themselves inferior. They may express low self-esteem or want to belittle themselves so as not to disturb or appear superior to others. While humility can, for example, help make us aware of our privileges, excessive humility can also mean that we feel the need to demonstrate it performatively, as evidenced by "statements of privilege" in online activism. This, ironically, feeds a dynamic of trying to score points, rather than actually helping other people or using one's resources and privileges to promote greater social justice.[28] This can be seen as a pathological form of disempowerment, which does not really contribute to creating the strong forms of solidarity useful for changing the world.

In the same vein, *critical thinking* is an intellectual virtue necessary for better understanding the mechanisms of domination that shape the world, thus helping to transform it. A lack of critical thinking leads to the vice of naïveté—a form of blissful blindness that unwittingly contributes to the suffering of others, alienation, and the destruction of the world. Except in certain circumstances (such as childhood), innocence should be considered more a vice than a virtue. Critical thinking is a disposition that constantly seeks to break through the veil of innocence, to turn toward the darkest aspects of the social world, not to contemplate disaster passively or cynically, but to find ways to change and overcome it. In *Minima Moralia*, Adorno clearly expresses this existential attitude of vehemently rejecting innocence:

> There is nothing innocuous left. The little pleasures, expressions of life that seemed exempt from the responsibility of thought, not only have an element of defiant silliness, of callous refusal to see, but directly serve their diametrical opposite … there is no longer beauty or consolation

---

27 Pascale Devette and Jonathan Durand Folco, "Retrouver la réceptivité: Écoute, attention et résonance pour débattre dans un monde polarisé," *Éthique publique* 22, no. 1, 2020.

28 Noémie Marignier, "Les énonciations de privilèges dans le militantisme féministe en ligne: Description et critique," *Argumentation et analyse du discours*, no. 17, 2017.

> except in the gaze falling on horror, withstanding it, and in unalleviated consciousness of negativity holding fast to the possibility of what is better. Mistrust is called for in face of all spontaneity, impetuosity, all letting oneself go, for it implies pliancy towards the superior might of the existent.[29]

Of course, this does not mean that we should sink into total pessimism or spend all day contemplating how gloomy the world is. We must cultivate critical thinking with "unalleviated consciousness [… of the] possibility of what is better," good intellectual discipline, and a dose of humility and openness, so as to avoid losing our way. The vice of excessive critical thinking is particularly dangerous and amplified in the age of algorithmic capital, with the widespread slide into conspiracy theories. This vice does not generally stem from a lack of education, knowledge, or rational thinking, but rather from an overdose of critical thinking, investigations, independent online research ("do your own research!"), over-rationalization, and sometimes wild and unfounded associations. This is an exaggerated desire to make sense of things, which may sometimes cloud our judgment. Obviously, conspiracy theories existed before the digital era. However, today they are spreading at breakneck speed with the help of algorithmic capital and social media. The platforms may themselves benefit in some ways from these bursts of unrestrained passion, even as—paradoxically—they sometimes try to combat them, often as a result of public pressure. The line between sharp critical thinking and outright conspiracy theory is not always so easily drawn. Conspiracy theories certainly have their share of ill-intentioned, misinformed, and dishonest proponents. But we also sometimes see critical, honest, intelligent people who—seeking to expose injustice, the abuse of power, and privilege—veer in the wrong direction.

Love, rage, and political projects

Finally, we must briefly mention three closely related emotional and political virtues: care, mutual aid, and the complex virtue of "love and rage." Care is at once an active concern for the needs of others, a relationship of interdependence, a form of mutual aid, and a set of practices that enable us to take care of others. Care, which is sometimes reduced to a

---

29 Adorno, *Minima Moralia*, p. 25.

simple interpersonal relationship between a mother and her child or the work of caring for sick or elderly people, has a much broader dimension, related to socially reproductive labor. A more politicized reading of care, which takes into account its unequal distribution in society (particularly the burdens on women, migrants, and racialized people, and so on), prevents us from falling into an overly naïve or romantic view of it. It can also be seen as a kind of relationship with the world: "a generic activity that includes everything we do to maintain, perpetuate and repair our world, so that we can live in it as well as possible."[30] Care as a virtue is a disposition to show concern for others and identify their needs (caring *about*), to take charge of those needs (taking care *of*), to provide appropriate care (care *giving*), to receive care or help from others (care *receiving*), and to care for the world together (caring *with*). According to Joan Tronto, each moment of care is linked to a set of values or virtues: attention (caring *about*), responsibility (taking care *of*), competence (care *giving*), responsiveness (care *receiving*), trust and solidarity (caring *with*).[31]

A lack of care manifests itself in indifference (not caring about others), while an excess tends to lead to a form of "codependency," a sometimes unhealthy tendency to care too much about others, forgetting or neglecting oneself, which paradoxically undermines the quality of a relationship. In the case of algorithmic capital, what prevails is the tendency toward carelessness, particularly due to the usual vices of capitalism. In driving competition, the quest for constantly higher performance, and the maximization of personal interest, it leads to a neglect of care.[32] Care relationships are essential in the fight against injustice and in overcoming the exhausting exercise of virtues (what have been called "burdened virtues") among oppressed and exploited groups, who must face situations of extreme precarity, violence, and internal tensions. The Care Collective, which authored the *Care Manifesto*, thus seeks to cultivate these virtues within activist groups so that the fight for the good life does not automatically lead to exhaustion, burnout, and other mental health issues. It likewise lays the foundations for a politics of care, which is a broader social project.[33]

---

30 Joan C. Tronto, *Un monde vulnerable: Pour une politique du care*, Paris: La Découverte, 2009, p. 143.

31 Joan C. Tronto, *Caring Democracy: Markets, Equality, and Justice*, New York: New York University Press, 2013, pp. 34–5.

32 On the ways that neoliberalism and market fundamentalism are hostile to care practices, see chapters 2 and 5 of Tronto, *Caring Democracy*.

33 The Care Collective, *The Care Manifesto: The Politics of Interdependence*, London: Verso, 2020.

This brings us to the virtue of *mutual aid* (or solidarity), which allows alliances to be built between social struggles, the working classes, and oppressed groups. This must avoid both the excesses of sectarianism (linked to rigid radicalism) and consensualism, which seeks to obscure real divisions by always seeking the lowest common denominator (we are all human beings, after all!). To foster this disposition toward mutual aid, it is necessary not only to take concrete actions of mutual support in times of crisis, but also to cultivate a form of love. This involves a love of justice and humanity, in the sense once associated with "charity" (*agapê*). This feeling must also be accompanied by righteous anger, a feeling of rage and indignation toward institutions and systems that dehumanize people and destroy life. Excessive anger takes the form of irascibility or hot temperedness, while a lack of due anger results in indifference to real evils. The right balance is difficult to find, and virtue ethics has recognized this for a very long time. As Aristotle put it, "Any one can get angry—that is easy—or give or spend money; but to do this to the right person, to the right extent, at the right time, with the right motive, and in the right way, *that* is not for every one, nor is it easy."[34]

The same applies to love, which does not here mean *eros* or love for a single person, nor even friendship (*philia*), camaraderie, or patriotism, but a more expansive sense of love that transcends the usual boundaries of belonging. An excess of love nevertheless leads to the ingenuousness, innocence, and naïveté that we criticized earlier. To avoid this trap, a healthy combination of love and rage—two opposing emotions—can sometimes counterbalance each other and form emancipatory synergies. Lama Rod Owens explains how unprocessed anger must be explored, respected, and better channeled in order to heal a lack of love and embark on the path to liberation.[35] While Aristotle saw a virtuous amalgam in the balance between anger and gentleness, the dynamic tension between love and rage can help us remain sensitive to injustice and motivate us to fight domination while not forgetting the role of gentleness, care, and respect for human dignity. This, too, is needed in order to avoid falling into hatred and pure thirst for destruction.

The critical theory of virtues, as here applied to the world of algorithmic capital, thus opens up further avenues for reflection. To continue exploring them, however useful this may be, could easily go beyond the

34 Aristotle, *Nicomachean Ethics*, 1109a–1109b.

35 Lama Rod Owens, *Love and Rage: The Path of Liberation Through Anger*, Berkeley, CA: North Atlantic Books, 2020.

scope of this thesis. Its presentation here in an outline form is preliminary—and necessarily incomplete. More in-depth work remains to be done to identify, in practice, the social, moral, and relational conditions that will enable us to lead a good life within, against, and beyond capitalism. However, we would like to mention a twofold objection that could be made to our argument, and which deserves proper attention. How can the virtues of which we speak—courage, respect, justice, humility, critical thinking, care, mutual aid, and love and rage—effectively combat the vices of algorithmic capitalism? Is this not just wishful thinking, a vain attempt to cultivate ethical dispositions that will give us a clear conscience while the world continues its race toward disaster? From this point immediately arises the second part: Even if virtues really could serve the effort to change the world, how can they be cultivated, taught, and disseminated in societies dominated by the vices of algorithmic capitalism? Since virtues do not come from nature, but from socialization, family, school, and (both traditional and social) media, how can we develop virtues opposed to the current system?

We think that both objections are valid. Virtues are a necessary but insufficient condition for social change. They facilitate practices, modes of organization, and collective actions aimed at transforming the established order, but institutions also play a major role in regulating practices and dispositions within society. Ethics, "inner change," or a "revolution in attitudes" are not premises that will magically lead to social revolution. Ethics is only one way of identifying practices and dispositions oriented toward good living. It must be achieved through a large-scale struggle, a comprehensive reorganization of institutions, and thus the practical materialization of a political project.

The role of institutions is thus essential: Even though human beings strive for fulfillment and virtue, they are imperfect creatures, and social conflicts remain inevitable. Neither beasts nor gods, individuals can exercise virtues to a certain degree, but we cannot assume that everyone can be spontaneously and continuously virtuous in their lives. Virtue is a complex and arduous process, "a beautiful and rare thing," as Aristotle put it—it is an experiment that leaves room for trial and error. This is why it is unrealistic and may even be dangerous to base a social model on the illusory assumption that all members of a community will be virtuous at all times. As Frédéric Lordon points out, social transformation cannot be conceived solely in terms of a society of "virtuosos."[36]

36 Frédéric Lordon, *Vivre sans?: Institutions, police, travail, argent*, Paris: La Fabrique, 2019.

This is why virtue is not enough alone. Laws, rules, institutions, and sanctions are also needed to ensure that individuals can live in a way that cultivates virtue. Institutions are of paramount importance, because they encourage virtues and discourage vices, promoting beneficial social habits that will become second nature to the virtuous, and rules to be followed by the nonvirtuous, under penalty of sanction for those who choose vice. Institutions both facilitate virtues and constrain vice. Laws, rules, and institutions thus have a dual role: to encourage individuals to act well and to discourage certain practices and behaviors. This obviously echoes the "nudges" familiar to algorithmic governmentality, which are combined with both positive and negative economic incentives, as well as the constraints, prohibitions, and obligations enacted by states. All these processes help to shape people's behavior, as do the social norms, customs, and expectations that already promote or inhibit certain practices.

But how can we know whether current institutions actually promote virtues rather than vices, and do not unknowingly reproduce certain forms of domination, for instance class inequality, sexism, or systemic racism? This is why we need a critical theory of virtues, drawing on the virtues of justice and critical thinking, as a normative compass for evaluating whether institutions promote the flourishing of the various individuals, groups, and communities concerned. Institutions are necessary in one way or another. Yet there is every chance that some of them need to be corrected, reformed, radically transformed, or even abolished if they do not meet the standards of virtue and justice—that is, the norms that serve as the spur for social critique.

Hence the constitutive incompleteness of ethics, which primarily concerns the study of dispositions that will promote or hinder the good life. These dispositions are inevitably influenced by institutions, with the effect that psychology is inseparable from an analysis of social, political, and institutional processes. Ethics precedes politics; it is prepolitical, but it naturally leads to politics, which reflects on the social and institutional conditions of the good life. "The goal of politics is happiness," Aristotle points out; and for him, the community or the *pólis* is what makes the good life possible. Politics is thus a kind of second-degree ethics, an ethics self-reflected in the horizon of a *pólis* that would allow it to fulfill its highest ideal. While the *pólis* brings together individuals and to meet economic imperatives related to their subsistence and the satisfaction of their needs, it also—and most importantly—forms a political community that reflects on the goals and purposes of this collectivity. It is not primarily motivated by the fact of living or of survival, but by the aspiration to live well.

How could the virtues discussed here guide a political project allowing us to build a society beyond algorithmic capitalism? How can practical reflection on the good life not only help individuals to "live better" with algorithms, or to code "virtuous robots," but also to overturn an unjust, oppressive, and alienating social order? Beyond the ethics that can be embodied in activist practices and collective actions, what is the desired social project—and what role do algorithms play in all this? Should a desirable postcapitalist world exist with or without algorithms? This is the focus of our final thesis.

Thesis 20

We need technosober democratic planning—reining in algorithms and tightening the limits of the digital world

We live in capitalism, its power seems inescapable—but then, so did the divine right of kings. Any human power can be resisted and changed by human beings.

—Ursula K. Le Guin

Technology, digitalization, and postcapitalist society: A complex issue

Replacing capitalism involves radically rethinking various parts of the system, such as property relations, the role of markets, resource allocation mechanisms, and labor, which are central to social life—and often taken for granted. Added to that is the need to rethink the institutional connections between economic production, the sphere of social reproduction, nature, political power, North–South relations, and so on.[1] In the era of algorithmic capital, we consider a collective and democratic reappropriation of algorithmic technologies a key dimension of any postcapitalist social project. This is, certainly, not the be-all and end-all

1 If we adopt Nancy Fraser's conception of capitalism as an institutionalized social order, then the socialist project surely needs to broaden its horizons. See Nancy Fraser, "What Should Socialism Mean in the Twenty-First Century?," *Socialist Register*, no. 56, 2020, and *Cannibal Capitalism: How Our System Is Devouring Democracy, Care, and the Planet and What We Can Do About It*, London: Verso, 2022.

of the project of human liberation. And simply changing the ownership of the means of production will not necessarily lead to a full transformation of the ways we produce, exchange, consume, and live. However, we contend that the rise of algorithms revives the debate between market and planning, thus creating a first possible lever for moving toward a world beyond capital and the "commodify everything" economy.

The idea of planning is bound to stir up memories of the Soviet Communist bureaucracy, and an economic model that presided over the construction of a great industrial power before it ran out of steam in the USSR's later decades.[2] This model clearly had serious limitations: sometimes major discrepancies between what was forecast in the plan and actual results; problems of wastage; considerable information deficits, worsening the gap between the USSR's notional production capacities and the products that ever got to consumers; political interference; the concealment of information under threat of reprisals; authoritarianism in the definition of needs without regard for individual preferences; and the sorts of standards imposed by a centralized bureaucracy.[3] Here we will not go into detail about the many factors that ultimately undermined the Soviet economy. But it is also important to understand that the planning of that era could not draw on the informational power offered by contemporary information technology. Of course, the socialist regimes of the past century did recognize the potential of IT for economic planning purposes. Project Cybersyn, for example, launched in Chile immediately after Salvador Allende took power in 1970, aimed to put cybernetics at the service of socialism.[4] Soviet economists and computer scientists had also begun exploring various computerized planning projects since at least the 1950s, but the technical limitations of the time and the conservatism of the bureaucratic elites greatly limited extensive exploration of these possibilities.[5]

In the era of Big Data and machine learning, new possibilities for economic planning are emerging. Platforms such as Amazon already anticipate future consumer needs by combining massive extraction of personal data, predictive behavioral analysis, supply chain optimization, and a logistical prowess that would have been unimaginable even just two

2 Alec Nove, *The Soviet Economy*, London: Routledge, 2011 [1961].

3 Ibid.; see also the detailed summary by Cédric Durand and Razmig Keucheyan, "Planifier à l'âge des algorithmes," *Actuel Marx*, no. 65, 2019, p. 86.

4 Eden Medina, *Cybernetic Revolutionaries: Technology and Politics in Allende's Chile*, Boston: MIT Press, 2011.

5 Durand and Keucheyan, "Planifier à l'âge des algorithmes," p. 87.

decades ago. While algorithmic technologies and their new applications in planning are primarily implemented within large capitalist firms such as Amazon or Walmart, there is—in theory—nothing to stop collective initiatives, states, or even workers' councils or citizens' assemblies from using the same technologies to pursue noncapitalist economic planning. Leigh Phillips and Michal Rozworski even identify the potential for a "People's Republic of Walmart." With this provocative phrase, they point to the possibility of viable and effective socialist algorithmic planning, which would take over planning and logistical technologies like the ones today used by the US retail giant.[6] Debate is thus (again) underway, this time on the effectiveness and desirability of algorithmically based economic planning. But is all this just chasing after a techno-optimist mirage? In what ethical and political conditions could algorithmic planning be useful for building a postcapitalist economy centered on satisfying human needs? What pitfalls must we avoid? What normative approach should be adopted when we evaluate these new technological possibilities?

Beyond technophilia and technophobia: Technosobriety

First, we need to steer clear of two rival perspectives which, while opposed, also share a certain form of technological determinism, a point we mentioned in Thesis 1. On the one hand, there is a techno-optimistic and technophile approach—for example, the one dear to accelerationists, which considers that capitalism is itself developing the practical and technological conditions for its own transcendence. All that is needed, therefore, is to accelerate this dynamic, exacerbate the system's contradictions, and reappropriate the means of algorithmic production as quickly as possible. This would mean removing capitalists from the equation so that we can build an economy of abundance that is already within reach. In Thesis 2, we mentioned Aaron Bastani's *Fully Automated Luxury Communism*, which represents this current of thinking with panache and ambition and, despite its limitations, is anything but boring.[7] Bastani believes that a completely socialized and largely automated economy could bring about a communist society of leisure. There would be no

6 Leigh Phillips and Michal Rozworski, *The People's Republic of Walmart: How the World's Biggest Corporations Are Laying the Foundation for Socialism*, London: Verso, 2019.

7 Aaron Bastani, *Fully Automated Luxury Communism*, London: Verso, 2020.

insurmountable ecological limits, as the depletion of natural resources on Earth could be offset, for example, by asteroid mining. Despite its appealing aspects, we consider this Promethean vision to have a few important blind spots and limitations. These include the existence of real biophysical limits to infinite economic growth; the ethical necessity of looking after the interests of future generations; and the underestimation of the political problem of how, exactly, we are meant to get rid of the capitalists. Furthermore, Kohei Saito notes that this "modernist" vision, based on a materialistic vision of abundance, reinforces the consumerist ideal of the Western and imperial way of life:

> Fully automated post-capitalism propagates an alternative hope that everyone keeps driving electronic SUVs, changing smartphones every two years and eating cultured meat hamburgers. Such a vision of the luxury future obviously sounds attractive to many people in the Global North because ecological modernization assures them that they do not need to change anything about their extravagant lifestyle. This kind of abundant future appeals to the satisfaction of people's immediate desires without challenging the current imperial mode of living in the Global North. The problem is, however, that such a vision accepts too uncritically existing value-standards and consumerist ideals. It ends up reproducing the social relations marked by oppression, inequality and exploitation that are inherent to capitalism.[8]

At the other end of the spectrum is what could be called a technophobic (or hard-technoskeptic) stance, which rejects the digital world and its possibilities outright, as well as any approach, tool, or innovation that promotes technological and/or industrial development. Anarcho-primitivism, some forms of neo-Luddism, anti-civilizational approaches, and certain versions of degrowth thinking belong to this family of tendencies, which reject anything "high tech" or close to it.[9]

---

8 Kohei Saito, *Marx in the Anthropocene: Towards the Idea of Degrowth Communism*, Cambridge: Cambridge University Press, 2023, p. 160.

9 See, for example, John Zerzan, *Future Primitive and Other Essays*, New York: Autonomedia, 1999; Steven E. Jones, *Against Technology: From the Luddites to Neo-Luddism*, New York: Routledge, 2006. Some current tech critics and writers such as Paris Marx, Brian Merchant, and Jathan Sadowski use the term "Luddite" to describe their own position. It seems, however, from their work that their critique of technology is directed chiefly at *capitalist* technology, and not at technology per se. In this sense, their position and ours share a similar view, close to original Luddism. See recent works by these writers exploring, among other things, the topic of Luddism: Jathan Sadowski, *The Mechanic and the Luddite: A Ruthless Criticism of Technology and Capitalism*, Berkeley: University

We, too, believe it is essential to question the logic of infinite growth, to robustly oppose certain harmful technological developments, to focus on energy descent and digital sobriety, and to take seriously the ecological limits that impose real constraints on our modes of production, exchange, and consumption. And yet one essential question remains: What should we do with existing technological infrastructures, innovations in machine learning, algorithmic technologies, and the possibilities that they offer in terms of computing, coordination, and organization? The question of rejecting algorithmic technologies (in part or in whole) cannot be resolved by a principled, absolutist, or a priori position rooted in a dogmatic conception of the good life, or an intransigent moral or philosophical posture. The use (or not) of algorithms in economic planning is above all a practical, strategic and ethical issue that must be assessed on a case-by-case basis, according to social, political and economic criteria. Our vision of a postcapitalist world is neither technophile nor technophobic, but rather *technosober*.

Defining Technosobriety

We conceive of technosobriety as an ethico-political principle that first and foremost questions and criticizes technosolutionism: Technosobriety is based on principles of moderation and caution, simplicity, and measure in the production and use of technology. It is not just a question of reducing the technological infrastructure, but rather of developing technologies that respond to real issues and that do not cause or worsen social and environmental problems. Technosobriety is not simply about individual use of technology but weighs in on the entire life cycle of technologies, from production to use to waste. A first technosober principle is to submit the ensemble of technological production, distribution, and consumption to normative criteria (social, ethical, and ecological) other than profitability or economic efficiency. Only technologies that solve socially recognized and democratically defined problems without degrading humans or the environment deserve our attention, time, and resources. Technosobriety is neither techno-optimistic nor technophobic. It is instead a redefinition of the meaning of progress and innovation

of California Press, 2025; Brian Merchant, *Blood in the Machine: The Origins of the Rebellion Against Big Tech*, New York: Little, Brown, 2023. On the history of resistance to technology and technoscience, see also the remarkable François Jarrige, *Techno-critiques: Du refus des machines à la contestation des technosciences*, Paris: La Découverte, 2016.

toward social ends. If we grant that technology is not a neutral force, and that it always aligns with some social interest, technosobriety means aligning technological development with the interests of the majority and with an intent to better the conditions of the most vulnerable among us. In this sense, technosobriety is a "technocritic" position that considers technology in its social and ecological context. It aims at developing technology along principles of reduction of energy consumption, moderation in the use of digital tools, durability and reparability of devices, climate justice, justice between Global North and Global South countries, freedom, and equality. It aims ultimately at a reappropriation by communities of decision-making processes regarding technological innovation.

While technosobriety is a general principle to be applied to technological development as a whole, it also issues direct and practical policy propositions. Some technosober policies are indeed very simple, and many are already being discussed in the literature. Drawing inspiration from authors such as Ruha Benjamin, Michael Kwet, and Evgeny Morozov, we can circumscribe digital consumerism by banning online advertising and targeted advertising, rendering illegal any and all forms of planned obsolescence, rendering mandatory the right to repair, and developing electronic waste recycling infrastructure as a priority. While we are at it, we should make interoperability between devices, software, and digital platforms mandatory as well.[10] These very simple propositions, taken together, can have a snowball effect. If implemented, they would pretty much rein in the power of the GAFAM companies and drastically reduce the ecological footprint of technology, which is not bad for a start. It would also take away most of the incentives to collect data on users and surveil them. If we really want to keep surveillance technologies, why not use them to surveil the police and governments instead? Why not use algorithmic surveillance to prevent fiscal criminality and the use of tax havens? Such uses would correspond better to the principle of aligning technologies with the interests of the majority and the common good.

The fundamental technosober question that can be applied to technological innovation could be summarized as follows: Is this technology indispensable in this sector, and does it maximize positive and minimize negative outcomes for the environment, the public, and the most

10 Based on a proposal by Cory Doctorow, *The Internet Con: How to Seize the Means of Computation*, London: Verso, 2023.

vulnerable? The answer to that question must come from democratic deliberation among users, workers, and those affected by that technology. And, if the technology is not indispensable, we should systematically privilege low-tech and eco-friendly solutions. We can imagine an evaluation grid with multiple criteria, subjected to democratic deliberation, and including scientific opinions of experts in a plurality of fields, civil society organizations, and the general public. We must make use of collective intelligence to assess the desirability of technological innovations instead of leaving these decisions to a handful of capitalists who are not incentivized to promote the common good, but to make profits. We think the principle of technosobriety should not direct the development only of AI, but of technology in general.

As such, a postcapitalist political project cannot be limited to a rejection of technology or its uncritical celebration. Rather, it should examine the possibilities of using these technological tools within certain limits and pursuing an energy descent with a view to building an ecological, democratic, and postcapitalist society. But how would a technosober postcapitalist world work? Can we think of possible ways to organize economic life that align with this principle? Would there be any role for algorithmic technologies in such a world?

To best answer the question of the proper postcapitalist use or dosage of algorithmic technologies, we will anchor it within the issue of algorithmic planning. On that basis, we will assess under what conditions such planning could contribute to the democratization of the economy, within ecological limits. We will begin by revisiting the "socialist calculation debate," following on from the contributions of Evgeny Morozov and others, before discussing the potential applications of these technologies to economic planning. We will then raise two main objections, one concerning the antidemocratic nature of algorithmic automation, and the second on the enormous environmental impact of the infrastructure that these algorithms require. Finally, we conclude with responses to these two objections, arguing for *limited algorithmic planning*, coupled with ethical principles of ecological self-restraint (energy descent based on convivial degrowth), public debate on the definition of needs, and decentralized forms of democratic participation. We believe that algorithms could, on these conditions alone, serve the interests of human flourishing.

The unexpected return of the socialist calculation debate

What new possibilities has the advent of Big Data and deep-learning algorithms opened up for economic processes? As Morozov points out, the socialist calculation debate may usefully be revisited in light of algorithmic technologies that enable new interactions between knowledge, prices, and social coordination.[11] The classic version of this debate in economic theory concerned the possibility of centralized economic planning in the absence of price signals. Neoclassical economists, such as Hayek, denied such a possibility. They contended that without the possibility of rapid feedback on consumers' ever-changing preferences, real-time adjustments between numerous economic actors, and sufficiently accurate information to ensure efficient resource-allocation, economic planning was doomed to failure, at least compared to the adaptive nature of interactions in the free market. However, in light of recent technological developments, could current algorithmic technologies prove superior to the conventional system of market prices in coordinating production, distribution, and consumption?

Big Data's possible impact on market mechanisms has not gone unnoticed among mainstream economists. Viktor Mayer-Schönberger and Thomas Ramge believe that Big Data could well replace capital, companies, and pricing systems with "data-rich markets" and horizontal interactions between individuals, thus driving a "reinvention" of capitalism.[12] One recurrent criticism of the market is that prices are only limited sources of information, reducing complex phenomena to purely quantitative, one-dimensional indicators that provide incomplete knowledge about what is being exchanged. Economic transactions are also subject to various uncertainties and inefficiencies, in particular social and environmental externalities. However, Mayer-Schönberger and Ramge argue that algorithms enable a relaunching and enrichment of market mechanisms. This could be done using a variety of highly useful qualitative and numerical data points, so that data, rather than prices alone, could coordinate market actors more optimally.

These data-rich markets (let us call them "algorithmic markets," since they would include algorithmic coordination) could convey much more accurate information than prices on traditional analog markets. Take

11 Evgeny Morozov, "Digital Socialism?," *New Left Review*, nos. 116/117, March/June 2019, pp. 33–67.

12 Viktor Mayer-Schönberger and Thomas Ramge, *Reinventing Capitalism in the Age of Big Data*, New York: Basic Books, 2018.

the example of a platform such as Airbnb, where the price of a good or service is combined with detailed information, photos, reviews from other users, comparisons, and a host of other data. Apps such as Tinder and Facebook have matching algorithms that connect multiple users based on common interests and compatible profiles without any need for price signals. Smart algorithms combined with an abundance of personal data make it possible to adjust supply and demand in real time, coordinating a multitude of actors based on highly detailed knowledge of their preferences. This makes it possible to automate calculations using a host of criteria, thus creating a process with impressive powers of prediction. In this sense, a data-based algorithmic system could be considered intrinsically superior to a price-based one both in terms of the quality of information and the coordination of economic activities.

However, Morozov points out that this conclusion is somewhat simplistic and even misleading. Economists such as Hayek were well aware that before engaging in commercial exchanges, economic actors can draw on a wealth of tacit (or experiential) knowledge, using the dynamics of competition to adjust in real time to market fluctuations, even if prices convey only fragmentary information. Big Data simply broadens the spectrum of available information, combining it with the dynamics of competition, the market, and prices, so that there is no break between the price signal system and data-based coordination; the two combine seamlessly, with producers and consumers simply having a more accurate picture of the goods traded on the market.[13]

Morozov's point is important because in the socialist calculation debate, the main issue is not simply whether the market or a central authority is the best vehicle for collecting or processing data. For Hayek, it was particularly important to understand the mechanisms of "discovery" and the dynamic updating of data by local actors, since these processes do not function through prices alone. However, many figures on both the right and the left were convinced that the debate was primarily a matter of determining the best mechanisms for communicating information with as little distortion as possible. Socialist planning, taking the premises of neoclassical economics for granted, thus gave way to market design strategies, as if the market were the only way to circulate information and coordinate a multitude of actors without too much friction.

It is worth emphasizing, however, that the main challenge associated with Big Data and algorithms lies in their ability to provide highly useful

13 Morozov, "Digital Socialism?," pp. 42–7.

sources of feedback to coordinate different social actors. The market is one such feedback mechanism, and so is the reason why algorithmic markets are one way of coordinating businesses and consumers. But we can also envisage nonmarket forms of coordinating social actors, which engage in different uses of data and algorithms, for instance, those based on principles of redistribution or reciprocity. Instead of promoting market uses of algorithms, why not devise new forms of social coordination outside of market parameters?

Rethinking digital feedback infrastructure

One possible form of nonmarket algorithmic coordination is already being implemented in the social credit system mentioned in Thesis 12, enacted to varying degrees and in various forms across Chinese jurisdictions since around 2017. It is a system of rewards and punishments based on surveillance, personal data extraction, and mutual rating systems. It applies to individuals as well as government agencies and businesses, and it then coordinates access to certain resources, especially financial ones. It is mainly used to rate companies and government agencies, in an effort by the Chinese authorities to strengthen public confidence in the market and local governments, following numerous corruption scandals. However, the system has a bad reputation outside China mainly because of its application to individuals. "Good" behavior (paying bills on time, not getting arrested, donating blood, volunteering, speaking positively about the government on social media, and so on) improves one's rating, while "bad" behavior (fraud, playing music too loudly, jaywalking, parking in a prohibited area, any form of crime) lowers it. On an individual level, rewards for good social credit vary, ranging from discounts on electricity bills to automatic approval for government services, or better interest rates on credit. Punishments tend to take the form of exclusions: bans on travel, on access to high-level positions, or on sending children to the best schools. This is thus an example of algorithmic technology being used for a nonmarket form of social coordination, but in a centralized and authoritarian way. Yet apart from this approach—in various respects an undemocratic, technocratic, and alienating one—are there other possibilities for nonmarket, nonauthoritarian forms of algorithmic coordination?

Morozov offers three options that deserve attention: (1) solidarity as a discovery process; (2) the design of nonmarket mechanisms; (3)

automated planning. The first option refers to the resolution of complex social problems, not all of which can be solved through competition. The exchange of information, experiences, and tips and solutions by a multitude of actors via platforms that promote communication and deliberation is one example of this nonmarket coordination. A range of YouTube videos aimed at sharing know-how in cooking, DIY, renovation, repairs, crafts, makeup, childcare, philosophy, music, political education, and language learning shows that this potential is already being realized in the interstices of the current system. A free social network such as Quebec's En commun, set up by Projet collectif in 2023, creates an interface that stimulates collaboration, the animation of hundreds of communities of practice, and the transfer of knowledge between a host of sectors, initiatives, and activities aimed at the common good.[14] Algorithms and/or platforms can here serve as a means of communication, of disseminating information, and even of deliberation as people solve various complex problems of general concern.

"Digital feedback infrastructures" could promote this type of individual and collective learning process without the need for mediation by prices or markets. Following Morozov's idea, we could envisage systems for matching problem-posers and problem-solvers, who could coordinate freely, without barriers or patents, thereby promoting knowledge-sharing and problem-solving beyond the commercial sphere.[15] Wikipedia already works like this, as do "hackathons," where intense collaborative work aims to develop new software or computer programs in a sprint involving several people. Author Yochai Benkler adds to this with his ideas of P2P collaborative production and of commons, which he presents as a third way between the market and centralized planning made possible by the advent of the Internet.[16] Cooperative and collaborative approaches are sometimes (and even often) more effective than competitive strategies for the purposes of knowledge discovery and problem-solving. It would be feasible (and even beneficial) to design and implement institutions, mechanisms, and incentives that encourage such practices instead of relying solely on cut-throat competition.

14 "*En commun* is a suite of ethical, free, accessible and collaborative platforms that connect knowledge and people to create a more equitable and environmentally friendly society. This suite aims to support cross-sector collaboration and the mobilization of knowledge among citizens, managers, professionals, researchers, entrepreneurs and anyone else working to develop innovative responses to the social and environmental challenges facing Quebec"; see projetcollectif.ca.

15 Morozov, "Digital Socialism?," p. 59.

16 Benkler, *The Wealth of Networks*.

The second option, designing nonmarket mechanisms, is a more intricate issue. It involves combining strategies to reduce complexity (through standards and procedures aimed at making a system more internally consistent) with processes to "amplify regulatory variety" (aimed at adapting a system to a more complex external environment by giving it maximum flexibility). The general idea is that social coordination mechanisms should not be designed in a rigid manner, based on a simple opposition between uniform bureaucracy and market freedom, since there exists a variety of institutional means to ensure social coordination. Libraries, museums, art galleries, universities, traffic controls, and barter systems all involve various models of regulating social practices. Algorithms can be used here to promote the viability, evolution, and functioning of these systems without resorting to a centralized technocratic approach or a data-enriched market logic.

Finally, the third option brings us back to the issue that particularly interests us: algorithmic planning. Even back in the 1990s, authors such as Cockshott and Cottrell argued that growing computing power would make it possible to overcome the technical problems inherent in economic planning.[17] It is also worth bearing in mind that, for several decades now, multinationals have been coordinating value chains and just-in-time production processes within their organizations and supplier networks, eliminating market uncertainties and formulating their own logic of planning. Planning is thus already part of a range of processes that are nonetheless associated with market capitalism. Moreover, recent technological innovations, such as anticipatory shipping, can be useful on the consumption side and could be combined with other innovations on the production side, such as 3D printers and other technical tools that promote a certain decentralization of production. Many possibilities for postcapitalist, decentralized planning and production are already here, working in the interstices of capital.

According to Morozov, these technological innovations have rendered obsolete the need for centralized economic planning, and instead point to a transition toward a form of "guild socialism," as Daniel Saros also suggests.[18] According to this perspective, workers' councils, conceived as "self-managed workers' start-ups," could display their products and services in a general catalog, similar to Apple's App Store or Google Play.

---

17 W. Paul Cockshott and Allin Cottrell, *Towards a New Socialism*, Nottingham: Spokesman, 1993.

18 Daniel Earl Saros, *Information Technology and Socialist Construction: The End of Capital and the Transition to Socialism*, New York: Routledge, 2014.

Consumers, for their part, would register their needs for the next production cycle, indicating their preferences and the quantities required. The use of Big Data and predictive analysis supported by algorithms would make it possible to fine-tune the play of supply and demand. Because there would be no imperative to maximize profits, producers would have an interest in selling off their stocks, even if it meant giving them away for free in some cases.[19] Special incentives, standards, or constraints could be added to such a model, which would promote a measure of restraint and more ethical and environmentally friendly consumption practices.

This type of planning system is already partially implemented in large multinational companies such as Amazon, Walmart, and Honda. The challenge for socialist algorithmic planning would be to promote a democratic reappropriation of these algorithmic technologies. It would seek to integrate them into a new economic system no longer focused on infinite growth and profit maximization, but on the optimal satisfaction of human needs, based on as little labor as possible. But is it enough to simply take current algorithmic technologies as they are and change their form of ownership? In our view, the issue is more complex than that.

The challenges of socializing the algorithmic means of production

Morozov's proposed solution for building "digital socialism" does not consist of simply taking the current technological machinery and replacing capitalist leaders with the state or a centralized planning office. As Marx pointed out in his writings on the Paris Commune: "The working class cannot simply lay hold of the ready-made state machinery, and wield it for its own purposes."[20] Morozov instead proposes that "social coordination can be mediated by a whole ecology of mechanisms, including law, democratic deliberation, decentralized 'radical bureaucracy' and feedback control, as well as the price system."[21]

But how can we choose the best mechanisms and strike the right balance between them? The important thing is to consider not only the criterion of economic efficiency, but also these mechanisms' ability

---

19 Morozov, "Digital Socialism?," p. 64.
20 Karl Marx, *The Civil War in France*, marxists.org.
21 Morozov, "Digital Socialism?," p. 66.

to solve complex social problems. This centrally poses the issue of democracy. Of course, it will not be possible to replace a model based on competition and profit maximization with a collaborative mode of production without addressing the control of this "feedback infrastructure," which we will here refer to as "algorithmic means of production." Two questions then arise: Can algorithms be democratically controlled? And what about this infrastructure's already-enormous environmental impact? Let us examine each of these questions in turn.

With regards to the issue of democratic control, one of the problems with machine-learning algorithms is that they are particularly opaque. They are "black boxes," and even experts do not know precisely why the algorithm arrives at one result rather than another. This is the problem of "explicability" discussed in Thesis 10, a major issue in AI ethics. Given their extreme technical sophistication, it seems difficult to democratize the production and fine-tuning of algorithms on a large scale. Furthermore, this technical opacity makes it impossible to ensure that algorithmic workings and results are comprehensible to the average person. It is hard to imagine democratic deliberation about something that hardly anyone understands. This is a significant challenge.

A first possible solution to this problem of explicability lies in democratizing the organizations that produce and use algorithms. The companies that currently design these systems, such as OpenAI, Alphabet, Uber, Tesla, Meta, Amazon, Apple, and Microsoft, are non-democratic, private organizations that rely on patents, trade secrets, and often closed ecosystems to maintain their lead in fiercely competitive markets. These companies compete for the best engineers, computer scientists, data scientists, and mathematicians in order to stay ahead of their competitors, and thus they have little interest in publicly sharing the recipes for their innovations. A review of the history and inner workings of OpenAI—originally created as a nonprofit with a mission to produce AI that benefits all humanity—shows that economic pressures have pushed this organization to turn into a private, for-profit company reliant on highly opaque practices.[22]

Instead of relying on competition and private ownership, a model promoting open innovation, free software (or algorithms), and collaboration with universities and public administrations would be better suited to fostering R&D in the public interest. This would have its benefits, even if the ins and outs of the functioning of particular algorithms

22 Karen Hao, *Empire of AI: Dreams and Nightmares in Sam Altman's Open AI*, New York: Penguin, 2025.

remained opaque to the general public. Democratization would then focus on the ownership of algorithms and the control of the organizations that use them, which would be transformed into cooperatives, nonprofits, or public companies that would be more accountable.

This is an interesting potential solution, but it faces certain obstacles. It should be noted that research in the field of AI already operates in a largely collaborative manner. Several publications are freely available; free programming training courses are on offer; Google uses Android, which is open-source software; and platform cooperatives are already emerging in the collaborative economy, without, however, being able to engage in sustained competition with their capitalist rivals. The GAFAM companies already widely exploit the wealth and creativity of knowledge in the form of digital commons. All this means that algorithmic capital operates in a hybrid logic between collaboration and competition, commons and the commercial appropriation of innovations. Yet this form of hybridization is obviously not enough to socialize or democratize algorithms, nor to solve the problem of explicability in itself.

What is more, since training algorithms requires large amounts of data, data collection and storage become crucial issues. Even if the means of algorithmic production were freely accessible, their usefulness and effectiveness would be greatly reduced if data were largely concentrated among a handful of private digital platforms. One solution would be to ban companies from directly appropriating personal data and to create social utility trusts responsible for data governance, as we discussed in Thesis 18.[23] The aim would be to pool data and make it accessible to companies and governments under certain conditions. Morozov's calls to "Socialize data centres!" and "Socialize the means of feedback production!" are thus prerequisites for any form of socialist economic planning that wants to make use of algorithms. But once we have socialized data in the form of social utility trusts and abolished private ownership of the means of algorithmic production through a mix of nonprofits, co-ops, and public firms, how can algorithms contribute to democratic economic planning? How can algorithms—that is, systems of automated recommendations or decisions—be combined with democratic decision-making about what to produce, consume, and distribute among the members of a community? Do we not run the risk that algorithms will be used and managed in a technocratic manner by individuals who are best-positioned in certain key

23 Leblanc, "Gouvernance des données."

sectors—with democracy nothing more than an indirect mechanism for rubber-stamping economic decisions? In short, what would be the role of democratic tools or spaces in an economic system in which a large number of transactions and decisions were automated and inaccessible to those most directly concerned?

Four models of postcapitalist economics

To examine this question, we can draw on some of the postcapitalist economic models that have been developed in recent decades. They have been usefully summarized by Audrey Laurin-Lamothe, Frédéric Legault, and Simon Tremblay-Pepin.[24] Here we are dealing with a schematic examination of their principles and how they could integrate algorithmic technologies, and not an in-depth or exhaustive discussion of the models concerned.

The Cockshott and Cottrell model centers on a central planning office that plays a key role in the development of economic plans, thus replicating the classic model of Soviet-style centralized, nationwide planning.[25] The democratic aspect comes into play mainly through the random selection of individuals for citizen committees and other public regulatory bodies, but the major economic decisions are largely centralized. Add to this the problem of algorithm explicability and the enormous concentration of data required to calculate the plans, and the democratic element of such a system seems limited. It remains largely subject to a technocratic, algorithmic logic.

Albert and Hahnel's model of participatory economics proposes a more decentralized system of planning.[26] A key democratic institution in their model is the council, which is made up of groups formed in workplaces and communities. These councils, combined in a federal structure, make decisions on production priorities and consumption needs. This model thus offers more substantial democratic deliberation.

---

24 For a more detailed overview of these issues, see Audrey Laurin-Lamothe, Frédéric Legault, and Simon Tremblay-Pepin, *Construire l'économie postcapitaliste*, Montreal: Lux, 2023. See also Frédéric Legault and Simon Tremblay-Pepin, "A Brief Sketch of Three Models of Democratic Economic Planning," *CRITS Research Note* no. 2, 2021, pp. 1–26.

25 Cockshott and Cottrell, *Towards a New Socialism*.

26 Michael Albert and Robin Hahnel, *The Political Economy of Participatory Economics*, Princeton, NJ: Princeton University Press, 1991. See also Pascal Lebrun, *L'économie participative: Une alternative contemporaine au capitalisme*, Montreal: Lux, 2014.

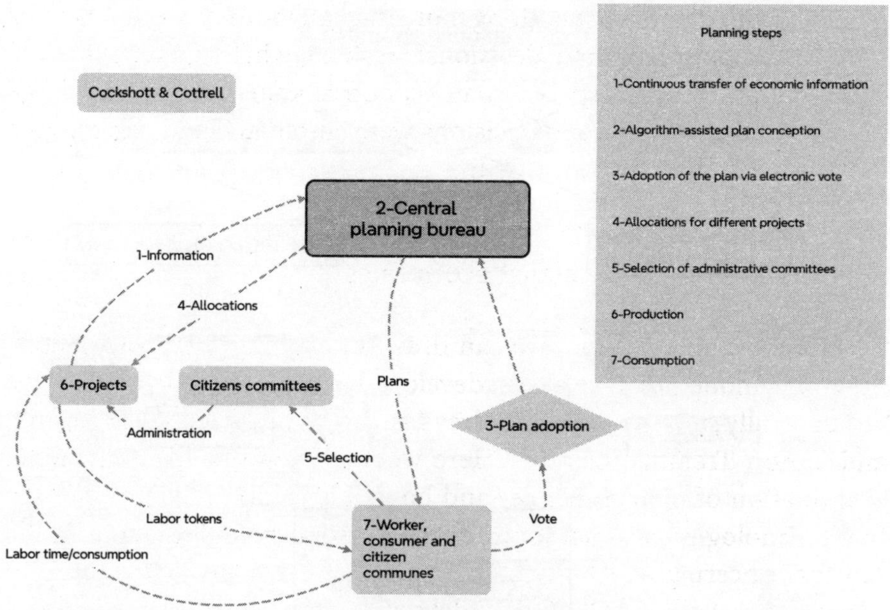

Figure 20.1 Cockshott and Cottrell's model of democratic economic planning
Source: Adapted from Frédéric Legault and Simon Tremblay-Pepin, "A Brief Sketch of Three Models of Democratic Economic Planning," *CRITS Research Note* no. 2, 2021, pp. 1–26

The algorithms are incorporated at the level of the Iteration Facilitation Board (see Figure 20.2) and aid the matching of production capacities with forecast consumption, as plans are developed via a dynamic, quasi-price-based system. Again, the main problem is explicability: This matching and the other coordination processes run through a black box whose operations are hard to understand, even for experts. However, democracy requires a certain level of transparency and trust that is difficult to reconcile with purely automated decision-making systems. This does not mean that algorithms are inherently incompatible with democracy, but AI cannot be the sole cornerstone of a democratic system for coordinating all local economic micro-decisions within a large society.

Devine and Adaman's model seems to incorporate more space for deliberation and democratic negotiation.[27] Production units send information to "negotiated coordination" bodies made up of multiple stakeholders, which send the information to a planning commission, which in turn sends these scenarios to an elected representative assembly and a second chamber (called the Chamber of Interests and

27 Fikret Adaman and Pat Devine, eds., *Economy and Society: Money, Capitalism and Transition*, Montreal: Black Rose Books, 2002.

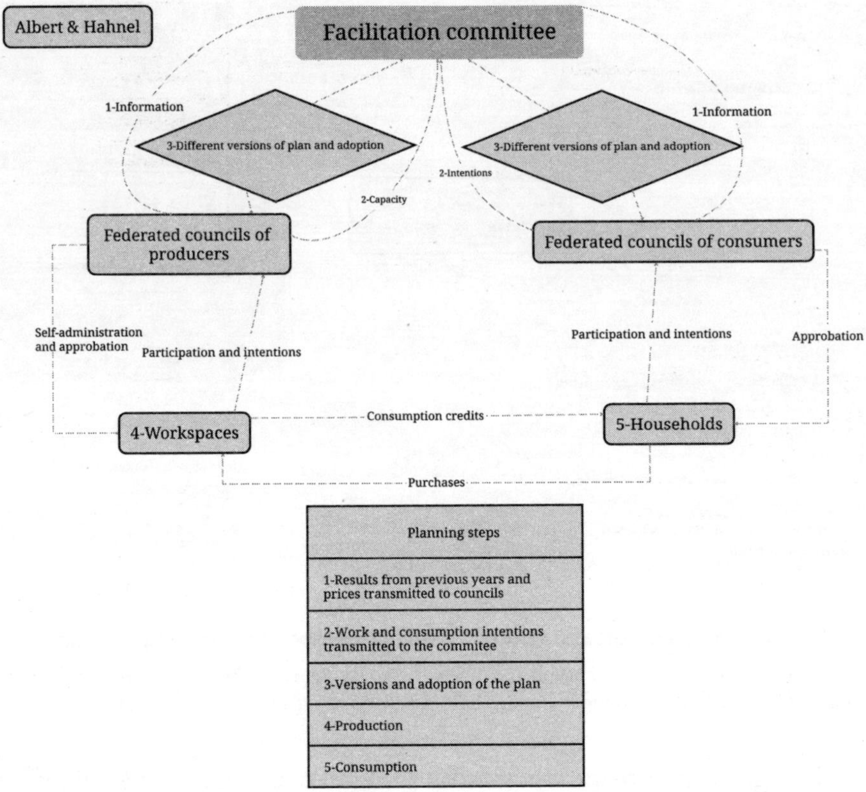

Figure 20.2 Albert and Hahnel's model of participatory economics

Source: Adapted from Legault and Tremblay-Pepin, "A Brief Sketch of Three Models of Democratic Economic Planning"

composed of members of civil society), which finally adopts a plan. Here, the role of algorithms seems much less central, but the chain of democratic decisions is more complex until formal adoption by a representative assembly. Algorithms play a subordinate role, but we can see that feedback mechanisms are reduced as well, with the preferred mechanism of social coordination taking the form of representative democracy. This can create challenges in terms of coordinating supply and demand, accuracy, and the possibility of making dynamic adjustments to the plan.

The same may be said of the model proposed by Legault, Theurillat-Cloutier, and Savard.[28] Their model for a postcapitalist economy is a variation on Devine and Adaman's decentralized democratic planning, in which companies send their data to sectoral committees, which in

28 Legault, Theurillat-Cloutier, and Savard, *Pour une écologie du 99 %*.

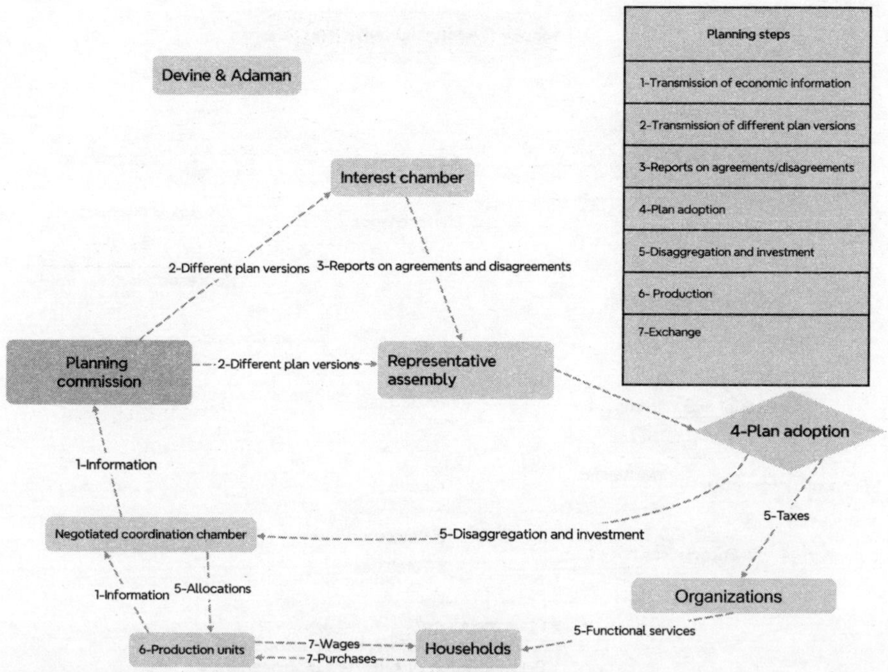

Figure 20.3 Devine and Adaman's model of decentralized democratic economic planning

Source: Adapted from Legault and Tremblay-Pepin, "A Brief Sketch of Three Models of Democratic Economic Planning"

turn send their data to a planning committee that develops scenarios for plans. A representative assembly is then responsible for adopting the plans by submitting them to a referendum. There are, of course, many democratic spaces, and the use of algorithms is concentrated at the planning-committee level, which would be responsible for calculating the plans based on data from the sectoral committees. Although political authorities have the final say, these decisions do not simply result from democratic deliberation but also reflect the balance of power between social groups and the Assembly, and potential tensions within this body. The economy is subordinate to political power, but in the same way as a government that makes decisions based on an "interface" controlled by a group of specialists from the planning committee. It is surely possible to use referendums for the eventual adoption of plans, but such a mechanism does not seem to be the best means of providing feedback to adjust productive capacities to the differentiated needs of individuals. Ultimately, as with all other models, the problem of explicability is a thorn in the side of any effort to democratize the economy using algorithmic technologies.

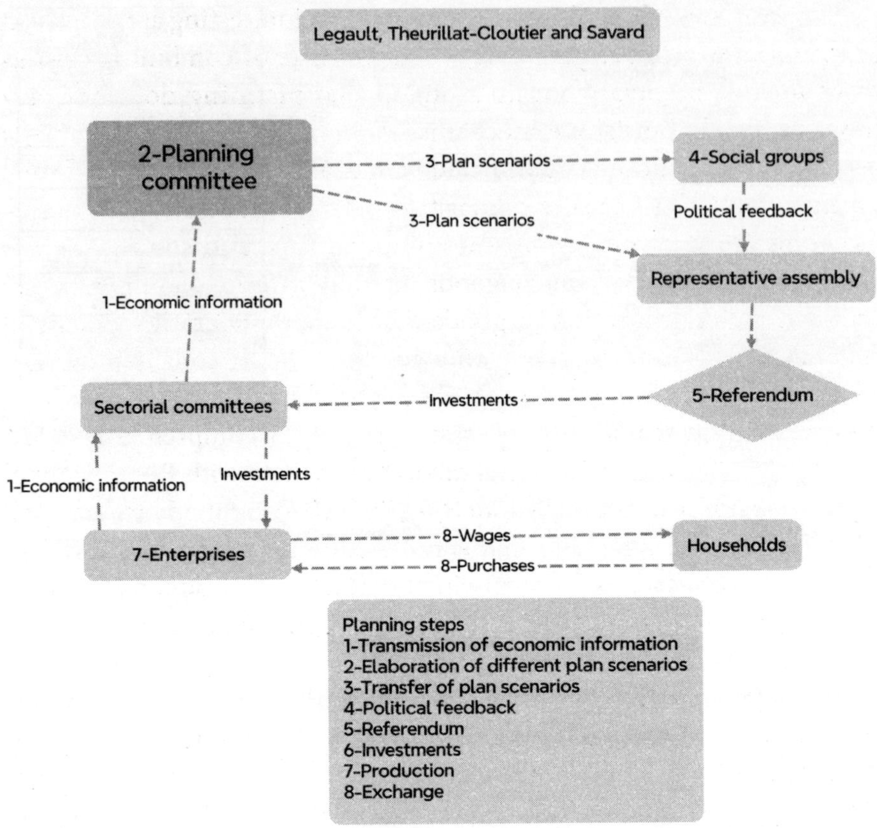

Figure 20.4 Legault, Theurillat-Cloutier, and Savard's model of decentralized democratic planning

Source: Legault, Theurillat-Cloutier, and Savard, *Pour une écologie du 99 %*

In short, despite their innovations and interesting features, there seems to be no perfect model for democratic economic planning (with or without the help of algorithms). One of the main problems is that they aim to base economic decisions entirely on planning (rather than on the market), incorporating a mix of democratic discussions and automated decisions at different levels. The central role of planning as the main mechanism for social coordination makes the processes somewhat cumbersome and heavily reliant on algorithms, experts, and the bodies that control these processes. The use of algorithms would undoubtedly promote greater economic efficiency, and even with the introduction of democratic spaces here and there, planning would remain centralized to a certain extent and subject to the problem of explicability. In general, what these models promise in algorithmic efficiency, they lose in democracy and transparency, and vice versa.

How can these difficulties be overcome? An interesting approach can be found in Morozov's proposal, which consists of combining various mechanisms of social coordination, so that planning does not play the central role but is one mechanism among others. For example, a postcapitalist economy could include a plurality of forms of economic organization (the social and solidarity economy, cooperatives, nonprofits, municipally owned companies, public firms, and so on), use the price system for certain niche goods, use algorithmic planning for the production and allocation of goods that meet democratically defined collective needs, and then rely on a combination of collaborative P2P production and commons to achieve the ideal of a "free association of producers" that would not necessarily require a comprehensive plan for all economic activities. In his theoretical debate with Robin Hahnel on postcapitalist economic models, Erik Olin Wright advocated such a hybrid socialist system, which is not a comprehensive model whose components are so closely intertwined that it must be adopted en bloc. Instead of a model aimed at building a tightly knit social unit under the umbrella of planning, it advocates the creation of an open ecosystem where the market can still play a certain limited role, but where various nonmarket coordination mechanisms govern collaborative production, the allocation of labor, and the satisfaction of needs. In this open system, algorithms are called upon to play a differentiated role depending on the coordination mechanisms used, and so do not form the core of the economic system. Thus, democratic mechanisms remain central, and explicability presents less of a problem because it does not affect all economic decisions and sectors.

The technosober imperative—and moving beyond an "all-digital" outlook

Even a well-designed, democratic postcapitalist system would still face the problem of ecological limits. This issue is central, since machine-learning algorithms are energy-intensive technologies, requiring the extraction of massive amounts of data, and thus data centers, telecoms infrastructure, and the extraction of natural resources to produce the devices on which our digital interactions rely. As we showed in Thesis 15, the global hardware required by the digital world and AI is far from sustainable in the long run. Rather than envisaging sustained growth in this sector linked to the increased use of data and algorithmic technologies for economic planning, it thus seems essential to focus on a collective principle of technosobriety, with a view to ecological self-restraint that allows us to keep the environmental footprint of digital infrastructure within the planet's biophysical limits. This means that we need to *stop* the endless accumulation of data and the unlimited development of algorithmic technologies, which require ever more energy and resources in a self-perpetuating cycle. The utopia of data centers based on renewable energies and the hoped-for gains in eco-efficiency do not, among other things, take into account the rebound effect, which is why the reformist horizon of the Green New Deal is also insufficient in the long run. It is necessary to *degrow* the overall size of technological infrastructure and shrink the place that the digital world occupies in our lives.

Here we draw on the principle of ecological self-restraint, corollary to technosobriety, which implies energy descent—that is, an overall reduction in energy production and consumption at the local, national, and global levels.[29] No sustainable solution to the climate crisis is possible without such a reduction. Next, technosobriety demands legislation to ban planned obsolescence, focusing on less energy-intensive technologies and promoting moderate use of digital devices at work, in our leisure time, and in our daily lives, as we mentioned earlier. This is not just a matter of individual choices, such as opting for the simpler life, or an ethic of disconnection aimed at improving our lifestyles and mental health. It is a choice for society as a whole: values, standards, habits, and social practices that need implementing in order to push for a world where digital and algorithmic technologies play a lesser, marginal, peripheral, and limited role. It means reducing them to the level at which they are collectively considered necessary and optimal. This does not mean that it is necessary to exclude all use of advanced technologies, but that they should be limited to ethical uses, and that user-friendly or intermediate (low-tech) technologies that promote greater autonomy, resilience, and reparability, and have a smaller ecological footprint should be prioritized. Reducing screen time, shrinking the digital world, and forging encounters and solidarity outside the predictable parameters of algorithmic capital count among the preconditions of technosobriety and the good life in our time.

Furthermore, this technosober ecological self-restraint cannot be achieved without the creation of democratic structures and processes aimed at the collective satisfaction of basic human needs. The concept

29 Rob Hopkins, *The Transition Handbook: From Oil Dependency to Local Resilience*, Totnes, UK: Green Books, 2008.

of "needs" is rich and poses a complex problem, and we cannot here outline a complete theory of needs that includes all its objective, subjective, intersubjective, and political dimensions.[30] Fortunately, some thinkers are beginning to explore the complex relationships between the definition of needs, the struggle over the interpretation of needs through public deliberation, and democratic planning.[31] It is thus essential to discuss which basic needs should be met first, and then consider the best means of satisfying other wants and desires that can contribute to the fulfillment of individuals, groups, and communities. For example, algorithmic production can be used to calculate the resources needed to meet anticipated demand for certain basic goods and services (food, clothing, furniture, drugs, computers, housing, and so on), in order to save as much as possible on the use of natural resources and human labor time, and to manufacture the most sustainable goods possible. The philosopher André Gorz uses the metaphor of the "heteronomous sphere" to refer to that part of the economy geared toward "the production of necessities" and which should be collectively managed to reduce human labor time to a minimum and ensure that needs are effectively met.[32] It is conceivable that algorithms could play a strategically important role in this sphere, while also being subject to democratic control processes through consumer councils, producer committees, and regional and national planning bodies. The important thing is that economic production should be guided by a principle of self-restraint and by the norm of "sufficiency."[33] Theoretical work on "sufficiency" and the capabilities approach advanced by Martha Nussbaum and Amartya Sen could be mobilized in this sense.[34]

Second, the establishment of this technoeconomic infrastructure based on democratic (and partly algorithmic) planning should not aim to control all economic processes. Its goal is to produce the essentials,

---

30 Ambre Fourrier, "Les besoins: Une notion à redéfinir?," *Notes de recherche du CRITS*, no. 3, 2022.

31 Nancy Fraser, "Triple Movement?: Parsing the Politics of Crisis After Polanyi," in Marian Burchardt and Gal Kirn, eds., *Beyond Neoliberalism: Approaches to Social Inequality and Difference*, London: Palgrave Macmillan, 2017; Ambre Fourrier, Jonathan Durand Folco, and Simon Tremblay-Pepin, "Redéfinir démocratiquement les besoins pour planifier l'économie," *Politique et Sociétés* 42, no. 2, 2023.

32 André Gorz, *Farewell to the Working Class: An Essay on Post-Industrial Socialism*, London: Pluto Press, 1982.

33 André Gorz, *Éloge du suffisant*, Paris: PUF, 2019.

34 Axel Grosseries, "Qu'est-ce que le suffisantisme?" *Philosophiques* 38, no. 2, 2011, pp. 465–91; John M. Alexander, *Capabilities and Social Justice: The Political Philosophy of Amartya Sen and Martha Nussbaum*, New York: Routledge, 2016.

so that the "autonomous sphere" can grow. This is the sphere where free time, creativity, and the flourishing of human faculties form the ultimate principle of a variety of social, economic, political, and cultural activities. As Gorz pointed out, such an economy does not aim to maximize profits, but to "save working time as much as possible in order to maximize available time."[35] In this logic, the incentives for productivity do not come from blind productivism linked to imperatives to produce relative surplus-value, but instead from the prospective goal of more free time. This could also feed decelerationist tendencies, in the interest of countering societal acceleration and temporal alienation. This autonomous sphere values the role of artisan production, the free manufacture of beautiful and useful objects, sporting and artistic activities, the creation of freely accessible computer programs, care work, mutual aid and solidarity networks, and other forms of self-expression. Most important, it values them outside of any reference to the criteria that define coordination by the market, the state, or algorithmic regulation.

This means, first, not basing our entire postcapitalist economic system on the heavy infrastructure of AI and algorithmic planning. The latter must play a limited role, combined with other feedback and social coordination mechanisms geared toward self-production, free time, collaboration, free association, community, care work, and creative self-expression. It is even possible to avoid an overly dualistic model, pitting a heteronomous sphere based on algorithms against a largely artisanal autonomous sphere, and instead favor a three-tiered social organization. A certain portion of production could be planned by making strategic use of algorithms to meet collective needs that are publicly recognized through democratic processes, whether at the regional, national, or international levels. Second, the mass use of the social and solidaristic economy, commons, and other forms of collaborative production would make it possible to meet needs and desires on a smaller scale. Finally, the autonomous sphere would be the main "informal" social space where all the desires that make for a good life can be expressed. As Gorz says: "The back-and-forth between heteronomous work, optional microsocial activities and autonomous personal activities is the guarantee of balance and freedom for everyone."[36]

Of course, much more could be said in developing a more specific vision of a free, rational, democratic, and technosober society that promotes the development of individual and civic virtues, as well as personal

35 André Gorz, *Les chemins du paradis: L'agonie du capital*, Paris: Galilée, 1983, p. 121.

36 Ibid., pp. 125–7.

and collective flourishing. The problem of necessity will not be magically solved by the advent of algorithms, and the mounting climate crisis warns against any notion that the promises of material abundance and boundless technological progress will guarantee our well-being. We believe, then, that democratic economic planning must aim to ensure optimal production of what is necessary based on a collective definition of priority needs, technosobriety, and the creation of free time. But this production of what is necessary is itself subordinate to a higher social purpose—what Marx called the "realm of freedom":

> Freedom, in this sphere, can consist only in this, that socialized man, the associated producers, govern the human metabolism with nature in a rational way, bringing it under their collective control instead of being dominated by it as a blind power; accomplishing it with the least expenditure of energy and in conditions most worthy and appropriate for their human nature. But this always remains a realm of necessity. The true realm of freedom, the development of human powers as an end in itself, begins beyond it, though it can only flourish with this realm of necessity as its basis. The reduction of the working day is the basic prerequisite.[37]

This is a horizon to be achieved—a project for a "society of good living." Reaching this future demands the overthrow of algorithmic capitalism.

37 Karl Marx, *Capital*, Vol. III, London: Penguin, 1991, p. 959.

Conclusion

In this decisive historical moment, let's get organized!

Listen to the forest that grows rather than the tree that falls.

—Anonymous

A guiding thread for interpreting our times

We have come to the end of our investigation. In this book, we have sought to develop an intellectual and theoretical toolkit to explore a whole series of phenomena that have been sweeping through our societies for the past twenty years, but whose common thread is not always so obvious. In Thesis 1, we asked:

> What connects such diverse phenomena as the rise of social media, the extraction of rare earth metals, new forms of online entertainment, the fragmentation of our schedules, hyperconnected forms of socialization, the fortunes of Bezos, Musk, and Zuckerberg, US–China tensions over Huawei and TikTok, or the ubiquitous spread of screens? What if we also work in the Brexit vote, teleworking, emotional robots, ChatGPT, the vagaries of cryptocurrencies, military conflicts in Congo, insurance discounts for "good behavior," polarized public debate, targeted political assassinations by drones, the climate crisis, binge-watching, suicides at Foxconn, the semiconductor shortage, influencers, Nvidia's record market capitalization, or same-day delivery?

We can now say that what connects all these phenomena is the advent of algorithmic capital.

We have dated this shift to around 2007–8, although it is worth remembering that major sociohistorical transformations never happen overnight, but instead take on their full dimensions over the long term. In other words, many elements of algorithmic capital appeared before 2007, others are developing in this very moment, and still others lie ahead of us. But it is often at pivotal moments and during major historical crises that new configurations of social order emerge, sometimes taking a few years to become fully apparent. Thus, in the context of the most recent global financial crisis, the dominant logic of the capitalist system underwent what we consider to be a profound change, and which we have conceptualized as a change in the stage of capitalism or its regime of accumulation. As we have seen, this change first crystallized in a new model of economic valorization based on data and algorithmic technologies—a model that, according to Zuboff, was "discovered" first at Google, but is based on a series of political, cultural, and social factors. This new form of economic valorization quickly spread throughout the digital world before its logic colonized the economy and society as a whole. We should not be mistaken about the nature of this change: Capitalism is still the dominant system on a global scale. But we are now entering a new episode in its history.

This was the guiding principle that led us to analyze the various aspects of social life examined in this book. Seeing the world in this way, through the lens of algorithmic capital, allows us to make sense of contemporary social transformations, to understand them as part of a moving whole—an open, complex, yet coherent historical process. In technical terms, we believe that the concept of algorithmic capital has great heuristic value: It contributes to discovery and understanding by organizing reality into an intellectual framework that makes it more intelligible.

The many facets of algorithmic capital

At the end of this overview, we can confidently speak of algorithmic capital—not simply as an economic phenomenon, a growth sector, or a revolutionary technology, but as a multifaceted social relationship that is creating a world of its own. It is a new form of social relations, subjectivity, power, and behavioral regulation. It implies shifts and reconfigurations in the institutionalized social order of capitalism. Throughout our research, we have often been struck by the rapid pace of its development and the multiple spheres of society affected and reconfigured by its logic (of extraction, coordination, prediction, and automation). We are now in a better position to sum up its many facets.

Economically, the accumulation of algorithmic capital is based on a certain reconfiguration of typically capitalist processes of valorization, developing mechanisms that integrate human experience beyond waged work as such in its circuits of valorization. Algorithmic capital is based not only on the exploitation of labor, but also on the extraction and datafication of human experience, the production of data commodities, and the close intertwining of the accumulation of capital with algorithmic power. The rise of captive screen time is central to this, but algorithmic capital also reconfigures labor itself, monitoring, automating, fragmenting, and outsourcing work. Platforms, a new form of market/company hybrid, now dominate economic life, and the production of goods and services is increasingly carried out using algorithmic technologies. Companies in various sectors—agriculture, mining, finance, energy, manufacturing, communications—are reinventing themselves as data firms and producing connected objects and goods. They are algorithmizing their production processes or adding data extraction and trading to their activities. In this sense, algorithmic capital does not only concern the online world, the digital economy, or Big Tech, but is also spreading to activities in the primary, secondary, tertiary, and domestic sectors, transforming all economic activities to varying degrees. In this economic form, the accumulation of capital is now effectuated through, and mediated by, algorithms and algorithmically organized social relations.

Social reproduction, as we saw in Thesis 6, is also subjected to imperatives and pressures coming from algorithmic capital through various phenomena such as the increasing use of AI in the spheres of healthcare and education, the "platformization" of domestic work, and the introduction of connected objects in the home. We have also seen that the social relations of reproduction are also affected by algorithmic technologies, since they have varied impacts on sexuality, affective labor, care, and even the metabolism of affects in the household.

Politically, algorithmic capital wields a new form of power. While Zuboff speaks of "instrumentarian power," we have explored the concepts of algorithmic regulation, governance, and governmentality. The terminology in this field is abundant, but in each case there seems to be a process of data extraction and analysis linked to predictive analysis, recommendation, incentive and/or (semi-)automated decision-making systems that can be integrated into a myriad of devices and spheres of human activity. This has given rise to a new mode of regulation of social practices and institutions, giving certain actors immense power over individuals and populations. The spread of algorithmic technologies in society has also launched an important tendency for the automation of systemic injustices, such as the reproduction in mechanical forms of male domination, racism, and heteronormativity. Algorithmic capital is also penetrating the state, the public administration, and the police, contributing to the erosion of democratic life. In the international arena, the rise of algorithmic capital coincides with the onset of a structural change in the international order, characterized by the decline of American hegemony and the rise of a powerful hegemonic rival, China. The Sino–American rivalry in the fields of AI and algorithms places the rest of the world in an asymmetrical position of dependence on the two powers (apart from Russia, which retains a degree of digital sovereignty). North–South relations are also changing. This is not because they have been turned upside down, but because a new algorithmic dimension has been added to the historical dynamics of colonial and neocolonial exploitation and domination.

On a *socio-cultural* level, the rise of social media and now generative AI are undoubtedly the most drastic and significant changes of recent decades. This has had important consequences. Social networks have enabled a vast quantitative growth in social relationships and allowed users to engage much more proactively with information. But the quality of social relationships, public discussions, socialization processes, and identity formation has suffered serious damage. The omnipresence of algorithmic technologies and the datafication of society have profoundly transformed fundamental relationships within communities: Relationships with space, objects, other people, time, nature, and oneself are increasingly mediated by the material and extractive infrastructure of algorithmic capital. This algorithmic mediation is no trivial concern, but one that has a profound structural role, in the sense that it produces, configures, or reorients social relations and personality types, such as the "connected self," according to its own logic and objectives. Its long-term consequences for the social fabric and psychological health remain to be determined. Meanwhile, the values associated with Silicon Valley are strongly imbuing the dominant algorithmic culture and are even being re-created as a model for territorial development, in constant dialectical interaction with dominated and/or nonhegemonic cultures.

The recent rise of Gen AI is rearticulating basic processes not only of cultural production, but even of the formation of meaning. The jury is still out on the impact such a technology will have on human thinking and culture.

On the *ethical* front, we have explored the types of subjectivity and world that algorithmic capital tends to reproduce. The outcome of this research has proved rather bleak, as we observe the development of new forms of alienation in the formation of algorithmic subjects and the worlds they inhabit. We have also proposed a critical theory of virtues in order to explore the moral dimensions of contemporary lived experience, with the various individual and collective vices produced or exacerbated by algorithmic capital. This process of alienation and reification of social relations is not complete or total. However, it is a real trend, a multidimensional and contested process that we must think about as a whole in order to resist it, attack it, and eventually overcome it. We have also analyzed algorithmic capital as it reconfigures the relation between society and nature along the lines of an accelerated and unsustainable extractive industry. As such, for us, algorithmic capital is a polysemic concept with multilayered meaning, referring simultaneously to a logic of accumulation, a form of mediation of social relations, a form of power, and the configuration of a relation between society and nature.

Of course, all these dimensions overlap in lived experience and social life. And ultimately, we are well aware that this ambitious examination of our subject remains incomplete, with some strong elements, and others that still need much more refinement, correction, and further research! Yet we hope to have demonstrated the expansive logic of algorithmic capital and its constant tendency to constitute and reproduce a world of its own. We hope to have made a contribution to the growing critical literature on contemporary capitalism and technology. As we stated at the outset, we believe that, in our day, one cannot understand contemporary technology without a critical understanding of capitalism, and one cannot understand capitalism without a critical understanding of technology.

A decisive historical moment

Does the concept of algorithmic capital explain everything going on in the contemporary era? Of course not. But as Marx described in his time, the dynamics of capitalism have something totalizing about them; they spread and diffuse ever further into society. It is a dynamic that always aims toward complete dominance, even if it can never quite achieve it. We do not believe that algorithmic capital is omnipresent, like the God of the great monotheistic religions, but it does drive an all-encompassing logic within society. This is an important point to recognize. We are even less prone to believe that algorithmic capital is immutable, or a "natural" development of capitalism; it is a historically specific phenomenon whose beginning we have identified, and it will last for a time and then disappear to make way for something else. Although we cannot predict when or how it will end, we believe it is important to think about what comes next.

This has led us to explore possible ways of mitigating its negative impacts, but also of building a postcapitalist and technosober society. This also concerns the new fronts of struggle and resistance that are springing up all over the world, alternatives to algorithmic capital in the form of digital commons, free software, and cooperative platforms, or more or less radical legislative reforms that counteract some of the harmful effects of capital in our era. Some more radical perspectives even envisage the potential dismantling of capitalism. These are based on the socialization of digital infrastructures and platforms, support for noncapitalist alternatives, and decentralized economic planning based in part on algorithms, but above all on democratic processes, collective action, and a principle of technosobriety aimed at reducing the space of digital technology in our lives.

This does not mean that we are predicting that algorithmic capital will be the final regime of accumulation or the "highest stage" of historical capitalism. We are engaged in critical theory, not prophecy. But this exercise in critical theory, drawing on a series of contributions from a myriad of researchers, aims to shed light on our historical moment and to devise tools for bettering our world, with the perspective of overcoming capitalism.

We are at a decisive moment in history, due to the convergence of two phenomena that are absolutely crucial for the future of human societies. First, we believe that algorithmic capital poses a real threat to political freedom, democracy, social cohesion, cultural diversity, and the possibilities for a good life, both individually and collectively. Its prolonged rise risks leading—without decisive collective action to regulate or reverse it—to an even more authoritarian, unequal, hateful, accelerated, "nudged," and standardized world. We see the current moment as one of potential rupture, either down a technoauthoritarian, or even technofascist, road, or a democratic postcapitalist one. Hence our effort to contribute to the thinking about democratic and noncapitalist paths.

Second, algorithmic capital is building its power in a context where climate crisis is already threatening the survival of human societies as we know them. The problem is that algorithmic capital and the climate crisis are likely to intensify each other in a vicious cycle of dynamics harmful to human and non-human life. This is also why we speak of a decisive historical moment. Algorithmic capital exacerbates the climate crisis through its vast energy footprint and its unsustainable exploitation of natural resources. It also presents itself as a solution to the climate crisis, in a worrying exercise in greenwashing and tech-washing, so that the exacerbation of the climate crisis creates what we believe to be an ill-advised craze for algorithmic technologies. Technological solutionism has never been so strong, aiming for a compensatory utopia seeking humanity's salvation by means of apps, AI, robots to manage natural disasters, and even trips to Mars. Left to its own devices, this cycle seems to be leading to a kind of "fully automated scarcity capitalism"—an authoritarian, automated, and ecocidal model of capitalism. Hence the importance of our moment and the urgency of collective mobilization against exploitation, oppression, and algorithmic domination, in the name of freedom, the good life, equality, democracy, and the balance of ecosystems.

For us, this research is only the beginning. Apart from a few articles that have already been published, this book represents an initial summary of our investigations, but it is far from exhaustive or flawless. As Marx pointed out, this means a work of "self-clarification (critical philosophy) to be gained by the present time of its struggles and desires. This is a work for the world and for us. It can be only the work of united forces."[1] Similarly, we believe that we must join the ongoing struggles that are only just beginning. These are, for certain, far from exhaustive or flawless, and present difficulties that may seem insurmountable. Overcoming algorithmic capital and mitigating the climate crisis may well be among the hardest tasks that societies have ever faced. But we must keep in mind the prodigious power that human beings have when they come together, determined and fighting. This is a power that has been demonstrated time and again in the history of anticolonial, abolitionist, feminist, labor, and other struggles that have changed—and continue to change—history. Without falling into naïveté or blind optimism, we

1 Marx to Ruge, September 1843.

should always remember that another world is possible and resist cynical, nihilistic, or resigned alternatives. This book is meant as a modest contribution to our collective understanding of this historic moment—and the possibilities for struggle.

Let's get organized!

As we emphasized at the start of this book, a critical theory of algorithms cannot remain confined to the realm of ideas, speculation, and academic research. It can certainly "shed light on the emancipatory potential, prospects for action and social alternatives present in today's world," but it must also translate into collective action and active participation in ongoing resistance movements. Thus, much more than a research program representing the "theoretical side" of human liberation, we need to articulate theory and practice through intervention aimed at changing the social world. In simpler terms, we must take action—and confront the powers that be. This requires organizing, networking, and bringing together the active forces in civil society, academia, and social movements that share a critical awareness of the social impacts of AI and the domination of algorithmic capital.

We are not offering any clear plan or manifesto. Nor are we proposing a detailed organizational structure. Should we create a new "algorithmic capital research institute," set up a collective that campaigns against capitalism and algorithmic governmentality, or infiltrate existing organizations? Perhaps we should build a broad, multisector coalition of social actors and organizations to produce in-depth critical studies and reports, make effective media interventions, and carry out direct actions and/or broad campaigns against capitalist platforms, the AI industry, and the "siliconization of the world"? We do not claim to have any ready-made answers. But this call for collective action is essential if we want to avoid getting stuck contemplating a looming disaster.

To further this work of organizing progressive forces interested in issues related to the digital world, data, and AI, but also the ecological crisis, injustices, and multiple forms of intersecting oppression, in Thesis 19 we identified a series of virtues that will help strengthen our resolve to change society and better unite our forces. Courage, respect, justice, humility, critical thinking, care, mutual aid, and love and rage are some of the moral and practical dispositions needed in order to build a radical, inclusive, and effective opposition to algorithmic capitalism and its defenders. We can be sure that they will not readily allow themselves to be overthrown.

Theoretical critique is only a first step, a moment of self-reflection on practice. Ultimately, it must be reintegrated into action in order to transform a world that is exacerbating and automating inequalities, alienating us and depriving us of our full human potential. This transition to action, however difficult and complex it may be, also brings joy. It is an experience of liberation, a strengthening of our personal and collective capacities, and generates the feeling of finally having a hold on the world and history, rather than passively submitting to the existing order. Faced with the seemingly relentless momentum of algorithmic capital and the world it brings with it, resistance and real utopias are scattered everywhere. It is our task to bring them together, without being afraid of bold experiments. As André Gide pointed out: "Many things seem impossible until we try them." All that remains is for us to roll up our sleeves and make the impossible happen.

Taking action, organizing, and actively fighting the established order is the best way to channel everyday frustrations, outrage at injustice, feelings of alienation generated by the digital world, and eco-anxiety, which affect a large proportion of our contemporaries. Philosopher Frédéric Lordon reminds us that the eco-anxiety that plagues much of the climate movement and young people is "one of those wonderful pieces of bullshit that the media system is so fond of. You take a real political problem, you put it through the media distiller, and you get psycho-societal bullshit from which all politics has been abolished."[2] Lordon particularly criticizes the tendency of neoliberal discourse to depoliticize and psychologize social issues as it lumbers individuals with the burden of changing things. If this fear risks paralyzing us, Lordon suggests that the way to avoid it is to remain laser-focused on the ultimate cause of the ecological disaster—capitalism—and the need to overthrow it. Lordon thus urges us to transform our relationship with the world faced with the current crisis: "Don't be eco-anxious, be eco-furious!"

This move from eco-anxiety to eco-rage could prove to be our salvation, similar to the virtue of rage discussed in Thesis 18. A study in psychology shows that "the experience of eco-rage indicates better mental health and stronger commitment to climate activism."[3] The same could be true of technological anxiety or the "AI anxiety" caused by the stress of social media,[4] generalized surveillance, the specter of superintelligence,

2 Frédéric Lordon, "L'éco-anxiété, une merveilleuse conner*e," lecture at the University of Lausanne, September 2022, available on YouTube, youtube.com/.

3 Marc-Aurèle Bary, "Pourquoi il faut en finir avec l'éco-anxiété," *Vice*, January 27, 2023.

4 Clémence Mary, "Après l'éco-anxiété, l'IA anxiété," *Libération*, June 19, 2023.

the risk of losing one's job to ChatGPT, the omnipotence of GAFAM, the blurring of truth and falsehood by deepfakes, and the awareness that AI will further increase humanity's carbon footprint. Faced with this unparalleled global crisis, the road of fear, anxiety, powerlessness, and indifference would only deepen our alienation faced with the rise of algorithmic capital. With a clearer understanding of the ultimate cause of these problems, we urge our readers to stop being afraid and to overcome our anxieties, through praxis. *So, let's stop being technoanxious—let's be technofurious!*

Index

Figure and tables are referenced by an f and a t along with the page number